A Complete Guide to Quality in Small-Scale Wine Making

To those whose wisdom has inspired me (jac): Flora May Dickson, pioneer journalist-editor who fought for rural communities (Grandmother, dec.); my parents, Grace and Jim Considine, viticulturists (dec.); Professor Bruce Knox (PhD supervisor, dec.); and Alan Antcliff (dec.) who taught me statistical analysis.

A Complete Guide to Quality in Small-Scale Wine Making

John A. (jac) Considine PhD (Melb), MAgSc (Adel), BAgrSci (Melb), Longerenong Dip AgSci

Professor Emeritus and Honorary Senior Research Fellow,
School of Plant Biology, Faculty of Science,
The University of Western Australia, Crawley, WA, Australia

Elizabeth Frankish BAgSci Hons (UTas)

Project Manager, Horticultural Food Safety,
Department of Agriculture and Food, Western Australia, Perth, WA, Australia

AMSTERDAM • BOSTON • HEIDELBERG • LONDON • NEW YORK • OXFORD
PARIS • SAN DIEGO • SAN FRANCISCO • SINGAPORE • SYDNEY • TOKYO
Academic Press is an imprint of Elsevier

Academic Press is an imprint of Elsevier
The Boulevard, Langford Lane, Kidlington, Oxford, OX5 1GB, UK
225 Wyman Street, Waltham, MA 02451, USA

First published 2014

Notices

Knowledge and best practice in this field are constantly changing. As new research and experience broaden our understanding, changes in research methods, professional practices, or medical treatment may become necessary.

Practitioners and researchers must always rely on their own experience and knowledge in evaluating and using any information, methods, compounds, or experiments described herein. In using such information or methods they should be mindful of their own safety and the safety of others, including parties for whom they have a professional responsibility.

To the fullest extent of the law, neither the Publisher nor the authors, contributors, or editors assume any liability for any injury and/or damage to persons or property as a matter of products liability, negligence or otherwise, or from any use or operation of any methods, products, instructions, or ideas contained in the material herein.

British Library Cataloguing-in-Publication Data
A catalogue record for this book is available from the British Library

Library of Congress Cataloguing-in-Publication Data
A catalogue record for this book is available from the Library of Congress

ISBN: 978-0-12-408081-2

For information on all Academic Press publications
visit our website at store.elsevier.com

Printed and bound in the United States
14 15 16 17 18 10 9 8 7 6 5 4 3 2 1

Contents

Acknowledgements

I wish to acknowledge my students, many of whom remain firm friends. It was a joy to go through the process of learning about the wine industry with you – thank you. I also wish to acknowledge the input of Bob Frayne who taught me and my students the craft of small-scale winemaking and who provided the early protocols. Bob, you were a fine teacher and winemaker and we all enjoyed the experience and support that you provided (and I thank your employer, the Department of Agriculture and Food, Western Australia, for making you available and for providing access to facilities). Another person who provided support in analytical chemistry, especially for HPLC and GLC of organic acids and sugars in wine and berries, was Greg Cawthrey of the School of Plant Biology, Faculty of Science, The University of Western Australia; and who was also responsible for providing basic training in Occupational Health and Safety in Laboratory situations. Thank you Greg.

I wish to thank those 'real' winemakers and wine scientists who allowed us into their inner sanctums and who gave us personal and inspiring insights into the art and craft of fine winemaking. Also, I thank those scientists within the Australian Wine Research Institute who gave insights into the higher levels of sensory science, especially Dr Leigh Francis. Winemakers who were especially generous with their time and support include James Talijancich (Talijancich Wines), Steve Warne (then of Evans & Tate Wines), Denis Horgan (Leeuwin Estate), Keith Mugford (Moss Wood), Will Nairn (Peel Estate Wines), Rob Bowen (then of Houghtons Wines), Bill Pannell (Picardy), and Kim Horton (Ferngrove Wines). I also wish to thank my research officer, Tony Robinson, for his energy and support in the practical aspects of the winemaking process.

The authors also thank the Institut Cooperative du Vin (ICV) for agreeing to the inclusion of their protocol for tasting berries (author Jacques Rousseau, see Chapter 7), the Faculté d'Oenology, Université Bordeaux Segalen, for permission to use data and methods from their website, www.bordeauxraisins.fr (Professeur Bernard Donèche), and Professor Roger Boulton (University of California, Davis) for providing a copy of his conductivity method for tartrate stability.

A number of people read parts of this text during the early phase of its writing and I would like to thank them; especially Dr Catherine Cox-Kidman (Treasury Wine Estates), Ross Pamment (Houghton Wines), Richard Fennessy (Dept. of Agriculture and Food, WA) and Anne Mazza (Mazza Wines). The authors, however, take full responsibility for the final text.

I would also like to thank my co-author, Elizabeth Frankish who, despite at the time being a busy CEO/Consultant, made time to provide my students with a strong industry-oriented insight into the science of microbiology and microbiological faults in wine.

Finally, I would like thank the editorial team at Academic Press for their professionalism and support during the production process: Thank you Nancy, Carrie, Charlotte and Edward.

jac

Preface

This text began as a manual for mature-age students taking a winemaking course, but without a requirement for prior studies in chemistry. There is a perception that winemaking is mainly a craft and while it can be taught as a craft this is with restraints and limitations. Winemaking is, however, also a science, and one grounded deeply in chemistry and microbiology (and engineering and physics). There was a period in the early renaissance of the table wine industry in Australia that it seemed to me, as a consumer, that art and craft were ignored and the emphasis was on producing technically sound wines (boring!). As the industry has matured, a shift has occurred towards wines that are well crafted, where the winemaker's 'art' is displayed, but in a background of sound science and technology. I thought this was a sound approach to developing a winemaker or a wine industry - get the science and technology right, and then the art and craft can occur with the opportunity to produce genuinely fine wines, consistently. Thus I set my students the task of producing technically sound wine as a basis on which to build their craft, and also to learn to appreciate the breadth of skills involved in contemporary winemaking. So they began with wine chemistry: wine analysis. That is, thus, the emphasis in this text. The winemaking processes provided are simple, uncomplicated, and suited to the novice or the researcher, but the chemistry is as complete as is practical for teaching, or a small winery.

The text can be approached in a number of ways. If you, the reader, are unfamiliar with the symbols of chemistry, then the text can be followed as a recipe and the analyses can be limited just to those that are essential (and they can be followed as a recipe without the need for deep understanding).

Chapter 1 provides an overview, an introduction, while Chapters 2 and 4 are essential reading. Chapter 2 concerns wine tasting. This is an absolutely central feature of winemaking. Any aspiring winemaker must develop their sensory skills to a level that may be compared with the physical skills of an elite athlete. This task should be disciplined and regular. Chapter 4 concerns occupational health and safety, especially related to handling laboratory chemicals, working in confined spaces, and using mechanical equipment. Chapters 5, 7 and 8 discuss and provide basic routines for the winemaking process and harvesting of fruit. Chapters 9 and 10 concern analysis of the parameters essential for making a sound wine, especially pH and sulphur dioxide. You may then delve into the other chapters as curiosity or problems require. The companion website for the book can be found at http://booksite.elsevier.com/9780124080812.

For the reader with a professional career in mind, or a deep curiosity, the remaining chapters provide a contemporary view of wine chemistry and of microbiology, with the purpose of developing deep understanding of the processes and working towards a command of the craft of winemaking: organic acids, tannins, oxidation, anthocyanins and many of the vast array of chemical substances that contribute overtly or discreetly to the sensory experience of wine. This material is suited to an introductory to intermediate level of study in wine science. More specialized texts such as those by Ribéreau-Gayon et al. (2006a,b), Boulton et al. (1996) or Jackson (2008) meet the needs of those seeking a more thorough exposition of the processes and chemistry. To learn the craft however, there is no replacement for experience - work with a fine winemaker.

jac

Chapter 1

Introduction to the Table Wine-Making Process

Chapter Outline

1. INTRODUCTION

Wine making is a blend of 'art' and of 'science'. It is complex and enduringly intellectually challenging. The endeavor of producing a truly fine wine, in successive years, from fruit that differs from one year to the next is a never-ending source of motivation and pleasure; how can the winemaker add value to this fruit? This book aims to guide the aspiring winemaker in the production of a technically sound wine. The challenge to you, the winemaker, is to go from that base to producing 'fine' wines consistently, and perhaps even exceptional wines.

Wines of all styles have a great deal in common, but producing a specific wine style may demand unique conditions. Broadly speaking, 'red' wines are exposed to oxygen during the primary ferment while 'white' wines are not (Figure 1.1). Red wine fermentation is carried out at a higher temperature (25–30°C) to aid extraction of tannins and color while white wines are fermented at 15–20°C to preserve volatile aromas. Red wine ferments include all berry parts, skin, pulp and seeds, but white wine ferments include the juice or 'must' pressed from the pulp only in order to minimize the extraction of bitter phenolics and tannins.

Red wines are double fermented: first with a yeast to process sugar to alcohol, and then with a bacterium to convert malic acid to lactic acid. The latter process reduces acidity, adds flavors that enhance red wines, especially in conjunction with oak, and helps to ensure stability against secondary fermentation in the bottle, even if the wine is not fully sterile.

White wines, with some exceptions such as Chardonnay, are yeast fermented only, then chilled and stabilized. Very careful filtration is required to remove all microorganisms to prevent fermentation of malic acid once bottled. This is a much quicker process and one that generally suits the dry, crisp, aromatic palate generally sought in such wines.

Elements common to both processes include hygiene, the fermentation of grape sugars, management of acidity and nutrient status, and certain steps to protect and stabilize the wine. The distinctions between the two processes are vital, however. It is possible to produce an acceptable red wine in your 'backyard' but challenging to produce a sound white wine under such circumstances. This is because of the much greater control that is required of oxygen status, hygiene, yeast nutrition and temperature.

Typically, a commercial yeast culture is added at the beginning to red and white crushed berries or musts. Wild yeasts that are present on the surface of grapes may do the job and may produce exceptional wine, but, then again, may result in an undrinkable wine. Few wild yeast taxa are able to complete fermentation to dryness, that is, to ferment all the available sugar. Unfermented sugar, termed residual sugar, impairs not only the palate of a wine but also its stability subsequently against in-bottle fermentation and contamination by spoilage microorganisms. Only commercial yeast genera (*Saccharomyces*) can complete fermentation reliably. These yeasts also occur naturally but are rare in the vineyard (however, they are common in a winery).

A Complete Guide to Quality in Small-Scale Wine Making.

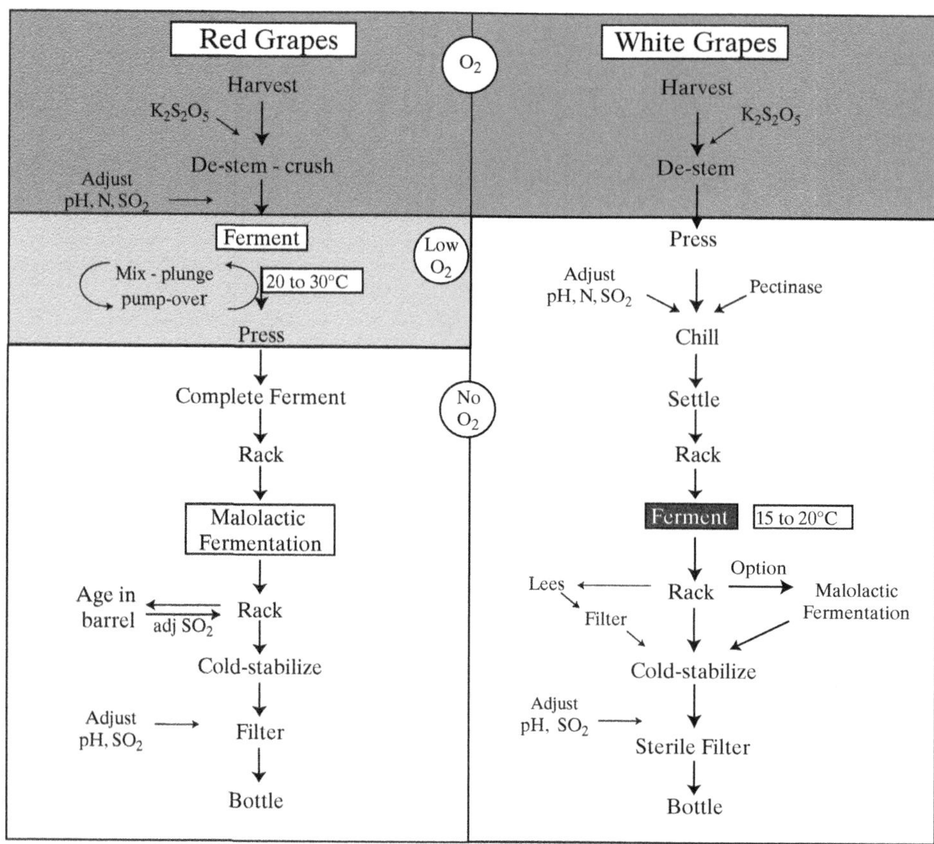

FIGURE 1.1 Outline of the wine-making process for red and white wines illustrating the flow of the processes and their distinguishing features. Note especially the regions that are fully aerobic, microaerobic and anaerobic. In red wine production, the anaerobic region may be managed microaerobically through a process known as microoxidation.

Sulfur dioxide (SO_2) is added at harvest, usually, or at crushing, to reduce the risk of oxidation and to suppress wild yeasts and bacteria that may spoil the wine. SO_2 is an essential part of modern wine-making practice. Its addition is especially important once fermentation has finished in order to reduce the risk of oxidation and spoilage. It is usually added as the salt, potassium metabisulfite ($K_2S_2O_5$), although it may be applied as a gas (SO_2).

Red wine making is an 'extractive' process that is carried out in the presence of the skins, seeds and possibly even some stems (Figure 1.1). It is less prone to oxidation because of the presence of high levels of antioxidants: tannins and anthocyanins. Indeed, some mild oxidation may be beneficial: it is part of the process. The skins, which naturally float and form a 'cap', are buoyed by carbon dioxide (CO_2) produced during fermentation. The cap is mixed with the must regularly (plunged or bathed by pumping the liquid over the top) to encourage the extraction of anthocyanins and tannins. Failure to do this will lead to the growth of spoilage yeasts on the surface. For a review of red wine-making processes see Sacchi et al. (2005).

Red wines and some white wines undergo a secondary fermentation that is accomplished by a bacterium, *Lactobacillus oeni*. This is termed a malolactic fermentation. It reduces wine acidity by converting malic acid to lactic acid. This may be especially important in wines produced from grapes grown in a cool climate and which may have a high level of malic acid at harvest. It also increases stability against spoilage and fermentation in-bottle by removing a remaining fermentable substance, malic acid. The bacterial fermentation may also broaden the spectrum of aromas and flavors. In large commercial wineries this fermentation is conducted in tanks but for premium wine and in small wineries it is often performed in small oak barrels.

White wine is sensitive to oxidation, especially after fermentation, is not extractive and must be stabilized by sterile filtration (Figure 1.1). Part of the stabilization process for white wine involves chilling before bottling to precipitate excess potassium bitartrate salts (as used in baking powder), which might otherwise form unsightly crystalline deposits in the bottle when refrigerated. White wines are also 'fined' with a clay (bentonite) to remove excess protein that might coagulate and form an unsightly haze should the wine get too hot during storage and transportation. Finally, they may be treated with copper sulfate to remove hydrogen sulfide (H_2S) 'rotten egg' gas that may be formed when starving yeast metabolize grape proteins (this may also be true of red wines but they usually contain more nutrients). Other fining processes to remove bitter tannins may include the use of natural products such as proteins from eggs, fish or gelatin, or synthetic products such as polyvinylpolypropylene (PVPP). These precipitate and are removed before bottling.

Aging in 'toasted' oak is a standard practice in the art of red wine making to broaden and balance the sensory characters of the final wine. Selection of particular oak sources and of the level of toast is a major part of fine wine making and an expensive aspect. Oak may also be used with some white wines (e.g. Chardonnay) and some Sauvignon blanc and Semillon wine styles (e.g. Fumé blanc).

The processes are simple conceptually. The art lies, however, in the science of analyzing the raw ingredients, in monitoring the process, and in the skill and judgment of the winemaker in managing the details of the process. Vital also are the skills of the viticulturist in matching cultivar to site and in devising appropriate vineyard management strategies—as well as luck in the season.

The final product is judged on its sensory appeal, and being aware of this throughout the process is at the heart of becoming a good winemaker. Once a wine becomes spoiled its value is limited or nil. The chemical and the quality assurance processes described in this text serve to assist in that goal, but do not guarantee it. Sensory alertness to 'off' flavors and aromas in vessels, pumps, tubing, etc., is vitally important as the human nose can detect some aromas down to picogram or even femtogram levels (10^{-12} to 10^{-15} g/L). As in all things, prevention is far better than cure: be alert and avoid problems. Care needs to be taken, however, to check one's palate outside the winery, for it is easy to become adapted to particular off-aromas (e.g. *Brettanomyces* is a problem requiring constant management in even the best of commercial-scale wineries).

2. GREAT WINES BEGIN WITH GREAT GRAPES

Quality is an issue that may make or break the smaller vineyard owner and determine the profitability of even the largest vineyard company. This is the case because the value of the fruit depends on the winemaker's judgment (and experience) regarding the value of the wine that may be made from that fruit. Fruit destined for super-premium wine may be valued at two- to 10-fold dollars per tonne above that for standard commercial fruit.

Defining grape quality and selecting appropriate maturity indices and quality measures are issues that are still widely debated among viticulturists and winemakers. Commonly, fine wines are associated with regions that inherently produce low vigor and low yields in a climate that enables full ripening under mild climatic conditions (Gladstones, 1992; Jones et al., 2010; White et al., 2006).

The best indicator of potential quality is history. This is the basis on which regional quality assurance labels are allocated, e.g. the Grand and Premier Cru classifications in France. Experience tends to be the benchmark on which site selection is determined. For example, the premium viticultural areas of the 'New World' in southern Western Australia were identified by comparison with the best viticultural regions in France and California. These regions had a long and reliable history and their climate and soils were used as benchmarks (Gladstones, 1965; Olmo, 1956).

Natural factors are predominant in determining wine quality: those that are based on location and the attributes of the vine itself, and its suitability for a particular location (Figure 1.2). This statement is not intended to diminish the importance of the human factor in viticultural practice and oenology. It is, however, indisputably the case that all great wine comes from particular regions. The French have coined a term for this, *terroir*; simply, the ecology of the site, its soils, aspect and climate (Figure 1.2) (Gladstones, 1992; White, 2003; Wilson, 1998). Whether a great wine is achieved, however, is influenced by the human factor, by chance and by season.

While the French terminology may seem to emphasize the physical environment, the soil and aspect, the factor that determines site suitability, cultivar selection and wine style is temperature (Gladstones, 1992; Jones et al., 2010). Grapevines, *Vitis vinifera* L., appear to have originated in the Near East and interbred naturally with local *V. vinifera* ssp. *sylvestris* populations in western Europe (Myles et al., 2011). Wine grapes are almost unique in agriculture in that the commercial cultivars that predominate today are those that arose naturally in historical times. They are adapted to Mediterranean and nearby temperate climatic zones.

The seasonal cycle of growth and dormancy, that is their phenology, the timing of bud burst, flowering, ripening and leaf fall, is tightly linked to an annual cycle of seasonal changes in daylength and temperature. Indeed, grapevine phenology has been used to gauge historical trends in climate (Chuine et al., 2004). From the viewpoint of quality, temperature during the ripening period, from veraison (softening) to harvest, is critically important; however, extremes at any period are detrimental (White et al., 2006).

Cultivar selection is included in the Natural/Enduring section of Figure 1.2 because selection of a cultivar and wine style is intimately associated with climate, and while humankind has intervened in its selection, it is essentially natural. Figure 1.3 illustrates the dramatic impact of climate on two important measures of wine quality. Table 1.1 represents one attempt at classification. A broader range is canvassed in other studies (e.g. White et al., 2006).

Determinants of Quality in the Vineyard

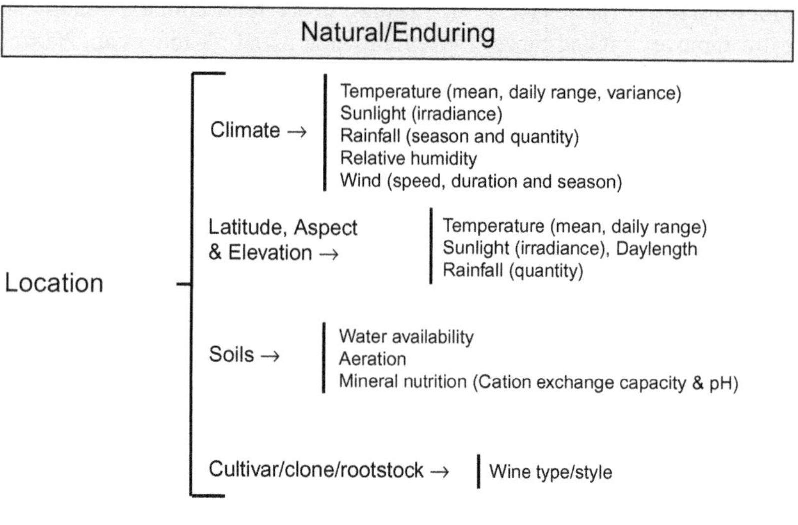

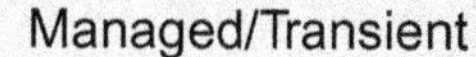

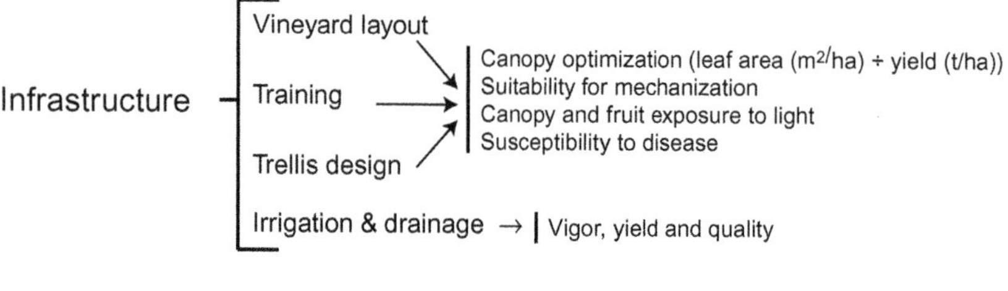

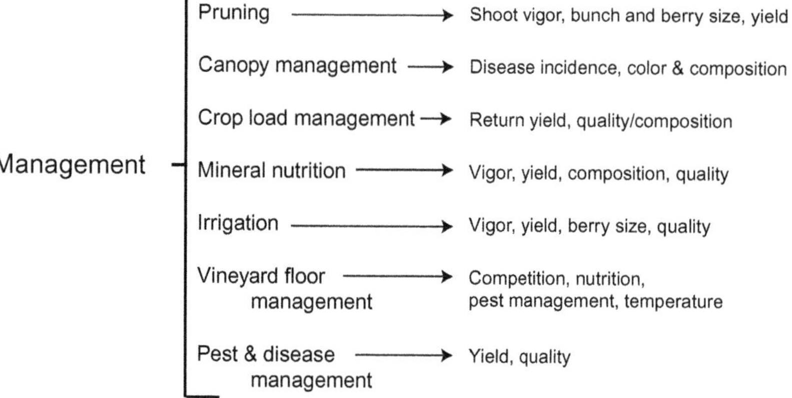

FIGURE 1.2 Outline of factors and activities that influence fruit quality for wine making in a vineyard.

While many elements are at work in determining quality, several of these are given expression through one particular aspect that seems to have a profound influence, namely vigor (Figure 1.2). Vigor is the expression of a plant's response to its environment: to water supply, nutrition and temperature, to consider three of the more important elements. Soil type and depth interact strongly with water and nutrition and play a substantial role in the expression of vine vigor.

Soils not only provide support for the vine and a medium for roots, but their physical characteristics determine water-holding capacity in the root zone, and their chemical properties control nutrient availability, quantity and balance. Viticultural soils are typically free-draining, open and shallow to moderate in depth, limiting the availability of water and nutrients, and are often rocky. However, these characteristics also expose the vine to risk of stress if water supply becomes deficient (a drought) or

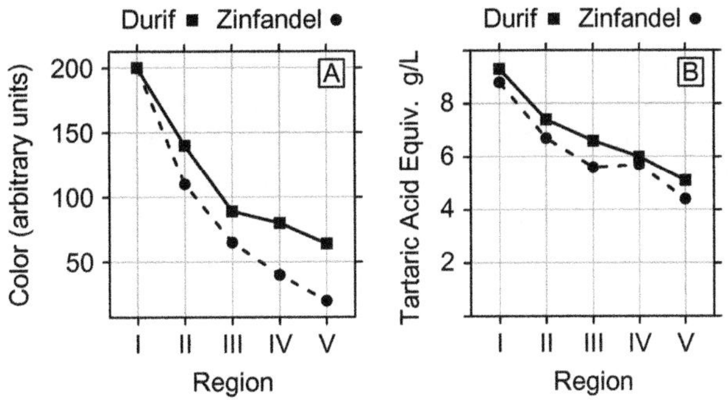

FIGURE 1.3 Impact of region on two measures of grape quality for wine making for two premium red grape cultivars. A: Color; B: acidity. *(Data from Winkler, 1963).*

TABLE 1.1 Site Classification for Wine Quality and Style as Indicated by Heat Summation for the Growing Period (April to October, Northern Hemisphere; October to April, Southern Hemisphere)

Region	I	II	III	IV	V
GDD (upper) F	2500	3000	3500	4000	4500
GDD (upper) C	1389	1667	1944	2222	2500
White cultivars	Aligoté	Aligoté	Chardonnay	Colombard	Doradillo
	Chardonnay	Chardonnay	Chenin blanc	Muscato fior d'Arancio	Muscat Frontignan
	Chasselas doré	Chasselas doré		Palomino	Palomino
	Gewürztraminer	Chenin blanc	Colombard	Ugni blanc	Pedro Ximénez
	Müller-Thurgau	Emerald Riesling	Emerald Riesling		Thompson seedless
	Riesling	Gewürztraminer	Riesling		
		Müller-Thurgau	Sauvignon blanc		
		Pinot gris	Semillon		
		Riesling	Sylvaner		
		Sauvignon blanc	Viognier		
		Sylvaner			
		Verdelho			
Red cultivars	Cabernet Sauvignon	Cabernet Sauvignon	Cabernet franc	Cinsaut	Carignan
	Pinot noir	Durif	Cabernet Sauvignon	Grenache	Souzao
		Grenache	Durif	Nebbiolo	Tinta Madeira
		Malbec	Grenache	Petit verdot	Touriga
		Merlot	Malbec	Sangiovese	Zante
		Pinot meunier	Merlot	Shiraz (Syrah)	
		Pinot noir	Nebbiolo	Souzao	
		Zinfandel	Sangiovese	Tinta Madeira	
			Shiraz (Syrah)	Touriga	
			Tempranillo		
			Zinfandel		

Note: Heat sums are calculated on the basis of temperature above a minimum of 10 °C (50 °F), termed growing degree days (GDD or Huglin Index). Compiled from Winkler (1963), Huglin and Schneider (1998) and Gladstones (1992) but interpolated to fit within Winkler's classification. Note that each author used a different method to calculate GDD, so the table is approximate only.

nutrition is neglected (e.g. White, 2003). Thus, it is not surprising that soils rank highly in the minds of those who search for quality in viticultural production and that contemporary viticultural practice is to add infrastructure to supplement natural supplies as necessary.

The outcome of excessive vigor is a dense leaf canopy that shades the fruit and buds and causes poor color development, overly large berries, low fruitfulness, bud senescence, and senescence of shaded leaves. This latter factor, in turn, leads to unpalatable, vegetative characters in the fruit (Bureau et al., 2000; Perez and Kliewer, 1990; Razungles et al., 1998; Ristic et al., 2007). A dense leaf canopy is also associated with poor color and tannin development (Cohen et al., 2012; Cortell and Kennedy, 2006).

The outcome of inadequate vigor is low yield and, if caused by water deficit stress, then unripe and seriously flawed fruit characters and unbalanced fruit chemistry (high pH) (R. Bowen, personal communication). If caused by nutrient deficiency, then the outcome is unbalanced fruit chemistry that affects not only the physiology of the fruit but also that of yeasts, which depend largely on the fruit for their nutrient supply. These are important areas of research, and the role of water and nutrient status has been studied and reviewed by a number of authors (Chapman et al., 2005; Matthews et al., 1990; Mpelasoka et al., 2003; Treeby et al., 2000).

The great vine regions of the world enable viticulturists to maintain vines readily in their 'Goldilocks' region: not too much water and nutrients and not too little—just right. Viticulture can and does succeed in other regions, but in those the viticulturist must intervene more intensely and, even then, there is only so much that can be achieved. Aspects that do not require constant intervention are those of planting density (and rootstock) and trellis design. It is essential that these are appropriate for the region and the cultivar (Figure 1.4).

The role of the viticulturist lies principally in the 'management' tasks, with the goal of ensuring that practices enable the attributes of the cultivar and sites to be fully and appropriately expressed. The winemaker, the end-user, usually plays a significant part in guiding viticultural practice, so it is not usually enough just to understand the wine-making process—the best winemakers also master viticulture.

Thus, in principle, a superior grapevine should be treated in a manner similar to that which would apply to a fine athlete: all the care and nutrition required; no more, no less. That is the essence of the science and art of

FIGURE 1.4 Grape vines known for quality fruit production showing a little of the diversity of vine training and planting density that characterizes particular vineyard production regions. A: Burgundy (high density, low vigor, vertically shoot positioned, VSP); B: Châteauneuf-du-Pape (low density, bush vine); C: Coonawarra (mechanically hedged, low density); D: Margaret River (low density, VSP).

viticulture. The ease with which this is achieved is the art of site selection.

2.1 Measuring Quality

Few hard and fast rules can be provided to measure quality, and the industry relies heavily on experience. Viticultural researchers have yet to adequately analyze the maturation process at a level that is fully meaningful for winemakers because the wine-making process itself is required to unmask the flavors and aromas. Advances in molecular biology and biotechnology could see this situation change quite soon. Sensory assessment or taste remains the best guide; chemical measures are important but those currently available are ultimately inadequate (but nonetheless are vitally important). The definitive test of fruit quality is wine quality: how much would a discerning public pay for this wine (especially if 'blind' tasted, with the label covered and perhaps transferred to a nondescript container).

Small-scale ferments can provide an economical and effective way of assessing the differences in the wine-making quality of fruit from different vineyards or from different parts of a vineyard, or the impact of changing a management system. However, it will take an expert to judge the finer differences and note that ferments may vary markedly, by chance, one from another. Therefore, fermentation trials should be replicated (see Chapter 12).

Regional differences are important in determining quality and style. Guidelines for a particular cultivar growing in a cool region may vary significantly from that growing in a warm region. Ripeness for end-use is important. For example, grapes for light, ready-to-drink and rosé-style red wines are picked earlier (less mature) than those intended for full-bodied premium, 'investment' or 'gift' wines (those worthy of aging). The practice of blending wines of differing maturity or from diverse regions to obtain a wider spectrum of sensory characters (or to obtain year-by-year consistency, as with Champagne) provides a further complication. Sometimes, blends are made at the wine-making stage (cofermented, common in Europe), while Australian practice is to blend before bottling, once fermentation is complete. This requires separation of batches and following them through the process independently until bottling. Blending requires an exceptionally well-trained palate to capture the diversity of varietal and site differences to maximize the value of the wine and achieve the winemaker's goals.

2.2 Measuring Maturity

Ripeness is the first guide to quality. As a grape ripens, sugars accumulate and organic acids, notably malic acid, decline. Thus measures of sugars and acids are the primary measures of ripeness. Note, however, that tartaric acid is the major acid in the ripe berry. It does not decline. The loss of titratable acidity is due largely to metabolism of malic acid. Malic acid loss is cultivar dependent as well as temperature dependent: the higher the temperature, the greater the loss (Figure 1.3).

Sugar content of the grape is generally measured as Brix (degrees) or as Baumé. Brix is convenient to measure in the field as only a small volume is required. It is an estimate of the weight of dissolved sugar in a solution (g/100 g). Baumé is an estimate of the volume of alcohol that will be present after fermentation and is measured as the specific gravity (density) of the solution relative to pure water. It is the standard laboratory method for fruit and fermenting must in the winery.

The commonly accepted rule of 10° Baumé equaling 18° Brix is systematically incorrect for red wines and may vary for white wines depending on the climate of the region of production (Ough, 1992). The European Economic Union uses a value of 16.83 g of sugar per liter per percent alcohol (Ribéreau-Gayon et al., 2006a). However, this too varies according to yeast strain and its conversion efficiency, which may be greater than a factor of 1.1 or less than 0.9 for 'low'-alcohol yeasts.

Rules of thumb that a winemaker may use for harvest decisions are:

- **Baumé**: normally 12.5–13.6° Bé or 22.5–24.5° Brix, may be lower for white wine
- **titratable acidity**: ideally 7.0–8.5 g/L (measured as tartaric acid equivalents) but may be as low as 4.5 g/L if from a warm area and, if legally permissible, tartaric (and/or malic or citric) acid can be added to adjust pH 3.3–3.6
- **yeast-assimilable nitrogen (YAN)**: 110 mg/L (red) or 165 mg/L (white)
- **ammonium:** 50 mg/L (red) or 140 mg/L (white)
- **anthocyanins:** 1.1–2.0 mg/g berry fresh weight (red) (note that actual level must be determined by experience: level varies greatly from one season to the next, regionally, and by cultivar).

These values impinge mainly on the technical aspects of quality wine production—ensuring sufficient sugar to produce a good fermentation but not so much that there will be excess residual sugar at the end. The upper limit is about 14.5% alcohol because levels beyond this value are toxic to many yeast strains. Such levels may also affect the palate negatively and are undesirable for health reasons. Sugar levels are also a good indicator of flavor development (Figure 1.5). Nitrogen is important to ensure that the yeast cells do not starve during the fermentation, causing a sluggish, faulty, or incomplete ferment. Moderate levels are desirable as an excessive level of nitrogen, especially of arginine, can lead to the production

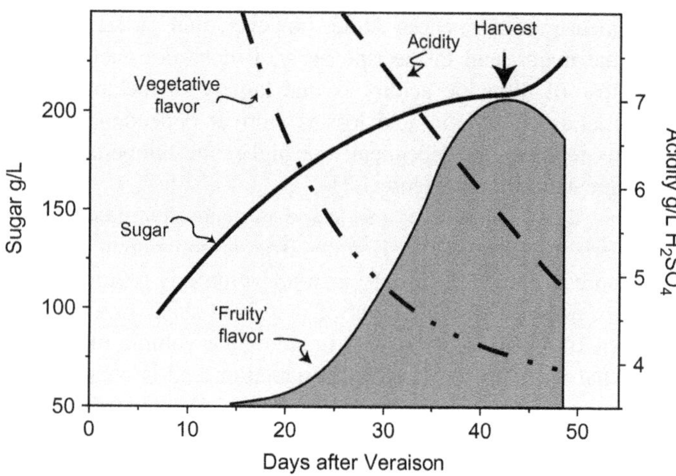

FIGURE 1.5 Generalized ripening behavior of Sauvignon blanc grapes showing the trends in the primary components of the berry from a wine-making viewpoint. The arrow indicates the ideal harvest stage. The rise in sugar from that point is due to shriveling of the berry. *(Redrawn from Ribéreau-Gayon, 2006a, with permission).*

of wine with levels of the potential carcinogen ethyl carbamate that exceed legal limits (Ferrari, 2002). This is another good reason not to liberally fertilize wine grapes with nitrogenous fertilizers.

Organoleptic quality is subject to climatic and vineyard management influences. The ideal vine has about 10–14 cm^2 of leaf area per gram of fruit, an open canopy with the fruit exposed or lightly shaded in hot climates and with restricted growth during the ripening period but green and photosynthetically active leaves. Vines exposed to excessive stress or conditions giving continuous growth are likely to produce fruit of inferior quality. The publication *Sunlight into Wine* provides a generalized vine ideotype (Smart and Robinson, 1991).

2.3 Tasting Grapes

Vine sampling should be systematic and focus on blocks of similar vines and soil–water conditions. Sampling based on soil or vine biomass maps can greatly facilitate this process (Bramley et al., 2005). A winemaker will assess fruit not only for sugar–acid balance but more importantly for its tannin and 'berry' characteristics (Rousseau, 2001). In red grapes, it is important to chew seeds and the skins and assess their flavor and 'ripeness'. As a berry ripens, the cell walls in the skin become softer and more easily macerated. Acids and anthocyanins and other phenolics are highest in the skin (Coombe and Iland, 2004). One should avoid chewing green seeds as this will be very unpleasant and can ruin the palate temporarily.

In white wines, there is usually less emphasis on skin characteristics and more on the subtle flavors of the flesh that give rise to the 'first run' or first pressings juice. Again, as the berry ripens, the flesh cells will become softer. Ideally, the berry should not be flaccid as this suggests that it is overmature and has lost water through the skin.

Seeds may be a good indicator of physiological ripeness, although this will vary somewhat from variety to variety. It may be most important in the early to mid-season varieties because veraison may start before the seed has fully matured. Seeds normally mature during the lag phase in growth prior to veraison—this may be too short in early varieties or delayed by cool weather or in overly vegetative vines. Immature seeds are an especial problem for red wine production because they are included in the primary ferment.

Immature seeds will be greenish, bitter and astringent to taste: they may cause burning of the mouth. Seeds may also adhere firmly to the flesh, although this too will vary from one cultivar to another. Loosening of the seed from the flesh is another indication of cell wall degradation that accompanies ripening and which improves juice yields. To judge just how important this process is, try pressing juice from a table grape cultivar. In these varieties, cell wall degradation is suppressed to provide a crisp flesh.

There is reliable anecdotal evidence that physiological ripeness may depend in part on management practices and circumstances between adjacent properties, and that the sugar-to-acid ratio may not be the best way of determining this. A fault causing considerable unease in eastern Australia is Shiraz (Syrah) shrivel, in which it appears that sugar accumulation ceases well before the target levels. Waiting for the sugar concentration to rise as the berry senesces and dries out (shrivels) leads only to lower flavor, lower acid and poor-quality wine: one should always taste the grape!

2.4 Postharvest Quality

Damaged and diseased fruit causes off-flavors and odors. Likewise, stored fruit, especially if harvested roughly or mechanically, undergoes senescence and deteriorates quickly in terms of both flavor and aroma and can result in stability problems. These processes are temperature sensitive and the cooler the fruit at harvest, the better. Therefore, harvesting is frequently conducted late at night or early in the morning. If fruit is to be stored, it should be at 0–2° C and preferably for no longer than 24 h. Fruit should be sprinkled lightly with powdered $K_2S_2O_5$ to minimize oxidation and microbial activity (this may need to be taken into account when making sulfur adjustments at pressing).

Winemakers are particularly concerned with 'material other than grapes' (MOG) as this will affect quality and the safe use of equipment. Excessive vegetative material will cause the subsequent wine to have a 'leafy' aroma and flavor; wire ties and pieces of wood will damage equipment, especially air bag presses, but possibly also destemmer–crushers. Recent developments in processing equipment that serve to sort and remove MOG and damaged berries include vibrating hoppers and destemmers (Pellenc, 2011) and magnets on harvesters. Hand picking eliminates many of these problems and allows for selection at the point of harvest.

3. AN OVERVIEW OF KEY ELEMENTS IN WINE MAKING

3.1 Fermentation

Humanity owes much to the work of Louis Pasteur, not only for the development of microbiology and the management of human and animal bacterial diseases but also for introducing science to the management of the process of wine making to ensure the production of reliable, unspoiled wines: 'Yeasts make wine, bacteria destroy it'; although we should really attribute the discovery of the role of yeast to Pierre Berthelot (Barnett, 2000). This view of bacteria has changed over time and we now recognize that a certain class of bacteria, the so-called 'lactic acid bacteria', can contribute to the development of high-quality wine by fermenting malic acid to form lactic acid. This type of fermentation can reduce acidity, add complexity to flavor and aroma profiles, and reduce the risk of unwanted secondary fermentation after bottling. However, bacterial infection of the primary ferment can poison the yeast and cause a 'stuck' fermentation, another good reason to use SO_2. Not all bacteria and yeast are good for wine. Management and selection have played a vital role in the development of the modern wine industry.

Three distinctive fermentations may occur:

$$\text{Sugar} \xrightarrow{\text{Yeast}} CO_2 + \text{Ethanol}$$

$$\text{Malic acid} \xrightarrow{\text{Lactic acid bacteria}} CO_2 + \text{Lactic acid}$$

$$\text{Sugar} \xrightarrow{\text{Acetobacter}} CO_2 + \text{Acetic acid}$$

Louis Pasteur studied all three and was particularly concerned with the prevention of spoilage (Barnett, 2000).

Four practices work together to prevent spoilage by undesirable microorganisms:

- **good hygiene**: maintaining all equipment in a high state of cleanliness using hot water and/or metabisulfite solutions or caustic washes neutralized with citric acid, or propriety cleansing agents
- **using good-quality fruit**, free from damaged or mold-infected berries and stems; if necessary, removing damaged fruit in the field, preharvest, and/or sorting postharvest
- **effective use of sulfur** as SO_2 or metabisulfite to control spoilage organisms
- **awareness** of the fragility of fermenting must and young wine and their sensitivity to damage by air (oxygen) and spoilage microorganisms.

Thus, the production of good wine relies on good viticultural practice in the field and quality assurance processes in the winery.

3.2 Composition Parameters

Sugar is an obvious requirement: no fermentable substrate, no alcohol, no beverage! However, a minimum level of about 10% is necessary. This limit presents a challenge to those who intend to supply a 'low-alcohol' product (not dealt with here). Alcohol acts as a preservative and helps to stabilize and preserve the wine. However, common yeast strains can only live in about 14% alcohol, although some selections tolerate somewhat higher levels. While this helps the winemaker in particular circumstances, the trend is toward lower alcohol levels in wine for health and palate quality reasons.

A sugar level higher than 14–15° Bé (25–27° Brix) may result in wine with an appreciable residual sugar because the yeasts die before they can complete the ferment. Residual sugar not only affects the palate but also results in a wine that may not be stable. Sweet wines with low alcohol require additional stabilizing additives. Nor do lactic acid bacteria (*Leuconostoc oenos*, syn. *Oenococcus oeni*) tolerate such high levels of alcohol, and therefore completing the secondary ferment may be problematic. Finally, a mix of sugar and lactic acid may result in volatile acidity, a highly undesirable outcome. High-alcohol or 'fortified' wines are produced by the addition of alcohol distilled from low-quality wine or winery wastes and the ferment is terminated while considerable residual sugar remains. That class of wine is not considered in this text.

Acids are important flavor agents and help to balance sweetness and other characteristics of the wine. The principal acids in wine are tartaric and malic acid. Both are dicarboxylic organic acids but only malic acid is fermentable. The product of the fermentation of malic acid is lactic acid (milk acid). Lactic acid has one carboxylic acid only and is thus about half as acidic as malic acid. In cool regions that have a high level of acidity at harvest maturity, malic fermentation offers a path to a balanced wine with less acidity without the addition of calcium (bi) carbonate as a deacidifying agent.

Acids are also important stabilizing agents in wine, minimizing the growth of spoilage organisms. pH is a measure of the concentration of acidic ions in a solution: the lower the number the higher the concentration of the acidic ions (H^+). This is a logarithmic scale; a wine with a pH of 3 has 10 times as many acidic ions as one of pH 4. At a pH below about 3.2, yeast and lactic acid bacteria struggle to grow. At a pH above about 3.7, other organisms can grow and the wine becomes bland and loses its 'brightness'. The loss of brightness, especially in red wines, is due to the dependence on pH of the color of anthocyanins, the pigments responsible for the red color. At a high pH (>7 neutral to alkaline) anthocyanins are either colorless or green.

Winemakers are interested also in the concentration of potassium (K^+) ions. These can replace H^+ ions, raising the pH of the wine and giving it an 'earthy' (muddy) flavor. However, K^+ is essential for normal functioning in plants and the transport of sugar to the grape berry (and into the yeast cell). The actual level that is acceptable varies with acid concentration.

3.3 Ammonium and Amino Acids

Yeasts depend on nutrients in the grape juice for their growth. One of the key elements for growth is nitrogen, which is used for the production of proteins (structural and enzymes). In grapes, nitrogen is found as ammonium ions (NH_4^+) and as free amino acids. Not all amino acids are metabolizable and one of the main amino acids found in grape juice, proline, is not. The principal assimilable amino acid is usually arginine.

Winemakers are concerned about these components because they determine whether there is sufficient nutrient to enable yeast growth throughout the fermentation and some are important precursors for yeast-derived flavor compounds. If there is insufficient nitrogen then fermentation may stop, leading to a stuck fermentation and high residual sugar (and a 'stinky' one owing to volatile, reduced sulfur compounds).

Additional nitrogen is usually added in the form of diammonium phosphate (DAP) to provide readily assimilable nitrogen and additional phosphorus, which is another essential element that may be deficient. However, increasingly, winemakers are adding more complex nutrients to ensure full flavor development and include a number of minor amino acids that are specific aroma precursors.

3.4 Sulfur

Sulfur in wine arises from two sources: (1) from the SO_2 that is added to manage the microbiology and oxidation–reduction state of the wine; and (2) from the decomposition of sulfur amino acids (cysteine and cystine). Metabolism of the amino acids and sulfates under anaerobic conditions may give rise to H_2S and other mercaptans which are objectionable. Development of a sulfurous aroma during fermentation indicates that the yeast is starving, metabolizing protein-based amino acids, and that the ferment may become sluggish or stuck.

Hydrogen sulfide is commonly removed by the addition of a small quantity of copper sulfate ($CuSO_4$). This forms the insoluble compound copper sulfide (CuS). It is highly effective at this but the addition of copper will also lower the concentration of vital sulfur-containing flavor compounds (thiols and mercaptans). Consequently, its use is best minimized or avoided if possible.

The level of SO_2 needs to be monitored carefully at the initial stages to ensure that the SO_2-tolerant yeasts and lactic acid bacteria can grow but other organisms (e.g. *Acetobacter*) cannot. It should not be so high that it bleaches the wine or is toxic to consume. The maximum limit that may be present in a wine is 250 mg/L, although most winemakers opt for minimum levels to reduce the risk of adverse effects in individuals who may be allergic to high levels of SO_2 (these people should drink aged wines or treat with a small quantity of hydrogen peroxide, available commercially in some bottle shops). SO_2 levels are usually higher in white than in red wines but decline as the wine ages in the bottle (Boulton et al., 1996; Iland et al., 2004; Ough, 1992). Winemakers quickly become adept at this analysis, which is repeated often throughout the process.

The issues raised in this introduction are those the reader should keep in mind when dealing with the details that follow, so that each section is seen in the context of the whole. It is important to choose the right product and the right process, and actively manage throughout. This means constantly paying attention to maintaining hygiene and an appropriate (and safe) environment, with the ultimate goal of achieving excellence. Chemistry and microbiology and technology are important, but the perception of the end-user is even more important. The ability of the winemaker to judge and communicate the path to that goal through thoughtful application of his or her human, sensory skills is what it takes to become a great winemaker.

Chapter 2

Flavors and Aromas in Foods and Beverages

Chapter Outline

1. INTRODUCTION

The objective of this chapter is to help the reader to develop an appreciation of the physiology of the interaction between a wine and an individual. Which sensory cues determine an individual's response? How can one make a consistent and reliable judgment of something that is arbitrary and individualized? This section comprises an introduction only. Other texts deal with the topic in far more detail and will be appropriate for those who wish to develop a professional level of expertise (Bakker and Clarke, 2011).

Taste, aroma and visual appearance are the primary means of assessing wine quality. Wine chemistry is a complementary tool that enables the winemaker to ensure quality and consistency: chemistry assists the winemaker to assess and resolve problems that may arise during the wine-making process. However important chemistry is to wine making, and this text is devoted substantially to chemistry, understanding the appeal of wine implies an understanding of elements of human physiology and psyche (Bartoshuk, 1993); wine is about people, not just chemistry. For a modern review of the chemistry underlying aroma in wine, see, for example, Roland et al. (2011).

A winemaker's ability is strongly dependent on his or her palate—so taste lots of wine and keep a diary. Practice putting your responses into words that can be applied consistently and communicated to others. Recall flavors and aromas from time spent in a kitchen because preparing and cooking food is a great way to develop sensory skills and language. This section is designed as an aid to that process. However, some care is required because of the level of alcohol in wine, so to minimize the risk of long-term damage to health, spit! Also, swallowing a wine will noticeably affect your response to the next wine. Even if you do spit, a lifetime of wine tasting may damage your teeth. Your dentist can advise you on how to protect your tooth enamel from leaching by wine acids if you taste many wines professionally and regularly.

As a winemaker or judge, it is important to be aware that people adapt to the aromas that surround them and quickly lose sensitivity to even the most malodorous of environments—it is the first perception that is critical (Figure 2.1). This also means that a winemaker can readily lose the ability to perceive a fault that might pervade a winery: *Brettanomyces*, for instance. Thus, it is good practice for a winery to bring in an outsider to assist with sensory analysis as this minimizes the risk of habituation, and for winemakers to visit other wineries. Likewise, perception is strongly influenced by previous experience and sometimes this can obscure judgment regarding flavors and aromas. So take care that the sequence of tasting does not inadvertently mask intensity of one or more important flavors or aromas; for example, a bitter flavor may mask the intensity of a subsequently tasted salty flavor (Bartoshuk et al., 1998).

Assessment comprises visual and organoleptic components (Table 2.1). Assessment is difficult because the elements are complex and we respond to each component and to the interaction of one component with another: aroma

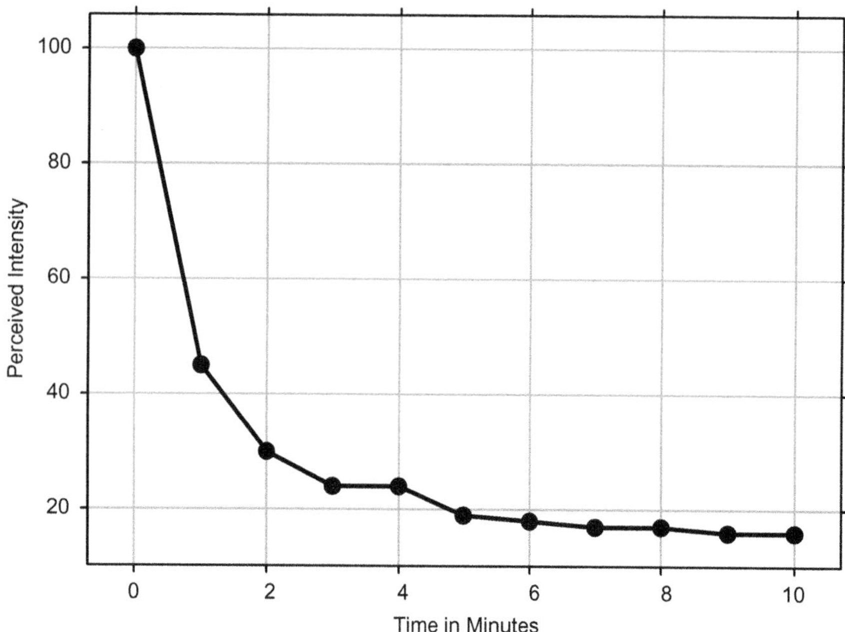

FIGURE 2.1 Graph of the decline in perception of intensity of 1 mg/L hydrogen sulfide over a period of 10 min. *(Data from Rankine, 1990)*

TABLE 2.1 Sensory Categories and Their Main Elements and Sequence of Assessment in Formal Wine Tasting

Visual (Eye)	Aroma (Nose)	Taste (Tongue)	Thermal (Mouth)	Tactile (Mouth)
Clarity	Floral	Acidity	Temperature (hot, cold)	Astringency
Color	Fruity	Bitterness		Drying
	Oxidized	Saltiness		Softness
	Pungent	Spiciness		Viscosity
	Vegetative	Sweetness		
	Woody	Umami		

interacts very strongly with taste, and even single components may stimulate unrelated receptors. Thus, while first impressions are important, it is also important to reflect and try to distinguish the individual elements of a taste/flavor sensation.

There is also a risk of making a poor assessment if cultural attributes of the marketplace are disregarded. Preference for sweet or what an Anglo-Saxon might regard as bland foods may need to be included in the assessment process depending on the cultural background of the market (Duffy and Bartoshuk, 2000). The flavors and aromas appreciated by different peoples of the world are often highly specific, culturally, although there are common links; sugar, for example, is generally liked by all.

An example of a cultural preference is that few people of European origin appreciate the delicacy of the flavor of the durian fruit because they are overwhelmed by its 'fecal-like' aroma when ripe. However, people such as the Thais who have been brought up with the fruit from childhood associate the smell with the pleasure of eating a fruit of great delicacy.

The task of making the comparison is made easier by the development of hedonic scales (Unpleasant versus Pleasant or Absent versus Present versus Strongly Present) ranked arbitrarily, usually from 0 to 9 or along an unmarked line.

The formal assessment of these attributes requires a panel of trained tasters operating under defined conditions. Such a panel comprises individuals who are interested, readily available, healthy, precise, discriminating and consistent in their evaluations, and who have been trained to recognize the important attributes. They come from all walks of life.

Regrettably, sensory skills deteriorate with age so it is best to learn these skills while you are young and to use young to middle-aged adults in tasting panels. Older people, with experience, remember, and the brain tends to fill in the gaps with what is expected, not necessarily objectively (Bartoshuk, 1989).

A professional taster may be able to discriminate between 2000 or more different compounds, whereas an untrained person can usually detect fewer than 1000. Sensitivity to particular compounds is also under genetic control (Bartoshuk et al., 1998) and this is also taken into account during the assessment process.

Conditions that are controlled in formal tastings include temperature, light, odor and distraction (Stone et al., 2012). Sometimes one needs to remove particular elements that may unduly influence a taster. For example, the product may be viewed under red light to minimize visual cues, allowing the taster to consider only the organoleptic aspects of a product.

Wine is a highly complex product with many thousands of known chemical components (Etiévant, 1991). It is known that the level and presence of these compounds is influenced by all conceivable influences: genetic, developmental, environmental and processing. The opportunity for variation is infinite. Complexity makes it impractical for wine assessment to be based solely on chemical tests (although chemical testing and fingerprinting are now used in Europe as a first level of discrimination and as a means of grouping wines to make the organoleptic analysis feasible on a national scale).

2. BASIC SENSORY ASSESSMENT

2.1 Taste

Those who spend time in a kitchen or who take a strong interest in cooking are usually skilled in this area for they are constantly tasting and smelling food ingredients and build a good vocabulary of sensory terms. There are five categories by which we formally or informally assess foods and beverages: sight, smell, temperature, touch (pain) and taste.

A list of the formal terms is presented in Table 2.1. Popular terms may not translate simply into these. The formal terms may be regarded as limiting because the senses interact and may flow into one another. For example, certain sugars may stimulate not only the sugar receptors but also the salty (reviewed by Bartoshuk et al., 2004). However, becoming aware of the interactions is critical to becoming skilled in perceiving and describing the finer details of the sensory attributes of wine (and food): it is the finer details that distinguish truly fine wines. The finer details may be perceived by the consumer at a subliminal (subconscious) level only, but are nonetheless understood by the brain at an emotional level, producing that feeling of heightened well-being that the finest wines and foods produce. The fine details of sensory perceptions are important whether the consumer can put them into words or not.

Some common responses to wine that reflect complex interactions are 'hotness' arising from high alcohol and falsely giving an impression of high temperature; 'pepperiness', which may be a blend of savory, umami and pain; 'mouth feel' as 'smooth/soft' or 'dry' (astringent), mediated through 'touch'; and viscosity, which may be sensed as 'thick/thin', creamy, high or low 'body'; while the terms 'fresh' or 'balanced' reflect the interaction between sweet and acid receptors but also have an element of 'feeling' or touch receptors. Becoming aware of each of the sensations that arise when eating or drinking is an art that requires much practice through reflection (cf. that of an elite athlete or artist). More complete lists of terms may be found in specialized texts (e.g. Bakker and Clarke, 2011; Peynaud, 1987; Rankine, 1990).

The sensory organs for visual and physical attributes are located in the eyes, tongue and mouth. Thus, astringency is assessed by the response of the lips, gums and the lining of the mouth (buccal cavity). In the case of wine and some fruits (e.g. persimmon), tannins react with the proteins in saliva and with the surface of the tissues lining the mouth and cause them to coagulate, giving a drying feeling. In the case of pineapple, kiwifruit, pawpaw and figs, an enzyme degrades these proteins (i.e. digests them) and causes a similar feeling.

The tongue is the primary sense organ for taste: sweetness, sourness, bitterness, saltiness and umami (Beauchamp, 2009). Umami is a savory flavor such as that of soy sauce which contains L-glutamate for which there are receptors on the tongue. While it was long considered that these senses were located at discrete areas on the tongue, this concept has been discredited: all parts of the upper surface of the tongue are sensitive to all tastes but thresholds and responsiveness to changes in concentration may vary depending on the density of the taste organs (Bartoshuk, 1993). Thus, the responsiveness of super-tasters to a doubling of a concentration of a test substance is about twice that of a non-taster or a medium taster (Bartoshuk et al., 1998).

There is a strong genetic component that is controlled by an allele T/t, where T is the dominant and t the recessive form. Individuals may be classified as 'non-tasters' (tt), 'tasters' (Tt) or 'super-tasters' (TT) based on the threshold of sensitivity to bitter or sweet substances (usually 6-*n*-propyluracil or quinine, although salt may be substituted as a guide (Bartoshuk et al., 1998). It is actually much more complex than this, with response being determined by the presence of three forms of T1 [T1R1 (umami), T1R2 (sweet), T1R3 (which works in tandem with R1 and R2)] and about 30 T2 (bitter), and the forms interacting to determine the nature of the response, pleasant as sweet or umami or unpleasant as bitter (Shi and Zhang, 2006). A serial dilution table is provided as an aid to determining genotype, taster, non-taster or super-taster (Table 2.2).

The density of taste buds varies nearly five-fold between individuals, from 33 to 156 per square centimeter, and largely accounts for the variation in sensitivity to bitter and indeed to other 'tastes' (Duffy and Bartoshuk, 2000). Each bud contains 50–100 taste receptor cells (Lindemann, 2001), but also pain and touch receptors. Thresholds for the basics tastes are provided in Table 2.2.

While the primary tastes are a consequence of food components reacting with specific receptors in the taste buds, spicy and savory flavors arise from a reaction between a chemical such as capsaicin, the active ingredient of chili peppers, and heat-sensitive nerves. They block the nerve and cause a sensation of 'heat'. Spicy and savory substances are mild anesthetics and may reduce sensation for a period but can also induce mild to strong physiological responses.

Studies of consumers and wine experts, including winemakers and judges, have demonstrated that super-tasters are more likely to be wine experts than either tasters or non-tasters (Hayes and Pickering, 2012). Super-tasters are also overrepresented among chefs (reviewed by Bartoshuk et al., 2004). It seems that super-tasters not only have a heightened response to taste but also are capable of discriminating among more flavors (reviewed by Hayes and Pickering, 2012). The difference in taste phenotype between experts and consumers is a matter of general concern to the wine and food industries: Will a consumer's preference mimic that of the artist, chef or winemaker? Possibly not.

2.2 Aroma

Taste and aroma interact strongly, with aroma usually predominating. A human has the potential to discriminate between millions of aromas through complex interactions at the receptor level in the nose (olfactory bulb) (Buck, 1993; Firestein, 2001). There is considerable genetic diversity, with individuals of African origin having about 10% more functional receptors than other groups of Americans; the diversity of sensory genes in humans is second only to that of the immune system (Menashe et al., 2003).

Aroma is perceived not only through inhalation, but more importantly through exhalation. Exhalation is important because the sensory organ, the olfactory bulb, is located at the base of the skull, toward the rear of the nasal cavity. It is directly connected to the brain and is an extension of it. Sense of taste is reduced profoundly by disorders and diseases that close the nasal passage, such as hay fever and colds. If your nose is pinched closed and your eyes are blindfolded, you may find it quite difficult to distinguish between a crisp apple and a potato! Thus, the capacity of the tongue alone is limited, although it is more resilient to aging than is the nose (Bartoshuk, 1989).

3. DESCRIPTIVE ANALYSIS

The concepts of descriptive analysis are presented here as a learning exercise rather than as a formal process such as may be undertaken in a quantitative descriptive analysis (QDA) (Stone et al., 2012). As a winemaker it is vitally important to train one's palate and to pay attention to the full scope of the sensory attributes of a wine and to its potential flaws. In the formal sense this usually involves not just familiarization with a range of flavors and aromas as is presented here. Discriminant analysis to determine

TABLE 2.2 Influence of Genetic Background on Taste of Three Bitter Compounds and Two Salts

Category	PROP		Quinine.HCl		Caffeine		NaCl		KCl	
	Threshold (M)	Magnitude	Threshold (M)	Magnitude	Threshold (M)	Magnitude	Threshold (M)	Magnitude	Threshold (M)	Magnitude
Non-taster	>0.001	10.0	0.00032	10.8	<0.01	7.8	<0.1	3.8	≅1.0	1.3
Taster	0.0001	8.8	<0.00001	12.5	≅0.0032	11.7	<0.1	10.1	≅0.32	3.4
Super-taster	<0.0001	15.3	<0.00001	15.8	<0.0032	13	<0.1	10	≅0.1	4.8

Note: Threshold = the average lowest concentration perceived by a subject; Magnitude = the perception the subject has of the highest concentration presented (on a scale of 1–20).
PROP = 6-*n*-propylthiouracil.
(Data interpolated from Bartoshuk, 1993).

sensitivity and repeated measures to assess the consistency of each participant's assessments are used as aids in selecting people for professional tasting panels. Sensitivity testing usually involves a series of duo–trio comparisons. These are also important techniques for sensory assessment during the wine-making and blending processes, where a winemaker may be working close to the lower limits of her or his sensory abilities.

Training related to Table 2.3 begins with exercises in which each participant assesses a range of samples of a wine that has been modified by the addition of a compound in which a single dominant sensory character is present (Noble et al., 1987; see also http://winearomawheel.com/). This may be done as a formal exercise in a class, among a group of friends or individually, and should be done over a period of separate classes, beginning perhaps

TABLE 2.3 Standardized Wine Aroma Terminology and Sensory Standards

Starter Set	Principal Term	2nd Level Term	3rd Level Term	Base Wine	Reference Composition (in 25 mL Unless Otherwise Specified)
✓	Floral	Floral	Linalool	White	1 mg (drop) linalool/100 mL white wine
✓			Orange blossom	White	Crushed orange blossoms
✓			Rose	White/red	1 mg 2-phenylethanol/150 mL white wine or crushed petals of 1 rose
			Violet	Red	Petals from 10 crushed violets
			Geranium	Red	Piece of ripped geranium leaf (10 mm × 10 mm)
✓	Spicy	Spicy	Cloves	White/red	Soak 1 whole clove for 10–20 min and remove
			Black pepper	Red	2–3 grains ground black pepper
✓			Licorice, anise	White/red	1 drop anise extract/50 mL wine
	Fruity	Citrus	Grapefruit	White	5 mL juice and small piece peel of fresh fruit
			Lemon	White	5 mL juice and small piece peel of fresh fruit
✓		Berry	Blackberry	Red	1–2 crushed fresh or frozen blackberries
			Raspberry	Red	1–2 crushed fresh or frozen raspberries
✓			Strawberry	Red	1–2 crushed fresh or frozen strawberries
			Blackcurrant/cassis	White/red	10 mL syrup from canned blackcurrants and 5 mL Ribena® (sweetened concentrate) or 10 mL cassis
✓		(Tree) fruit	Cherry	Red	10 mL brine from canned cherries
			Apricot	White	15–20 mL apricot nectar
✓			Peach	White	15–20 mL peach nectar or syrup from canned peaches
			Apple	White	Slice fresh apple, 5 mL apple juice
✓		(Tropical) fruit	Pineapple	White	2–4 mL freshly opened canned pineapple juice
✓			Melon	White	1 piece fresh ripe cantaloupe (20-mm cube)
			Banana	White	1, 10-mm slice fresh banana
		(Dried) fruit	Strawberry jam	Red	1 tsp strawberry jam
			Raisin	Red	5–8 crushed fresh raisins
			Prune	Red	1–2 mL prune juice
			Fig	White/red	½ fig or 5–10 mL brine from canned figs
		Other	Artificial fruit	Red	7–8 grains Tropical Punch Kool-Aid® or tropical fruit flavoring

(Continued)

TABLE 2.3 (Continued)

Starter Set	Principal Term	2nd Level Term	3rd Level Term	Base Wine	Reference Composition (in 25 mL Unless Otherwise Specified)
			Methyl anthranilate	White/red	2–5 mL Welch's® grape juice
✓	Herbaceous/ vegetative	Fresh	Stemmy	White/red	4 crushed grape stems
			Grass, cut green	White/red	1 shredded 20-mm blade of green grass
✓			Bell pepper	White/red	12-mm × 10-mm slice of bell pepper; soak for 30 min and remove
			Eucalyptus	White/red	1 crushed eucalyptus leaf
✓			Mint	White/red	1 crushed mint leaf or 1 drop mint extract
✓		Canned/cooked	Green beans	White/red	3–5 mL brine from canned green beans
✓			Asparagus	White/red	2–3 mL brine from canned asparagus
			Green olive	White/red	4–6 mL brine from canned green olives
			Black olive	White/red	4–6 mL brine from canned black olives
			Artichoke	White/red	2–5 mL brine from cooked artichoke
		Dried	Hay/straw	White	Several pieces of hay, finely cut (no wine)
			Tea	White/red	3–4 leaves of black tea
			Tobacco	White/red	3–4 flakes of tobacco (least aromatic possible)
	Nutty	Nutty	Walnut	White/red	1–2 walnuts, crushed (no wine)
			Hazelnut	White/red	1–2 hazelnuts, ground (no wine)
			Almond	White/red	1 drop almond extract/100 mL wine or 1–2 almonds ground (no wine)
	Caramelized	Caramel	Honey	White	5–8 mL honey
✓			Butterscotch	White/red	1 butterscotch Life Saver® and 1 cut Kraft Caramel®
✓			Diacetyl (butter)	White/red	1 drop butter-flavored extract/100 mL wine
			Soy sauce	Red	1–2 drops soy sauce
			Chocolate	Red	2–5 mL chocolate flavor or ½ tsp powdered cocoa
			Molasses	Red	1–3 mL molasses
	Wood	Phenolic	Phenolic	White/red	1 mg ethyl guaiacol
✓			Vanilla	White/red	1–2 drops vanilla extract
		Resinous	Cedar	Red	1 drop cedar oil or few shavings of cedar wood
✓			Oak	White/red	2–3 mL oak flavor (e.g. Oak Mor; Finer Filter Products, Newark, CA, USA)
✓		Burned	Smoky	White/red	1 drop smoky flavor extract/150 mL wine
			Burnt toast/ charred	White/red	1 small piece burnt wood in 200 mL wine
			Coffee	Red	2–4 grains ground coffee
	Earthy	Earthy	Dusty	Red	No standard available
✓			Mushroom	White/red	1 small mushroom, finely sliced/10 mL wine

(Continued)

TABLE 2.3 (Continued)

Starter Set	Principal Term	2nd Level Term	3rd Level Term	Base Wine	Reference Composition (in 25 mL Unless Otherwise Specified)
		Moldy	Musty (mildew)	White/red	Piece of mildewed cloth (no wine)
			Moldy cork	White/red	Pieces of moldy cork (no wine)
	Chemical	Petroleum	Tar	Red	1 drop roofing tar, let sit overnight in wine
			Plastic	White/red	Cut-up plastic tubing
✓			Kerosene	White/red	1 drop kerosene/150 mL wine
			Diesel	White/red	1 drop diesel/150 mL wine or 1 drop WD-40®/50 mL wine
✓		Sulfur	Rubbery	White/red	1, 10-mm × 5-mm piece of cut rubber tubing or bike tire (let sit several hours)
✓			Hydrogen sulfide	White/red	0.03 μM hydrogen sulfide in wine; or ⅛ yolk of hard-boiled egg (no wine)
✓			Mercaptan	White/red	0.08 μM ethanethiol in wine; or smell natural gas which contains ethanethiol
			Garlic	White/red	1, 5-mm × 5-mm piece crushed garlic in 150 mL wine; soak for up to 1 min and remove
✓			Cabbage	White/red	2–3 mL brine from boiled cabbage leaves
✓			Burnt match	White/red	Burn 1 wooden match, extinguish; when cool, add 150 mL wine
✓			Sulfur dioxide	White	250 mg/L sulfur dioxide
			Wet wool, wet dog	White	Small piece of heated wet wool (no wine)
		Papery	Filter pad	White/red	Soak 20-mm × 20-mm piece filter pad overnight in 100 mL wine
✓			Wet cardboard	White/red	Soak 20-mm × 20-mm piece cardboard overnight in 100 mL wine
✓		Pungent	Ethyl acetate	White/red	1 drop ethyl acetate/50 mL wine
			Acetic acid	White/red	2–5 mL vinegar/50 mL wine or 2 drops glacial acetic acid/50 mL wine
✓			Ethanol	White/red	10–15 mL ethanol/50 mL wine
		Other	Fishy	White/red	1 drop trimethylamine/50 mL wine or few grains anion exchange resin in hydroxide form (no wine)
			Soapy	White/red	Few flakes Ivory® soap or grains borax
			Sorbate	White/red	50 mg potassium sorbate
			Fusel alcohol	White/red	300 mg/L of 2-methyl-l-butanol and/or 3-methyl-l-butanol
✓	Pungent	Hot	Alcohol	White/red	40% v/v ethanol in wine or water
		Cool	Methanol	White/red	No standard developed
✓	Oxidized	Oxidized	Acetaldehyde	White/red	40 mg/L acetaldehyde or 5 mL sherry/25 mL wine
	Microbiological	Yeasty	Flor-yeast	White	No standard developed
			Leesy	White/red	No standard developed

(*Continued*)

TABLE 2.3 (Continued)

Starter Set	Principal Term	2nd Level Term	3rd Level Term	Base Wine	Reference Composition (in 25 mL Unless Otherwise Specified)
		Lactic	Sauerkraut	White/red	2–5 mL brine from canned sauerkraut
✓			Butyric acid	White/red	1 drop *n*-butanoic acid/100 mL wine
✓			Sweaty	White/red	1 drop isopentanoic acid/100 mL wine
			Lactic acid	White/red	No standard developed
		Other	Horsey	White/red	1 mg *p*-cresol/100 mL wine
			Mousey	White/red	0.5–1 mg 2-ethyl-3,4,5,6-tetrahydropyridine/L wine

(From Noble et al., 1987, with permission).

with examples from the range of classes (floral, fruit, vegetative, etc.) before moving to explore finer details within each class (e.g. the vegetative group).

While these standards may be readily prepared, commercial preparations are also available and some of these may be more extensive than the list provided here (e.g. www.winearomas.com, http://www.thegiftedman.com.au, http://www.wineenthusiast.com, http://pros.co.nz; especially spoilage standards).

The extensive list in Table 2.3 includes aromas that are normal and some that may indicate a 'flawed' wine if dominant, for example diacetyl. Likewise, 'kerosene' would normally be considered a flaw but may be the dominant and sought-after character in an aged Riesling wine (1,1,6-trimethyl-1,2-dihydronaphthalene, TDN). In general, the majority of the characters that lie within the chemical and microbiological categories are flaws.

In this process, the student, you, smells the 'enhanced' wine. It is good practice to waft the aroma toward your nose rather than 'burying' your nose in the vessel because it may be very unpleasant and overwhelm the senses!

The process should not be hurried. Each person should take care to learn to discriminate between the various samples within each group (floral, vegetal, etc.).

Do not swallow the wine, because to do so would impair your capacity to evaluate subsequent samples. You should drink water or refresh the palate with a bland biscuit and cheese.

Even an elementary appreciation of the complexity of a wine will greatly improve your capacity to enjoy wine, as well as and other fresh foods and beverages.

4. TASTING AND MAKING JUDGMENTS

This is a distinct process, one not to be confused with descriptive analysis. It is a form of rank analysis in which one assesses like products and ranks them in order of desirable characters.

In tasting wine it is necessary to choose a glass that is 'water' clear with no color present at all so that the apparent color of the wine is not altered. It is usual also for it to be shaped so that it narrows toward the top. An ISO XL5 glass is a common standard. Its shape allows the wine to be swirled and tends to concentrate the aromas. The glass must be absolutely clean as the meniscus should form normally and not be impeded by an oily surface. Do not wash glasses with a household detergent as these contain aromatic substances that may impair judgment; use metabisulfite or a commercial, odorless detergent. If available, rinse with distilled, deionized or rain water. Fill to no more than 25–30% of the volume so that the wine can be swirled to encourage aromas to concentrate in the headspace.

Temperature is important. Too cold ($<10°$ C) and faults may be masked and aromas diminished. Too warm ($>20°$ C) and aromas may be driven off, oxidation promoted, and the perception of sugar–acid balance and astringency altered (Peynaud, 1987). A cellar temperature of 15–18° C is good for both red and white wine, noting that body temperature in the mouth will raise the temperature beyond those values (Bakker and Clarke, 2011; Peynaud, 1987).

Most wine-tasting protocols judge first the color, clarity and apparent viscosity, then the aroma and lastly the taste or palate (e.g. Table 2.4). In a clean glass the wine will 'wet' the glass surface and produce a meniscus. If the glass is held at a slope, one can more readily assess color and, if present, browning due to oxidation (age), especially in red wines. Swirling the wine within the glass enables a judge to appreciate aspects of wine viscosity and the 'pearls' that run down the inner surface give an indication of viscosity, which is affected by alcohol concentration and by the presence of sugar alcohols (glycerol and mannitol among others).

TABLE 2.4 Example of a Wine-Tasting Score Card with a Total Score of 20, as Commonly Used in Australia

Name				Date	
Wine	Color & Clarity	Bouquet	Palate	Total	Comments
	3	7	10	20	
1					
2					
3					
4					
5					
6					
7					
8					
9					
10					

Note: Premium (Gold medal) 18.5–20; subpremium (Silver medal) 17–18.4; good commercial (Bronze medal) 15.5–16.9; commercial 14–15.4; subcommercial/faulty < 14.

One possible flaw in this approach is that aroma may be assessed more sensitively via the mouth, and on exhalation, rather than via the nasal vestibule (front opening). In addition, enzymes within the mouth, including those from bacteria, may release aromas (although the most important enzyme in saliva, an α-glucosidase, is inactive because most aromas are bound to a sugar by a β-arrangement and thus are not affected by that enzyme).

The taster should look for color, balance—sweetness/acidity, bitterness versus astringency—length of flavor, pleasant versus unpleasant after-tastes, viscosity, apparent 'heat' (alcohol) and aroma. While the taster may note the presence of known aromas, such as tropical fruits in Semillon, what counts is the overall impression, the taster's integration of the aromatic or taste attributes of the wine. It is important to taste many examples of wines made from your cultivars of choice in order to become familiar with the nature of the characters of that cultivar (blend) and the range of their expression. A good memory is a great asset. Take notes as this will help you to remember a wine and an association.

Color provides strong visual clues to the style and quality of the wine. White wines vary from pale straw to a deep straw or even to gold, perhaps with greenish tinges. In general, the darker the color of the wine, the greater its oxidation or age, or possibly it is a particular style. The wine may be made from overripe or dried fruit to achieve a 'late-pick', 'spätlese' or Vin Santo style wine, for example.

Red wines vary in color and color depth depending on cultivar, style, maturity, viticultural practice, acidity (pH) and age. Cultivar and wine-making practices have a major impact, with some, such as Pinot and Grenache, being less dense owing to a relatively low level of color-stabilizing tannins present compared with Cabernet Sauvignon, Shiraz (Syrah), Merlot or Petit Verdot.

Once you have learned to appreciate the range of characters it is appropriate to extend the exercise to wines selected to be distinctive, ranking each attribute on a scale of 0–9. In this process, you not only smell the wine but also taste it. Note the sugar–acid balance, the astringency, bitterness, if present, and the fore-taste, the middle-taste and the after-taste. After-taste is particularly important. Fine wines will 'live' in the mouth, have prolonged pleasant flavor and finish cleanly. Lesser wine will finish quickly ('short') and may finish with a bitter or somewhat distasteful flavor. Once this is done, you will be able to produce a rose diagram showing the distinguishing characteristics of each wine and each cultivar (Figure 2.2). Rose diagrams are used to bring together values from many variables into a single diagram. These can be overlaid and the sensory aspects of individual wines compared visually (you may change the axis labels according to your needs—there is no standard).

When purchasing and assessing commercial wines, one should be aware of the finer details of the wine-making practices adopted. Thus, in Australia, a wine may be labeled as a single cultivar, but contain as much as 15% of another. These additions can dramatically alter the characteristics of the wine; for example, a 1% addition of 'Orange Muscat' will completely alter the volatile aroma range, as will a small addition of Viognier as is common practice in the Rhône region of France. The winemaker should take every opportunity to visit other wineries, talk with their winemakers and taste their wines through all stages of production. Only then can one develop a truly educated palate.

Formal tasting protocols are now well developed for research on food and beverage preference (Bakker and Clarke, 2011). Some elements of these processes are appropriate for the winemaker, especially duo–trio tests that are designed to assist the taster in making fine distinctions between wines. These would be appropriate when making final adjustments to wine such as determining the amount of a fining agent to add, a pH adjustment or a blend (Peynaud, 1987; Rankine, 1990).

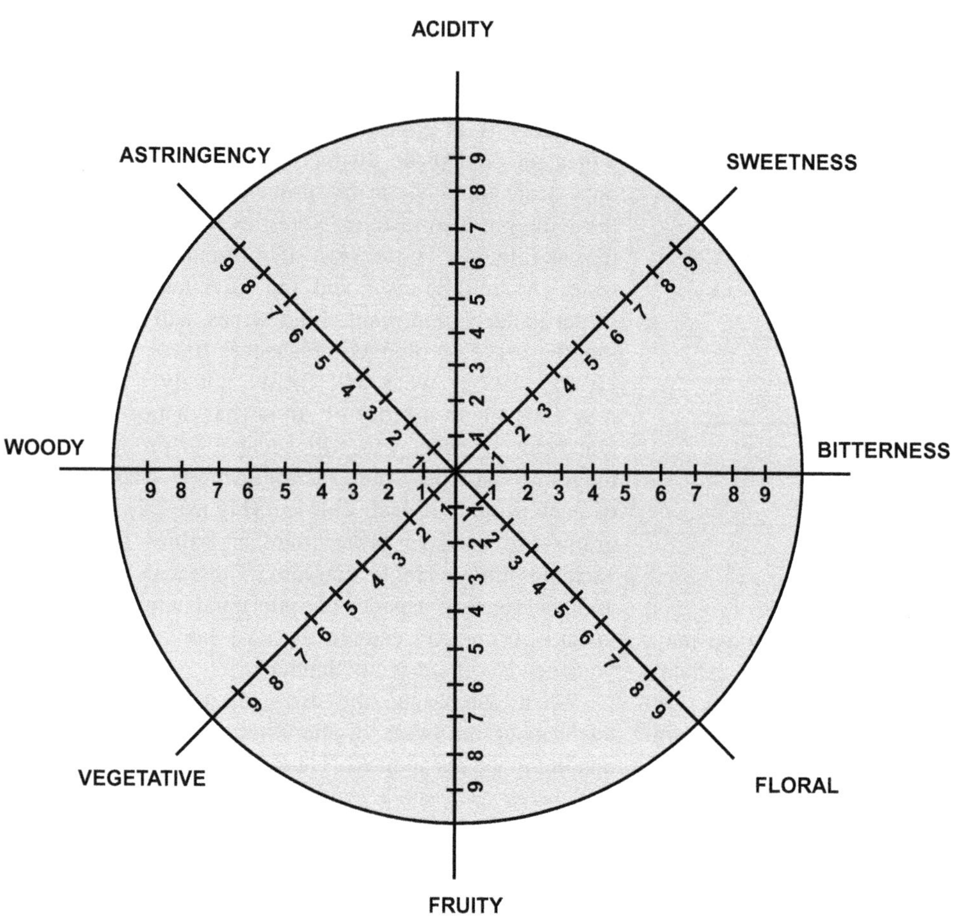

FIGURE 2.2 Example of a tasting rose. The judge ranks a wine in terms of its intensity of each character and marks each axis with that rank, 0–9 (absent to exceptional). Once completed, a line is drawn linking the points on each axis to produce a sensory rose that characterizes the wine.

5. OTHER TASTE AND AROMA TESTS

5.1 Discrimination Tests

Three forms of discrimination test are relevant to the winemaker: paired, duo–trio and triangle comparisons. In the paired comparison, the test is to ascertain whether the two sample wines differ in a particular way: 'Which sample is more sulfurous, more acidic or sweeter?' It concerns relativity. In duo–trio tests, one sample is known, a reference or control; one of the unknowns is the same as the control and one is unknown. The question asked usually is, 'Which of the two unknowns is closer to the standard or control?' The triangle test has no reference but two of the samples are the same. The question asked is, 'Which sample is different?' It is in fact a combination of three paired comparisons and makes that test more robust by forcing repetition, but at the cost of complexity (i.e. there are more combinations in terms of comparisons and order of assessment).

In practice, however, winemakers may not adhere firmly to these simple tests, despite their rigor and their statistical attributes being well understood (e.g. Stone et al., 2012). Practical wine making may involve ranking a series of samples in order of magnitude for a particular attribute such as bitterness or acidity and comparing these with a known control. Corporate wine making may adopt a more rigorous approach, especially where a particular product is produced annually, such as some well-known non-vintage champagnes. To further your knowledge on formal sensory tests see a text such as that by Stone et al. (2012), which also discusses the psychological and statistical aspects that are important in assessing experimental wines. These tests are also discussed in a more general wine context by Peynaud (1987) and Rankine (1990).

5.2 Mapping the Tongue

The density of taste buds on the tongue may be readily mapped using a dilute solution of a blue, food-grade dye. The taste-bud papillae are revealed as pink spots on a blue background. Typically, non-tasters have a sparse distribution pattern, super-tasters a tiled, dense pattern and tasters are intermediate. If you wish, you may test the now discredited theory that the tastes map to discrete areas on the tongue using a cotton bud or paper taper dipped on one of the following four solutions: 1 g/L tartaric acid, 20 g/L sucrose, 10 mg/L quinine sulfate and 8 g/L sodium chloride.

TABLE 2.5 Preparation of a Dilution Series for Determining Bitter or Salty Taste Threshold

Series Number	6-*n*-Propylthiouracil *M*	$\log_{10}M$	mg/L
1	**0.000032**	**−4.50**	**5.39**
2	0.00010	−4.00	17.0
3	**0.00032**	**−3.50**	**53.9**
4	0.00100	−3.00	170
5	**0.00316**	**−2.50**	**539**
	NaCl (*M*)	$\log_{10}M$	g/L
1	**0.01**	**−2**	**0.584**
2	0.032	−1.5	2.92
3	**0.10**	**−1**	**5.84**
4	0.32	−0.5	29.2
5	**1.0**	**0.0**	**58.4**

Note: Comprises two parallel, 10-fold dilution series (bold is set 1 and normal is set 2). Each set may be extended downward if necessary.

5.3 Taste Sensitivity Testing

5.3.1 Bitterness

Sensitivity to bitterness may be determined more exactly by tasting a closely spaced series of concentrations (Table 2.5). The relationship between concentration and taste magnitude tends to be non-additive; therefore, sensitivity tests use geometric series, usually doubling.

Use a taper (paper) and apply to the tongue of a partner to determine the threshold for sensitivity to the 'bitter' taste. Allow the taster to rinse his or her mouth with clean water between tests. Typically, people are either non-tasters (i.e. not very sensitive), tasters or super-tasters (Reed et al., 1995). An alternative is to use one of the higher concentrations [0.0018 or 0.0032 M 6-*n*-propylthiouracil (PROP)] and rank response from nil to intense (e.g. 1−9) (Bartoshuk, 1993). Quinine.HCl or NaCl may be used instead of PROP (Table 2.5). Because responsiveness may change more than threshold, the latter test may be more reliable. A reference response, such as a sound or tone, may be used to assist in ranking (Marks et al., 1988; Moskowitz, 1971) but this is not practical in most circumstances.

5.3.2 Sulfur Dioxide

Assessing sulfur dioxide (SO_2) is important because it is ubiquitous in the winery. Take care to determine whether individuals are either allergic or prone to developing allergic responses to SO_2 (see Section 2.1 on metabisulfites in Chapter 4). Such an individual may require urgent medical attention if exposed. This test is an extension of that in Table 2.3 and involves assessing a serial dilution of SO_2 in either wine or potassium bitartrate (1 g/L). Rankine (1990) suggests a two-fold series from 0 to 160 mg/L SO_2: 0, 20, 40, 80, 160, but other ranges may be devised. Note that if using wine about half of the added SO_2 will be bound, so adjust the levels accordingly. Assess, in random order, both the wine series and tartrate series so that the interaction with wine aroma may also be assessed. In conducting this test it is important that the taster wafts the aroma to the nose and does not immediately bury the nose in the glass! The higher of the levels may be intensely irritating even if the subject is not allergic to SO_2.

Chapter 3

Wine Chemistry

Chapter Outline

1. ORGANIC ACIDS, PH AND THEIR ROLE IN WINE MAKING

1.1 Introduction

Grapes and wine contain naturally occurring organic acids, formed by plant and microorganism metabolism. These play a key role in cell biology and in foods and beverages derived from living organisms, and in the safety of those foods and beverages. Plants contain the same basic organic acids as are found in animal cells and which play a key role in energy metabolism. A difference is that plants have a vacuole and frequently store particular variants as secondary rather than as primary metabolites: malic and citric acids are common secondary metabolites in higher plants. Tartaric acid is peculiar to a few species including *Vitis* and *Musa* (banana).

These 'secondary' organic acids may function as counter-ions—anions that balance charge and electrical properties across membranes, especially the tonoplast and the vacuole it contains. They may also function as anti-feeding agents and toxins (e.g. oxalic acid) or have other specialist functions. Thus, oxalic acid sequesters excessive calcium and helps to ensure that the calcium signaling pathways function correctly. Salicylic acid is a regulatory compound in plants but also serves as a useful pharmacological role in humans. It too is overproduced as a secondary metabolite in particular plants such as willow (willow bark is a natural source of 'aspirin').

A secondary metabolite is one whose function is not clearly related to primary metabolism, the production of energy and structural macromolecules. The presence of high levels of certain organic acids is one of the reasons that we can make wine from grapes, the others being the high level of fermentable sugars and the presence of astringent phenolics that provide 'mouth feel'. Winemakers routinely measure three aspects of the chemistry of fruit and wine: pH, titratable organic acids and sugar content.

1.2 Physical Chemistry

Acids are members of a class of compound containing one or more electrostatic bonds. They comprise balancing positive and negative ions. In an acid, the positive ion is hydrogen. Salts differ only in that salts of strong acids and strong bases (hydrochloric acid and sodium hydroxide) are neutral, i.e. when dissolved the pH is not altered. The release of hydrogen ions when an acid is dissolved causes the pH to become lower to a degree that depends on the strength of the ionic bond between the hydrogen atom and the remainder of the molecule. Salts of weak acids and/or bases may alter the pH when dissolved.

- **Acid:** Commonly, an acid is a substance that when dissolved in water ionizes to produce a proton (H^+) ion.

  Acids taste sour, turn litmus paper red and react with bases to produce a salt. Thus, hydrochloric acid, a strong acid, reacts with sodium hydroxide, a strong base, to produce sodium chloride (a neutral salt) and water:

$$HCl + H_2O \Leftrightarrow H_3O^+ + Cl^-$$
$$NaOH + H_2O \Leftrightarrow Na^+ + OH^- + H_2O$$
$$H_3O^+ + OH^- \Leftrightarrow 2H_2O$$

  Thus, by addition:

$$HCl + NaOH \Leftrightarrow NaCl + H_2O$$

- **Dissociation constant:** A value that defines the ratio of the ionization products with respect to the undissociated parent compound (for acids this is referred to as K_a).

Thus, using the example above: $K_a = [H^+] \cdot \frac{[Cl^-]}{[HCl]}$, where the square brackets indicate concentration (moles/L).

For weakly dissociating compounds the dissociation constants may be quite small and thus for convenience chemists have introduced the concept of the minus log of the dissociation constant (cf. pH) to provide positive values which are numerically small, $pK_a = -\log K_a$: the larger the value the weaker the acid (i.e. the more strongly the proton (H^+) is attracted to the compound in question).

Note that this is a logarithmic scale and an acid with a pK_a of 2 is 10-fold stronger than one with a pK_a of 3.

Acids with more than one proton have more than one pK_a value: citric acid is a tricarboxylic acid, i.e. it has three (−COOH) groups, each of which is increasingly weaker than the other, labeled 1, 2, 3 (Table 3.1, Figure 3.1). To understand how this affects pH we need to consider the concept of buffering:

- **Buffer**: A weak acid that moderates the effect of addition of an acid or base on pH. Weak acids are most effective in buffering change in pH within one pH unit of their pK_a.

Table 3.1 gives the dissociation constants for the principal acids in wine and juice. The list is far from exhaustive but tartaric and malic acids predominate. The impact of these on pH buffering is demonstrated in Figure 3.2, which shows the buffering range for each of the association constants superimposed on an idealized titration curve for a juice.

These acids are so widely known that their former names continue to be used, rather than the International Union of Pure and Applied Chemistry (IUPAC) naming convention. Under that convention, tartaric acid is known as 2,3-dihydroxydibutenoic acid while malic acid is referred to as hydroxysuccinnic acid and citric acid as 2-hydroxypropane-1,2,3-tricarboxylic acid.

TABLE 3.1 Common Classes of Acid Found in Wine and Their Dissociation Constants (pK_a) in Water

Acid	pK_{a1}	pK_{a2}	pK_{a3}	Form in Wine
Strong inorganic				
Hydrochloric (HCl)	≪1			Fully dissociated ions
Phosphoric (H_3PO_4)	1.96	6.7	12.44	
Sulfuric (H_2SO_4)	1.0	1.6		Fully dissociated ions
Sulfurous (H_2SO_3)	1.77	7		Bisulfite ion
Strong organic				
Citric ($C_6H_8O_7$)	3.14	4.75	5.4	Partly ionized
Formic (CH_2O_2)	3.77			
Lactic ($C_3H_6O_3$)	3.86			Partly ionized
Malic ($C_4H_6O_5$)	3.40	5.20		Partly ionized
Salicylic ($C_7H_6O_3$)	2.97			
Tartaric ($C4_3H_6O_6$)	2.95	4.25		Bitartrate ion
Weak organic				
Acetic ($C_2H_4O_2$)	4.76			Largely undissociated
Benzoic ($C_7H_6O_2$)	4.20			Largely undissociated
Butyric ($C_4H_8O_2$)	4.82			
Propanionic ($C_3H_8O_2$)	4.85			
Succinic	4.2	5.56		
Weak inorganic				
Carbonic (H_2CO_3)	6.37	10.32		Undissociated
Hydrogen sulfide (H_2S)	7.0			
Phenols				
Polyphenols	7–10			Undissociated

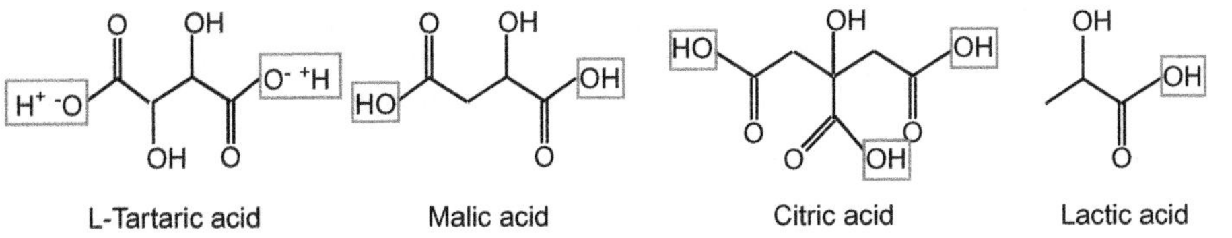

L-Tartaric acid Malic acid Citric acid Lactic acid

FIGURE 3.1 Structure of the four principal wine acids. In carboxylic acids, the ionizable hydroxyl is that which terminates the chain adjacent to the aldehydic oxygen (double bond) as in the tartaric acid example.

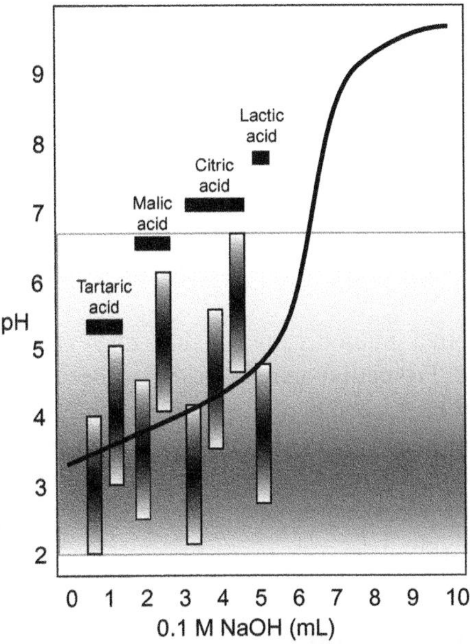

FIGURE 3.2 Graph of volume of 0.1 M NaOH plotted against pH for an idealized grape juice with the buffering range for each of the acid groups of the major grape (and wine) acids. The shaded zone indicates the buffering capacity of the juice observed.

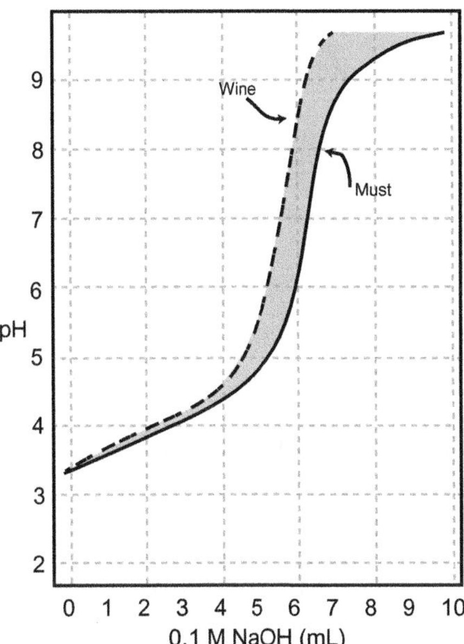

FIGURE 3.3 Comparison of the titration curve of the juice of berries and of the wine made from those berries following a malolactic fermentation. *(Data from Ribéreau-Gayon et al., 2006b, with permission).*

Observe that the graph turns between about pH 4.5 and 5.0, which is about the upper limit of the buffering associated with pK_{a2} of tartaric acid but below that of malic acid. The conclusion, in this case, is that there is little malic acid and certainly very little citric, otherwise the slope would have remained fairly linear until the midpoint of their respective buffering ranges had been exceeded.

The pH of a solution containing a weak acid and its salt is a function of the pK_a of the substance and the concentration of the ionized form (A^-) and the undissociated compound (H·Acid *cf. lower line of eqn 1* $[H{\cdot}A]$ known as the Henderson–Hasselbach equation):

$$\mathrm{pH} = \mathrm{p}K_a + \log\left(\frac{[A^-]}{[H \cdot A]}\right) \qquad (1)$$

Thus, pH is controlled by the association constants of the acids and their concentration. Adding another acid or base will shift the balance in one direction or the other. According to Ribéreau-Gayon et al. (2006b), it is not yet feasible to predict the pH of a wine or the impact of an addition because of this complexity, although progress is being made in this area.

Figure 3.3 shows that it is easier to change the pH of a fermentation at the end than at the beginning. The difference between this and Figure 3.1 is due, presumably, to the metabolism of malic acid to lactic acid, which buffers at a lower pH and reduces the overall acidity available as a buffer ([malate^{2-}] 2[H^+] vs [lactate^{1-}] 1[H^+]) as a contribution to the denominator of Equation (1). However, a winemaker will usually make the main adjustments as early as possible in order to maximize the 'integration' of the acid with the remainder of the products—it is easy to imagine the extremely complex interactions between the myriad ionic ions and compounds that exist in a juice/wine; it is not as simple as adjusting the pH of a standard solution.

Tartaric acid is present as a mixture of tartaric acid, potassium bitartrate and dipotassium tartrate. The pH and the concentration of potassium ions determine the actual

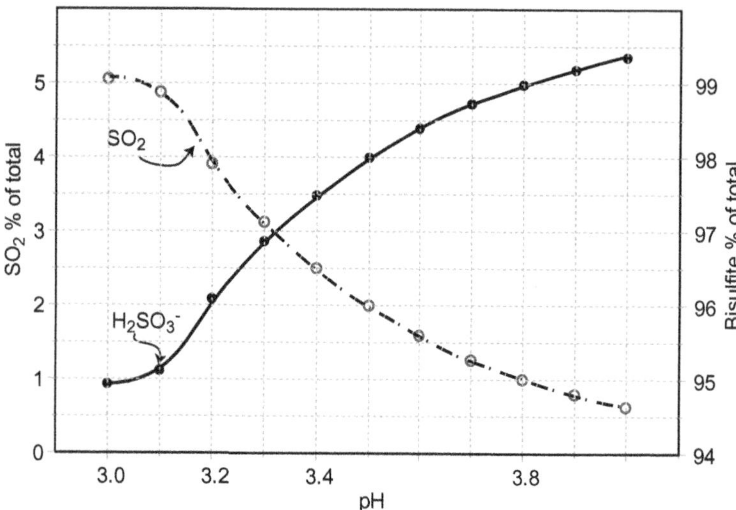

FIGURE 3.4 Impact of pH on the equilibrium balance of molecular SO_2 and bisulfite ion in water, i.e. no acetaldehyde to bind the SO_2. *(Data from Ribéreau-Gayon et al., 2006b).*

proportions. In fruit, it is present as a 'super-saturated solution', one that depends on the integrity of the complexes that hold it in solution.

In wine these complexes are modified or absent and so the salt crystallizes out over time, forming crusts on the bottom of the bottle. To avoid this problem, winemakers usually chill the wine to a temperature of −2 to −4°C and add finely ground potassium bitartrate crystals as a nucleating agent. For this reason, you cannot expect to recover all the tartrates if you freeze samples before carrying out analyses—if frozen they must be heated to redissolve the salts.

1.3 Preservative Role

Organic acids and salt were the mainstay of those who prepared food for storage prior to the invention and common use of refrigeration. Vinegar is about 1 mol/L acetic acid, has a pH of about 2.4 and is still widely used as a preservative. Wine has a pH that is above this but is still too low for growth of the vast majority of microorganisms; for example, *Escherichia coli* cannot grow at a pH below about 5 while *Streptococci* require a minimum pH of 6.5.

No disease-causing microorganism can grow at the pH of grape juice and few spoilage or other microorganisms can grow. Therefore, the content of organic acids as secondary metabolites confers an antimicrobial attribute to the juice. Only those microorganisms specifically adapted to low pH, so-called acidophiles, can grow. The yeast *Saccharomyces cerevisiae* is one of these, but even it struggles at a pH below about 3.2.

1.4 Impact on Sulfur Dioxide

While the importance of protecting wine against oxidation may be minimal at crushing, protection against wild, acid-adapted, microorganisms that may cause faults is important. Protection against oxidation becomes critically important at the completion of a ferment, especially so in white wines, but even in red wine. In all cases it is the molecular sulfur dioxide (SO_2) that is the active component and the concentration of this is strongly influenced by pH (Figure 3.4).

1.5 pH and Color

The acidity of a wine also has an important impact on color intensity and hue. Anthocyanins may exist in a number of forms (Mazza and Miniati, 1993). Only two of these are colored and one only with the color we associate with wine. This principle is familiar to growers of hydrangea, who alter the color of the flower bracts by changing soil pH (and thus cell vacuole pH); acid soils give pink bracts and alkaline soils blue. The pink is the cationic flavylium ion (plus a proton, H^+) and the blue is the quinonic form. The other two forms are colorless: the chalcone and the carbinol form. The change is dramatic over the pH range found in wine (Figure 3.5). As the pH rises, color density declines and the hue shifts toward purple (red plus blue) (Figure 3.6).

1.6 Role in Fermentation

While the sugars glucose and fructose are the primary fermentable substrates, malic acid can also be fermented, but to lactic acid rather than ethanol and by a bacterium (*Oenococcus oeni*, syn. *Leuconostoc oenos*, a *Lactobacillus*). This is especially important in red wines because they may not be sterile filtered before bottling. To avoid the risk of fermentation occurring in the bottle, this fermentation is carried out in the winery. Some white wines are also taken through this process but mainly for reasons of

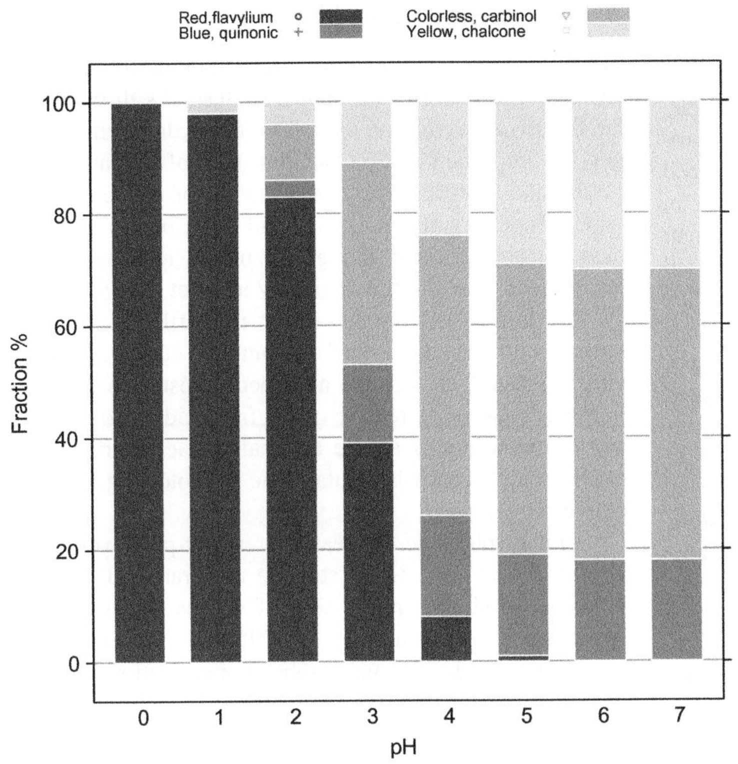

FIGURE 3.5 Influence of pH on the equilibrium of the four major forms of anthocyanins found in wine. *(Data from Ribéreau-Gayon et al. 2006b).*

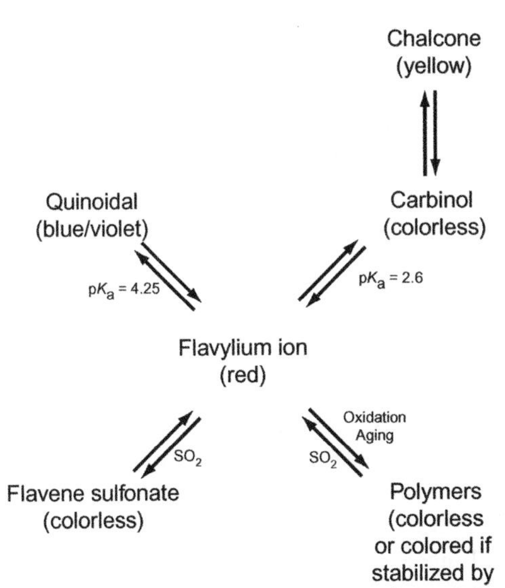

FIGURE 3.6 Protonation of anthocyanins and changes in electron charge distribution that influence pigment intensity (ε) through the formation or otherwise of a flavylium ion. *(Ribéreau-Gayon, 1964; Mazza and Miniati, 1993; Boulton, 2001).*

aroma (this is true also of red wines). As lactobaccilli struggle at a pH below pH 3.5, the pH of wine is maintained as near to this value as practicable and possibly adjusted up or down on completion.

1.7 Adjusting pH

As indicated previously, it is easier to adjust pH after the completion of the wine-making process than before, all things considered, but this should be reserved for fine-tuning before bottling. Any major change should be done at the beginning to ensure proper integration. However, the pH should be below the pK_{a2} of tartaric acid when a wine is cold-stabilized so that as the concentration falls with precipitation of solid potassium bitartrate, the pH falls rather than rises. For further information on this read the section on cold-stabilization of wine (Chapter 10, Section 8).

It is not yet possible to predict the final pH of a wine and so it is necessary to be conservative, another reason for making the adjustment prior to bottling rather than earlier (Boulton et al., 1996; Ribéreau-Gayon et al., 2006b). While winemakers learn from experience how much acid or alkali to add, the simplest way of determining the amount to add is to titrate a sample of the must or wine with a solution of known concentration of tartaric or other acid (often citric so as not to destabilize the wine prior to bottling, but taking care, as citric acid is fermentable).

1.8 Aroma and Flavor

Aroma in wine is usually present in an odorless form as conjugates of sugars, and these are present in quite high concentrations, certainly higher than the fruitiest wine. Many of the important odoriferous compounds in wine are formed during senescence of the berry, that is, they are breakdown products of other molecules. For example, the *nor*-isoprenoids, famously responsible for fruity aromas in Chardonnay but present in all wines, are formed by degradation of xanthophylls (related to carotene) in the chloroplast. This chemical is also the main ingredient of Boronia oil. Such breakdown products are detoxified in the plant by the addition (bonding) of sugars as they are exported from the chloroplast.

The sugar–product complexes are bonded by glycosidic bonds and these are susceptible to acid-catalyzed hydrolysis. This releases the aroma compound. It is primarily this reaction that leads to changes in aroma in wines as they age, although oxidation also plays a role. Examples are terpene and *nor*-isoprenoid sugar esters. Other important aromas arise from the degradation of proteins, peptides and amino acids.

On the other hand, acid-based condensation contributes to the formation of colloidal particles or aggregates of highly condensed tannins. These are brown and eventually reach a size at which they precipitate leaving an almost clear, red–brown liquor in aged wines.

2. PHENOLICS, DIVERSITY AND THEIR ROLE IN WINE MAKING

2.1 Introduction

Phenolics are essential ingredients in wine. They are a source of flavor, of physicality, of color and of reducing potential: wine isn't wine without them. Understanding and managing phenolics is an essential aspect of the production of fine wine. All phenolics are unsaturated cyclic structures and have hydroxyl groups (one to three) at the base of the ring (in the *ortho* and *para* positions). This feature lies at the core of their chemistry. Hydroxyls at these locations are readily oxidized forming highly reactive products known as quinones and, usually, hydrogen peroxide in the process (see Section 3.2 on oxidation, below). Also, the hydroxyl groups serve as attachment points for other molecules, principally sugars and organic acids, in processes known as glycosylation and esterification (Figure 3.7). This is their normal state in the plant.

Grapes contain two broad groups of phenolics: non-flavonoids and flavonoids. These are found especially in the skin and the seed but also in stems and leaves. Flavonoids are not usually found in the flesh but anthocyanins may be in a few red wine cultivars (Tienturier varieties, e.g. Pinot tienturier, Alicante Henri Bouschet). Oak is a source of distinctive phenolics termed the ellagic or hydroyzable tannins.

2.2 Non-flavonoid Phenolics

The non-flavonoid phenolics are important in all wine styles as the majority occur in the flesh. As discussed in the following section on oxidation, the hydroxycinnimates play an important role in the first step of oxidative processes and contribute to the straw color of aged white wine. They may also make a small contribution to bitterness but are generally thought to play a minimal role in comparison with the flavonoids (Vèrette et al., 1988).

The hydroxycinnimates occur primarily as esters of tartaric acid, principally caftaric but also coutaric and fertaric (Figure 3.7). Various non-specific esterase enzymes in musts lead to the release of the free acids, caffeic, coumaric and ferulic. Caftaric and caffeic acids are readily oxidized and react with glutathione to protect against oxidative browning.

Another phenolic of great interest but perhaps of little importance in wine is the stilbene resveratrol. This compound is produced in the skin of grapes in response to fungal infection (Jeandet et al., 1995). It is therefore present in red wines in much higher concentrations than in white but at levels that are unlikely to be physiologically effective as a protection against disease in consumers, although there has been great community and scientific interest in the prospect.

2.3 Flavonoids

Flavonoids and their many derivatives are essential in wine, especially red wine. In particular white wine styles they add astringency, bitterness and mouth feel, and contribute to color. Best known are the anthocyanins (Figure 3.7). They bring deep color to wine and subtle differences in color between the various forms assist in distinguishing one cultivar from another (Figure 3.7) as well as age, pH and oxidation state of a wine. It has been suggested that *Vitis vinifera* may be classified into five groups of cultivars according to the pattern of anthocyanins in the berries and type of substitution in addition to glucose at position 3 on the C ring (Wenzel et al., 1987). Malvadin and its derivatives generally dominate except in the 'Trollinger' group that includes Pinotage and Sangiovese, in which cyanidin tends to predominate. Figure 3.7 is a simplified view of anthocyanin structure. Acetates and coumerate esters are also common (reviewed by Mazza and Francis, 1995). Anthocyanins also participate in copigmentation reactions in wine, enhancing and shifting in hue according to pH and companion molecules (see below).

While anthocyanins confer color and help to stabilize procyanidins (tannins), it is the colorless to light yellow flavonoids that arguably have the greatest impact on wine, as anthocyanins are tasteless. The flavonoids are many

Hydroxycinnamates

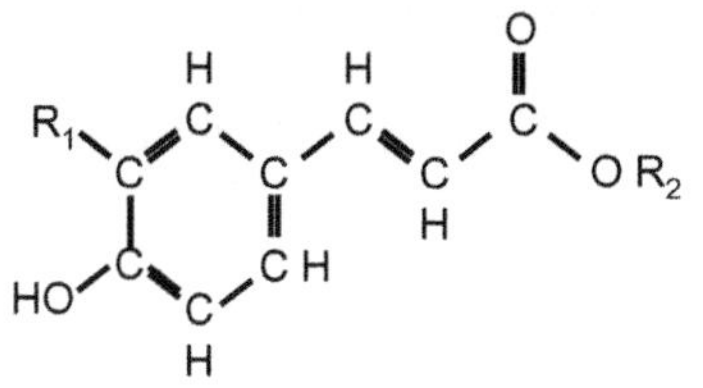

R_1	R_2, -H	R_2, -tartaric ester
-H	Coumaric acid	Coutaric acid
-OH	Caffeic acid	Caftaric acid
$-OCH_3$	Ferulic acid	Fertaric acid

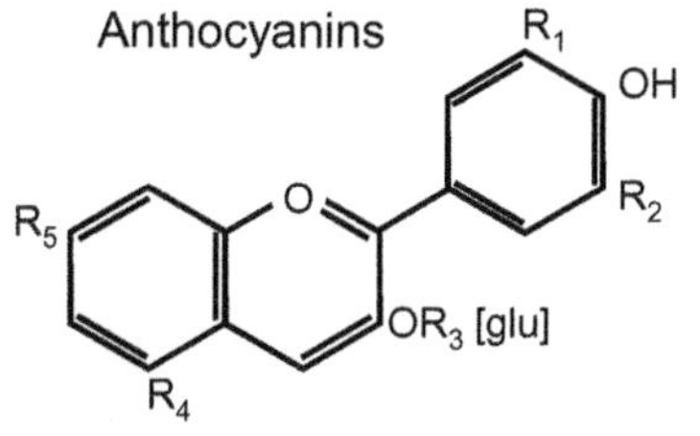

Flavanols [Catechins]

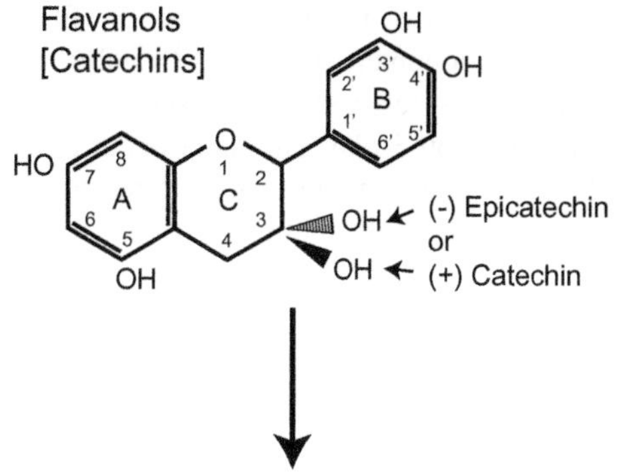

(-) Epicatechin or (+) Catechin

Procyanidins (tannins)

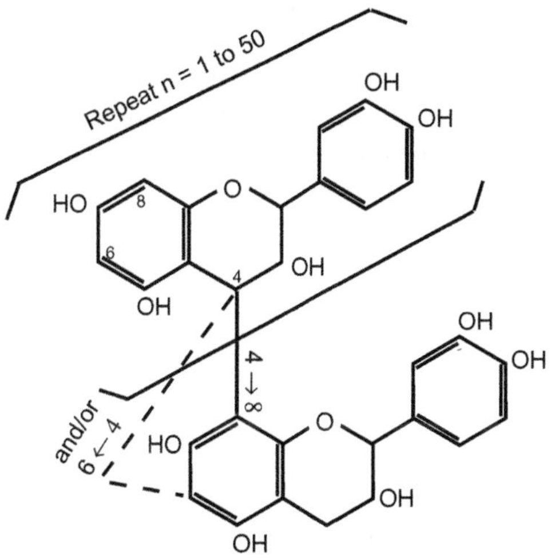

Repeat n = 1 to 50

4→8

and/or 6←4

Procyanidin	Arrangement
B1	(-) epicatechin 4→8 (+) catechin
B2	(-) epicatechin 4→8 (-) epicatechin
B3	(+) catechin 4→8 (+) catechin
B4	(+) catechin 4→8 (-) epicatechin
B5	(-) epicatechin 4→6 (-) epicatechin
B6	(+) catechin 4→6 (+) catechin
B7	(-) epicatechin 4→6 (+) catechin
B8	(+) catechin 4→6 (-) epicatechin

Flavonols

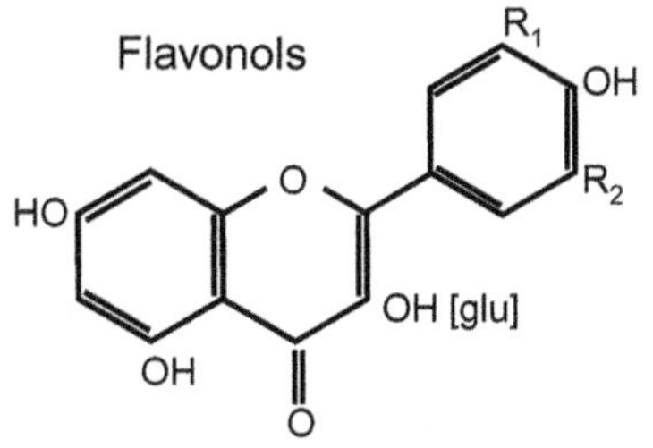

Flavonol	R_1	R_2
Kaempferol	-H	-H
Quercetin	-OH	-H
Myricetin	-OH	-OH
Isorhamnetin	-H	$-OCH_3$
Laricitrin	-OH	$-OCH_3$
Syringetin	$-OCH_3$	$-OCH_3$

Anthocyanins

Anthocyanin	Color	R_1	R_2
Cyanidin	Orange-red	-H	-OH
Delphinidin	Bluish-red	-OH	-OH
Peonidin	Orange-red	$-OCH_3$	-H
Petunidin	Bluish-red	$-OCH_3$	-OH
Malvidin	Bluish-red	$-OCH_3$	$-OCH_3$

Anthocyanin	Malbec	Merlot	Cabernet Sauvignon	Nebbiolo	Barbera
	%	%	%	%	%
Malvidin	94.2	81.3	83.3	24.1	40.4
Peonidin	1.4	10.3	2.5	54.9	7.5
Delphinidin	0.4	0.3	5.0	4.2	18.6
Petunidin	3.9	7.9	8.8	4.0	18.6
Cyanidin	0.1	0.2	0.4	12.9	10.1

FIGURE 3.7 The principal, naturally occurring, phenolics found in *Vitis vinifera* grapes and wines. The 'C' ring distinguishes the flavanols (saturated, no double bonds) from the flavonols (unsaturated with ketone at carbon 4) and the anthocyanins (unsaturated, two double bonds, no ketone). *(Table of fraction of anthocyanin types from Blouin and Guimberteau, 2000, with permission).*

variations within two themes: flavanols (flavan-3-ols), which occur as monomers, esterified to gallic acid, or as condensed polymers, referred to commonly as procyanidins, and flavonols (flavon-3-ols). The latter occur as glycosides ([glu]). The flavanols and flavonols differ biochemically and chemically and should not be confused.

2.3.1 *Flavon-3-ols*

The principal flavonols are quercetin, myricetin and kaempferol, but laricitrin, isorhamnetin and syringetin also occur in minor amounts (Figure 3.7). White cultivars possess mainly quercetin with low levels of kaempferol and isorhamnetin (Figure 3.8). They are synthesized pre-veraison, with quercitol glucuronide being predominant at veraison in red cultivars but declining with maturity (Castillo-Muñoz et al., 2007). They are yellowish with absorption maxima in the ultraviolet of approximately 255 and 370 nm. Flavonols occur mainly in the skin and are thought to be a response to ultraviolet light, possibly a protectant. Levels in wine tend therefore to reflect degree of exposure of fruit to light during maturation, and concentration is correlated with quality in red wine and with degree of skin contact in white (Ribéreau-Gayon (1964); Ristic et al., 2007; Ritchey and Waterhouse, 1999).

2.3.2 *Flavan-3-ols*

As monomers, the catechins or flavanols are a simpler group comprised primarily of two stereoisomers, (+)catechin and (−)epicatechin, with the difference being the arrangement of the hydroxyl at the 3 position on the C ring. Small amounts of (−)epigallocatechin also occur with an additional hydroxyl substitution at the 5′ position on the B ring (Waterhouse, 2002). They are important antioxidants, bitter and astringent factors and precursors of condensed or polymeric procyanidins (tannins). They occur in the skin, seeds and stems of grapes and are the most abundant phenolics, being as high as 1.1 g/kg in Pinot noir. Their sensory attribute is largely one of bitterness (Figure 3.9) (Arnold et al., 1980).

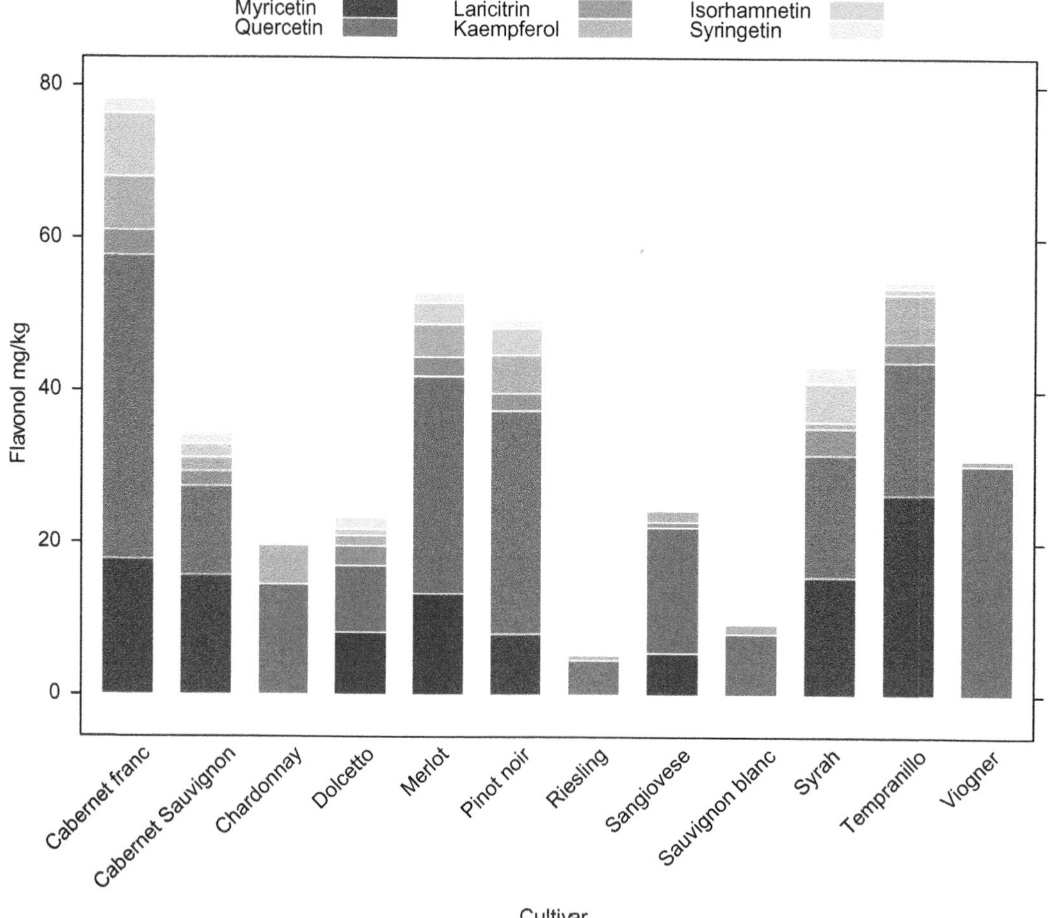

FIGURE 3.8 Flavonols from a selection of red and white cultivars grown in the one location, under one management system and harvested at similar maturity. *(Data from Mattivi et al., 2006).*

2.4 Tannins

Tannins are the most complex phenolics in grapes and wine. Terminology is therefore an issue and not always consistent. The term 'tannin' is a loose term. Any plant compound that has been used to tan leather falls into this category. These are, generally, large condensed molecules based on a complex of phenolics and capable of forming stable cross-links between proteins in skin. The condensed tannins are formed from catechins and epicatechins in the grape, but in the wine anthocyanins may participate in the condensation process. The hydrolyzable or ellagitannins are oak derived. Tannins are so called because they can cross-link proteins. This is thought to be the process that causes the mouth dryness associated with astringency (Arnold et al., 1980). In wine making, it is important to be more specific because the chemistry and therefore the management and activity vary markedly according to class.

Terminology is inconsistent and not necessarily logical given that the words used now have a Greek base meaning precursor to a dark blue color (procyan-idin/-ogen) while 'antho' pertains to flower (proanthocyan-idin/-ogen). The term procyanidin tends to be reserved for flavan-3-ols with a low degree of polymerization (dp) of 2–3, while proanthocyanidin is applied to higher orders of polymerization. Similarly with anthocyanins: the suffix -idin is used when referring to the base molecule, that without sugar or ester conjugates, while the contraction -in refers to the stable conjugated form.

- **Procyanidin**: A small polymer based on two or three linked catechin/epicatechin molecules (Figure 3.7). It occurs widely in nature and is present in all tissues of the grapevine but especially in seeds (Downey et al., 2003).
- **Proanthocyanidin**: A large polymer of linear or branched catechin/epicatechin molecules with a dp of 4–50 units, mainly in skin (Downey et al., 2003) and wine. In wine, the polymer may include condensed anthocyanins and non-flavonoid phenols.

2.4.1 *Grape-Derived*

The procyanidins (syn. proanthocyanogens) occur as dimers and higher polymers in the grape skin, seed and rachis (stalk). Dimers are named according to the manner in which they are linked: B1 to B8, B1–B4 tending to form linear, stacked, chains, while B5–B8 enable branching (Figure 3.7). The degree of polymerization seems to differ markedly between tissues in the mature fruit, with a mean dp of 5 and 40 being the mean upper levels for seed and skins, respectively, of Shiraz (Syrah) at maturity (Downey et al., 2003). This condensation was correlated with an increasing proportion of insoluble

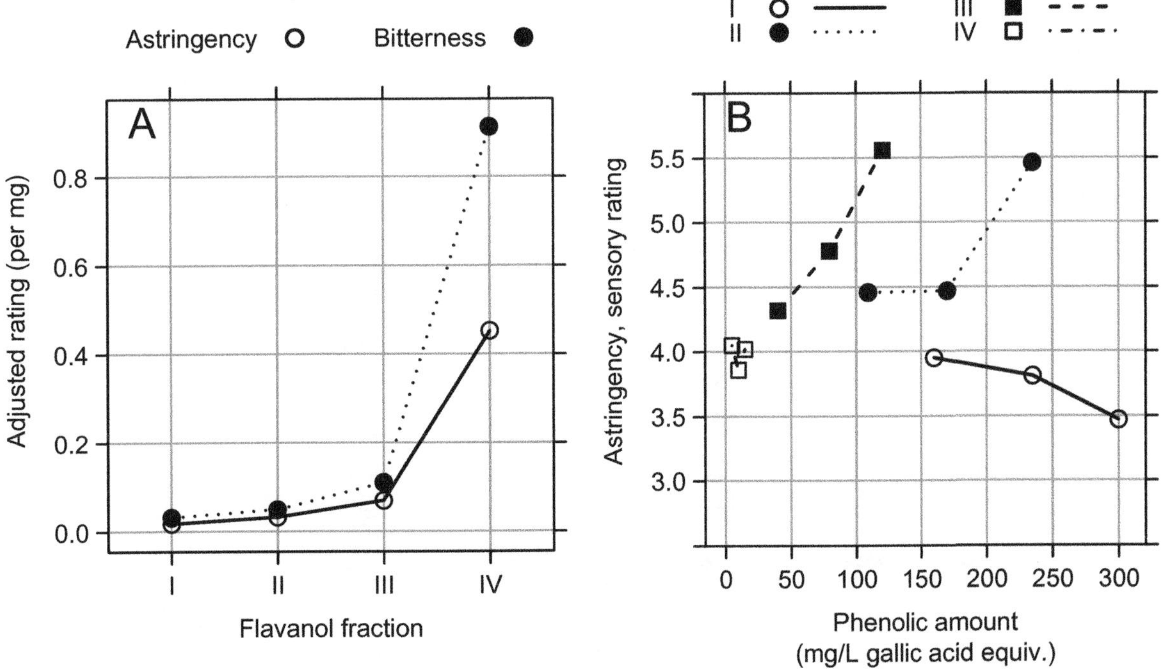

FIGURE 3.9 Sensory assessment of flavanols. A: Relative bitterness or astringency by formal sensory analysis; B: Dose–response curves for extracted seed flavanol groups. I = monomeric catechins; II = primarily dimers (B1–B8); III = mainly trimers and tetramers; IV = large, >5 or 6 condensation units. *(Data from Arnold et al., 1980).*

proanthocyanidins, especially in the skin. The proportion of these forms differs markedly between cultivars, organs and stage of fruit development (Figure 3.10, Table 3.2). In general, the levels decline sharply, increasing the palatability of flesh and seed. While it seems that certain of the dimers are more prominent in the red cultivars at harvest, the two examples, Cabernet Sauvignon and Merlot, used in the study are closely related and may not be truly representative. It is highly likely that sensory differences between wines of different cultivars reflect not only the absolute content but also the molecular arrangement.

In red wine making, the contribution of seed- versus skin-derived flavanols changes with time; those from the skin diminish in relative importance and those from the seeds rise (de Freitas et al., 2000). This is thought to be associated also with the rise in alcohol, which increases flavanol solubility.

Changes occurring through the first two years of aging in the bottle include a loss of two-thirds of the free anthocyanins, a rise followed by a loss of pyro- or conjugated anthocyanins and vitisins, a rise in vinylphenol and vinylcatechol anthocyanin adducts, a rise then a loss of direct condensation anthocyanin–catechin products, and a dramatic decline in the concentration of acetaldehyde-mediated anthocyanin condensation products (Alcalde-Eon et al., 2006). At 23 months, the pigments are dominated by acetone derivates, vinyl–catechol adducts, and epicatechin–anthocyanin condensation products (Figure 3.11).

Condensation of catechins has been shown to lead to reduced astringency (Weber et al., 2013), perhaps due to the formation of complexes with anthocyanins. Previous

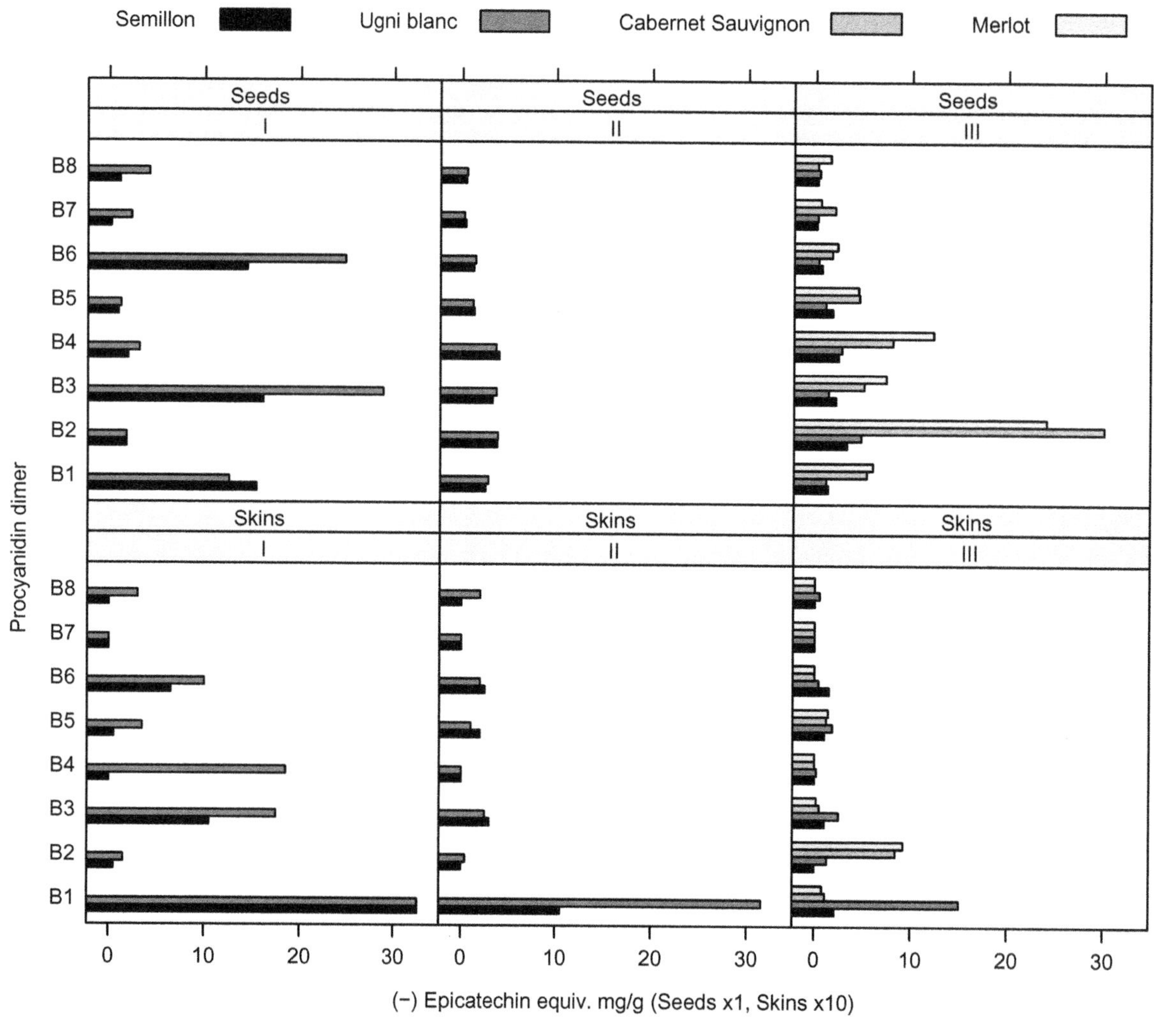

FIGURE 3.10 Development and varietal differences in procyanidin dimers present in seed and skin of two white and two red cultivars. The scale for the 'skins' is 10-fold that for the seeds (i.e. skin values were multiplied by 10 before plotting). *(Data from de Freitas and Glories, 1999; de Freitas et al., 2000).*

studies have shown that, on a weight for weight basis, astringency and bitterness increase with molecular weight (Figure 3.9A). However, solubility issues are probably at play and the concentrations achieved with the highest molecular weight fraction were very low compared with the other fractions. Fractions II and III revealed a marked, positive dose–response curve whereas the low monomer fraction declined and the high molecular weight showed no response over the range tested. Bitterness levels were similar and very high for all levels of all groups tested. The most recent study has confirmed this finding (Weber et al., 2013). Thus, the reduction in bitterness and astringency with wine age is probably due to precipitation and the formation of colloids by the highest molecular weight groups.

2.4.2 Oak-Derived

Oak-derived phenolics and tannins are quite different from those that occur naturally in the grape. They are based on polymers of ellagic acid glycosides but like the condensed tannins are complex (Puech et al., 1999). The impact of

TABLE 3.2 Flavanol Dimer and Monomer Content in Seeds and Skins of Four Grape Cultivars, mg/g Dry Weight (–) Catechin Equivalents at Harvest

	Seeds			Skins		
Cultivar	Monomer	Dimer	PAC	Monomer	Dimer	PAC
Semillon	11.7	214.7	23.7	19.1	0.6	19.1
Ugni blanc	11.8	191.3	39.0	2.2	82.0	113.9
Cabernet Sauvignon	46.8	52.8		0.5	1.1	
Merlot	33.7	53.8		1.0	1.2	

Note: PAC = proanthocyanidins.
(Data from de Freitas and Glories, 1999; de Freitas et al., 2000).

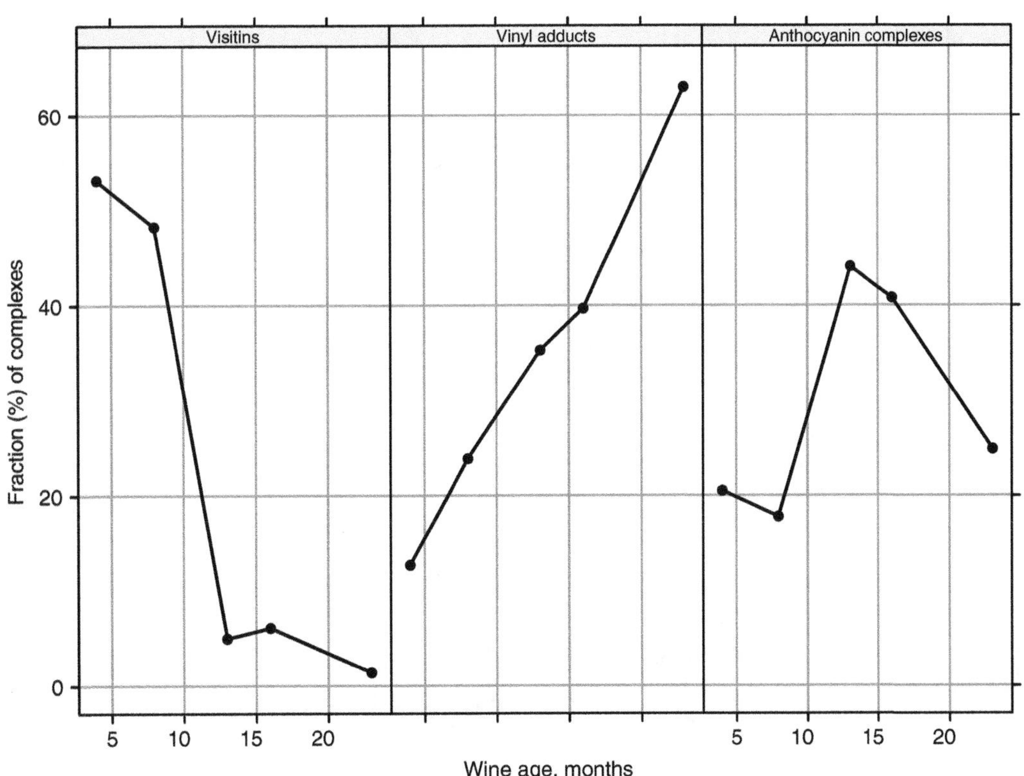

FIGURE 3.11 Changes in the principal anthocyanin complexes during aging of wine made from Tempranillo grapes. *(Data from Alcalde-Eon et al., 2006).*

oak tannins, as such, on sensory aspects of wine remains controversial. The gallic acid molecule contains three hydroxyl groups capable of reacting to form ester or glycoside linkages. Examples of some common oak-derived phenolics are depicted in Figure 3.12. Castalagin and vescalagin are isomers differing only in the arrangement of the hydroxyl at the terminus of the linear, aldose form of the glucose backbone (bold line). The hydroxyl groups are reactive, can be oxidized or esterified or form glycosidic linkages generating considerable diversity of composition and arrangements. In ellagic tannins, common to oak (*Quercus* spp.), the gallic acids are generally cross-linked, whereas in the gallotannins they are not. Gallic acid is also a minor component of grape seeds. Oak lactone, which exists as two isomers and is also known as 'whiskey' lactone, is a component of the aroma profile of beverages matured in oak, but is not a phenolic. Vanillin and guaiacol and 4-methyl-guaiacol are important sensory aspects of wines stored in contact with oak (Spillman et al., 1997).

2.5 Copigmentation and Matrix Effects

Change in color and color intensity are among the most striking changes that occur during fermentation (Figure 3.13). A substantial part of this is due to a solvent or matrix effect: the impact of alcohol is such that the

Gallic acid

Ellagic acid

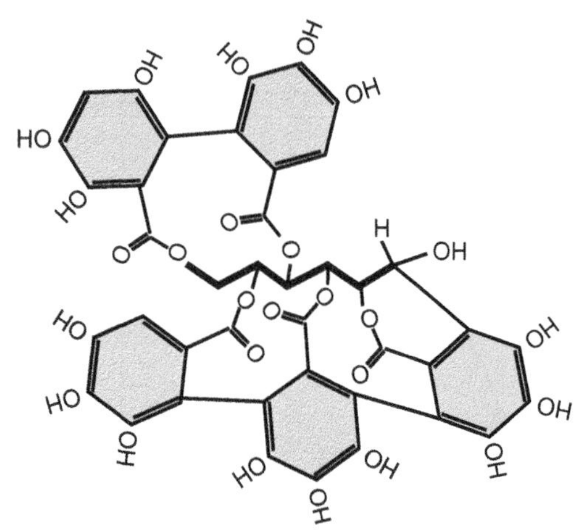

Ellagitannin
(Vescalagin/Castalagin)

Oak lactone

R [H} - Guaiacol
R [CH_3] - 4 Methylguaiacol

Vanillin

FIGURE 3.12 **Examples of the oak-derived heartwood tannins, phenolics and other secondary products found in wines stored in contact with oak.**

apparent anthocyanin concentration is reduced by nearly 40%. However, the actual reduction is less than 25%. The difference is due to the solvent, ethanol, reducing the molar absorptivity of the chromophore, the anthocyanin, probably by disrupting copigmentation associations (Boulton, 2001). The loss during fermentation is nearly double this value. Many factors contribute to this loss, including oxidation and condensation, while copigmentation may augment apparent color. These changes are accompanied by a marked change in hue or tint as the apparent color shifts toward the blue/purple part of the spectrum. Copigmentation reactions are responsible for as much as 50% of the color of young red wines (Boulton, 2001).

These interactions are also observed as a failure of solutions of anthocyanins (wines, juices, musts) to conform to Beer's law, which is important when using optical density to estimate concentration. This phenomenon has been ascribed to the formation of colloids but it is more likely to be due to interactions between anthocyanin glucosides and flavonoids which affect both the absorbance spectrum, generally shifting the absorbance maximum upward, and the absorbance coefficient. The coefficient may be increased by as much as two- to 10-fold for individual anthocyanins (reviewed by Boulton, 2001). The effect is due to stacking of molecules and is concentration, pH and counter-anion dependent (e.g. Figure 3.14). It is also affected by conjugation (and thus by oxidation). Therefore, one needs to be acutely aware of these phenomena when measuring color and anthocyanins in wine and must/juice and be careful to compare like with like.

The manner in which anthocyanin derivatives and phenolics interact is complex and thought to be important in anthocyanin stability (or lack of it in cultivars such as Pinot noir and Sangiovese). This is also considered to account in good part for the positive effect of cofermentation/blending even of white and red cultivars (e.g. Shiraz and Viognier) on wine color and stability, not just on aroma (reviewed by Boulton, 2001).

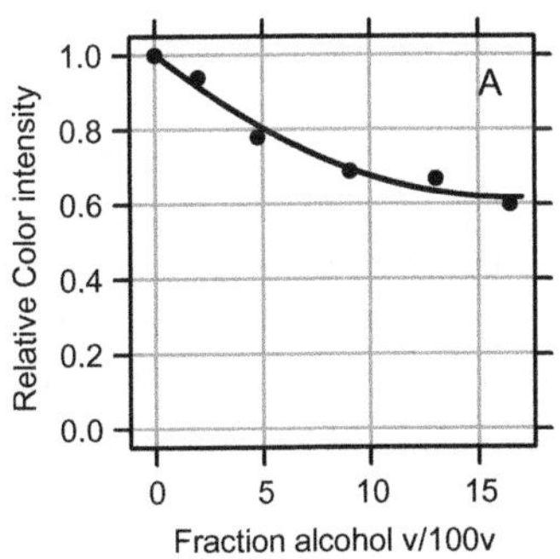

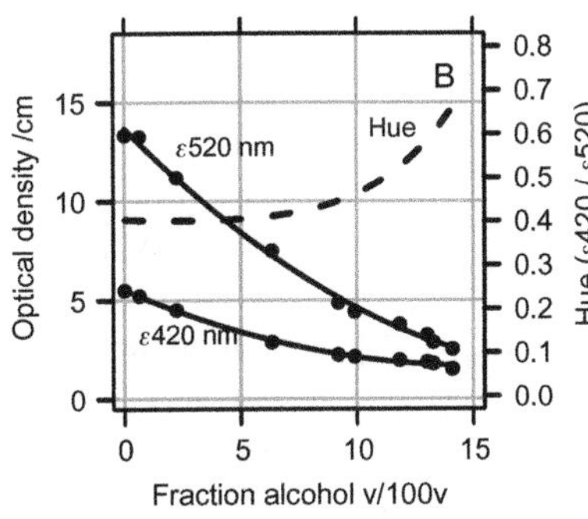

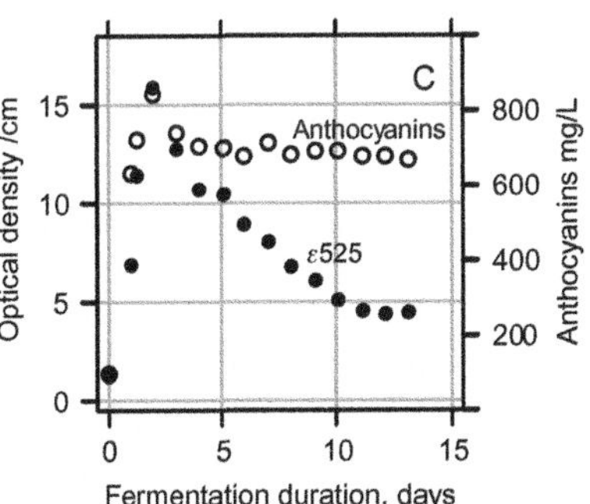

FIGURE 3.13 Impact of alcohol content on absorbance (au/cm) of Shiraz must (pH 3.9) that has been heated to 80°C to extract the anthocyanins and then pressed (thermovinification). A: Color intensity of must dosed with alcohol and measured after standing for 3 h ($\lambda^{525\ nm}$). The curve shows the proportional change allowing for dilution. B: Changes to the same must during fermentation showing the brown ($\lambda^{420\ nm}$) and red ($\lambda^{520\ nm}$) and the ratio of the two as an index of hue or tint. C: Fermentation of a similar must using traditional processes but pressed on day 2. The curves compare the apparent anthocyanin content if measured by optical density at 535 nm and the actual level measured chemically. *(All data from Somers and Evans, 1979, with permission).*

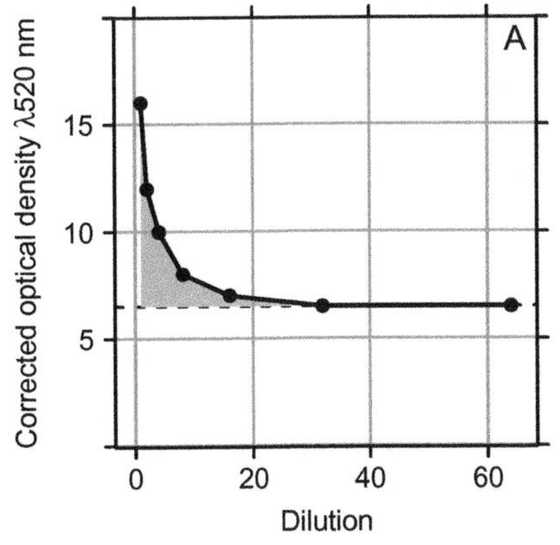

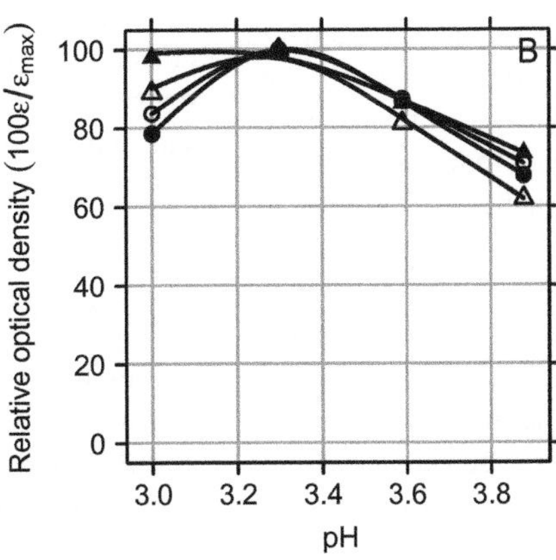

FIGURE 3.14 Effect of dilution (concentration) and pH on the observed extinction coefficient of anthocyanins in wine. A: The impact of dilution (binary series, 1, 2, 4, ... , 64); the shaded zone indicates the region affected by copigmentation; B: The effect of pH on copigmentation and thus on optical density and apparent anthocyanin content (cv. Cabernet Sauvignon). *(Modified from Boulton, 2001, with permission).*

3. OXYGEN AND ITS ROLE IN WINE MAKING

3.1 Introduction

Thoughtful management of oxygen status throughout the wine-making and aging process is one of the most important elements in the production of quality table wines, whether red or white (du Toit et al., 2006). Oxygen status influences yeast behavior, risk of infection by fault-causing organisms and all aspects of wine quality as judged by the consumer, including color, palate and aroma. It is not a matter of total exclusion, or of open access to air; it is about timing, degree of exposure according to wine style and managing exposure throughout the life of the wine. Even subtle factors such as choice of bottle closure and head space in the bottle can have marked effects on oxygen-related sensory attributes of wine stored in the bottle (Kilmartin, 2009; Skouroumounis et al., 2005).

Negative aspects associated with excessive oxygen during the wine-making practice include browning and loss of varietal flavors and aromas, especially those due to certain mercaptans (e.g. 3-mercaptohexanol, 3MH); development of sherry-like, oxidized, aldehydic aromas; and increased susceptibility to contamination by fault-causing microbes, notably surface-growing yeasts (*Pichea* spp.) but also *Brettanomyces* and acetic acid bacteria.

Factors associated with these risks include low free sulfur dioxide (fSO_2), the presence of copper (Cu^{2+}) and iron (Fe^{2+}) ions, laccase (an oxidase, usually of fungal origin, e.g. *Botrytis*) and polyphenol oxidase (PPO, syn. tyrosinase, naturally occurring in grapes). SO_2, in the context of oxidation–reduction reactions in wine, has been shown to increase oxygen consumption, thus reducing oxygen levels. This then leads to the regeneration of polyphenols. It inhibits PPO and reacts with hydrogen peroxide (H_2O_2) and thus prevents the formation of acetaldehyde from ethanol, which requires H_2O_2 (reviewed by Danilewicz, 2012).

Positive aspects of oxygen in wine include enhanced condensation of tannins, and thus reduced astringency and enhanced color stability and sometimes color intensity. It assists aging of red wines in particular, but also some white wines (e.g. production of 1,1,6-trimethyl-1,2-dihydronaphthalene, TDN, in aged Riesling wines); and the presence of some oxygen helps to eliminate reduced, sulfurous aromas (e.g. cooked cabbage and rubber), especially in white wines. Microoxidation is a modern practice to promote aging and softening of red wines while minimizing risk of fault-causing microorganisms; this method is cheaper and much more manageable than using oak barrels (reviewed by du Toit et al., 2006; Kilmartin, 2009).

Antioxidants are the antidote to oxygen and some of the most important are naturally occurring; for example, ascorbic acid is both naturally occurring to a small degree and a common addition; however, problems can arise through the formation of H_2O_2 and acetaldehyde in ethanolic solutions. Glutathione is also highly important, especially in musts, because it protects oxidative browning-sensitive phenolics. While anthocyanins and flavanols react and condense and brown in the presence of oxygen, the process, when carefully managed, provides greater stability and protects red wines against the negative effects of excessive oxidation.

The most common antioxidant added to wine is SO_2. One of its main effects with respect to oxidation is to inhibit the action of naturally occurring oxidative enzymes such as PPO (Singleton, 1987). Ascorbic acid is another commonly used antioxidant but is restricted mainly to white wines; even for those it can be technically challenging to use effectively.

Poor fruit and wine hygiene caused by the use of disease-infected fruit can have devastating effects on wine quality. Many fungi, and *Botrytis* in particular, contain resilient oxidative enzymes known as laccases. Preventing these enzymes from contaminating wine is critically important, and if that cannot be achieved then destroying, reducing or otherwise managing their activity is highly important.

While discussion is generally directed toward minimizing or avoiding exposure to oxygen, some exposure is necessary, even in the production of white wines. Wine is saturated with oxygen at about 6 mL/L (more at low, less at high temperature). However, it is a dynamic process and saturation is a rare state owing to the capacity of phenolics principally to react with and remove dissolved oxygen (reviewed by Boulton et al., 1996). The capacity of white wine to absorb oxygen is about one-tenth of that of red wines owing to the absence of flavanols and ellagic tannins (from oak). White wines that had absorbed 60 mL/L of oxygen were clearly brown and maderized. However, red wines generally improved up to and even well beyond this level of oxidation. Transfers and barrel aging add about 20 and 40 mL/L oxygen per annum (reviewed by du Toit et al., 2006).

Three aspects of oxygen in wine that require further consideration are browning, aldehyde formation and condensation reactions. For a more thorough consideration see one or more of the following: Boulton et al. (1996), du Toit et al. (2006) and Kilmartin (2009).

3.2 Oxidation

Somewhat surprisingly, oxygen has limited ability to affect oxidative processes in wine owing to the pairing and the sharing of electrons which are in a ground state in the O_2 molecule. Another source of electrons is required to give effect to the oxidative potential and usually this is a metal ion, copper or iron, either in solution or associated with an enzyme such as PPO (tyrosinase) or laccase that has copper ions as an essential part of its active sites.

Browning or yellowing of white wines generally increases with age and is accompanied, at first, by desirable, aged-related sensory attributes and then those of decay and senescence: old age and flabbiness! Aroma degradation has been noted to occur before color changes are visible (Oliveira et al., 2011) owing to the production of compounds such as methional (boiled potato) and phenylacetaldehyde (honey-like). Yellowing, then browning is not a simple chemical reaction. It requires the presence of an enzyme and/or a metal ion and oxygen and a substrate. In white and also in red wines, the phenols, caftaric and coutaric acids, are the primary substrates (Singleton, 1985). In grape juice, must, glutathione intervenes:

$$\text{Caftaric acid} + O_2 \xrightarrow{\text{PPO}}$$

$$\text{Caftaric quinone} \xrightarrow{+\text{ Glutathione}} S\text{-Glutathionyl caftaric acid}$$

This product does not brown, leading Singleton (1985) to conclude that grape musts possess a built-in system to protect against browning, at least to a degree. The product, *S*-glutathionyl caftaric acid, is stable and relatively colorless except in the presence of laccase, which is able to react with the glutathionyl derivative. Most of this reaction occurs during crushing and pressing and explains the relative resistance to oxidative browning at this stage of wine making. Browning, such as that seen in freshly cut fruit or in oxidative styles of white wine production, involves the reaction of quinones with amino acids and proteins. This may also occur to a degree in wine but the predominant emphasis is on the behavior of the hydroxycinnamic acids, the flavonols, and their derivatives. The higher the content of flavanols in white wines, the greater the tendency to oxidize due not to their inherent sensitivity but to the manner in which they interact with diphenols such as caffeic acid (Singleton, 1987) and form brown condensation products that are susceptible to further oxidation, producing colored polymers (Table 3.3).

pH is critically important as it determines the proportion of phenolate ions, the reactive state of a phenol, present in the wine; lower is better. At pH 4, the concentration of phenolate ions is about 10-fold that at pH 3 and therefore high pH wines oxidize rapidly and do not age well (Singleton, 1987).

It is the quinones of the dihydroxyphenols and trihydroxyphenols that are primarily responsible for the production of the yellow to brown pigments, although monohydroxyphenols such as coumaric acid may also be involved. Thus, caftaric acid (and its hydrolysis product, caffeic acid) and the flavonoids, catechin and quercetin, may be involved. The latter are more evident in white wines that have had skin contact, postcrushing, or have been heavily pressed. In these reactions metal ions play an important role (Schema 3.1).

In this schema, ferrous and cupric ions cooperate to reduce oxygen with a dihydroxyphenol as the substrate and in so doing are regenerated while producing two quinones

TABLE 3.3 Optical Characteristics of Example Reactants and an Example Product (Singleton, 1987)

Substance	cp	λ_{max} (nm)	ε	λ_{max} (nm)	ε	λ_{max} (nm)	ε	Color
Catechol	No	214	6300	292	2300	–	–	None
o-Benzoquinone	No	240	2400	280	400	430	20	None
Diphenyl quinone	Yes	400	69,000	–	–	–	–	Yellow

Note: cp = condensation product of oxidation; λ_{max} = absorption maximum; None = not colored; – = no data.

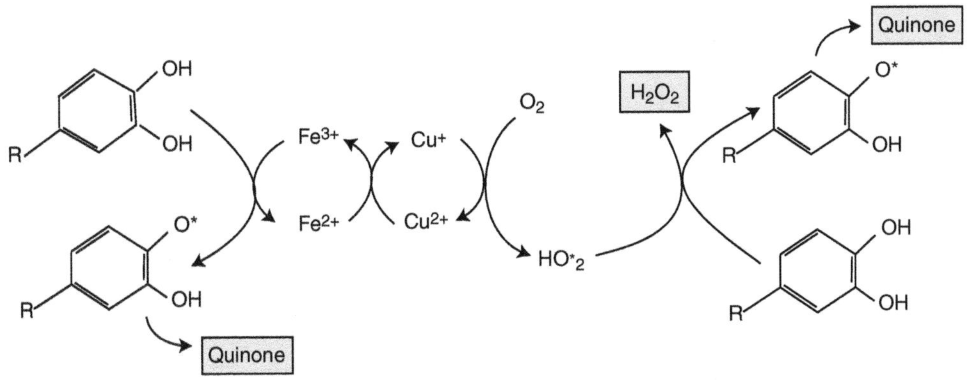

SCHEMA 3.1 Redox cycling of copper and iron while catalyzing the oxidation of catechols to produce quinones and hydrogen peroxide. *(From Danilewicz, 2007, with permission).*

and one mole of hydrogen peroxide. The peroxide may react subsequently with ethanol to produce acetaldehyde. Acetaldeyhyde in high levels is a fault but it is also an important player in the condensation of anthocyanins and other flavonoids. Copper and iron levels have been reported to be in the range of 0.08–1.39 mg/L and about 5.5 mg/L, respectively, levels on the whole that are quite sufficient to support oxidative processes in wine (reviewed by Danilewicz, 2007).

3.3 Aldehyde Formation

Two aldehydes have been shown to arise from oxidative processes in wine. The first, acetaldehyde, is well known and has been thoroughly studied (Figure 3.15). The second, glyoxylic acid, is less well known but may play an even more important role than acetaldehyde in condensation reactions. Aldehydes are considered to be intermediates in the age-related color and flavor changes that occur in wines (reviewed by Oliveira et al., 2011).

Acetaldehyde is the penultimate step in the biochemical formation of ethanol under anaerobic conditions by yeast. However, the primary source of this important chemical in wine is from the oxidation of ethanol by H_2O_2 produced by the oxidation of phenols, either biochemically in conjunction with PPO or laccase in musts or as catalyzed by copper and iron in wine. Acetaldehyde inactivates SO_2 by forming an adduct, and it also plays a role in the formation of condensation products of flavonoids. However, more than this, high concentrations of acetaldehyde cause aroma faults and are closely related to oxidative faults generally.

Acetaldehyde in wine is a product of the Fenton reaction, in which hydroxyl radicals ($OH^\bullet$) derived from the H_2O_2, formed during the first step in the oxidation of phenolic compounds, react with ethanol (Danilewicz, 2007):

$$H_2O_2 \xrightarrow{H^+,\ Fe^{2+}} OH^\bullet + H_2O +$$

$$CH_3CH_2OH \xrightarrow{O_2 \text{ or } Fe^{3+}} CH_3CHO + H_2O$$

Sulfur dioxide, however, being more reactive than ethanol, is the preferred partner:

$$H_2O_2 + H_2SO_3 \rightarrow H_2SO_4 + H_2O$$

A role for glyoxylic acid, a product of oxidation of tartaric acid in the condensation of flavanols, is a fairly recent observation (Fulcrand et al., 1997). These authors showed that when glyoxylic acid was formed from tartaric acid, in the presence of Fe^{2+}, it was far more reactive than acetaldehyde as a partner in condensation reactions. Furthermore, while the immediate dimers were colorless, those dimers served as precursors for more stable, yellow-colored polymers (Fulcrand et al., 1997). Copolymers may also form with both glyoxylic acid and acetaldehyde being active simultaneously (Drinkine et al., 2005).

$$HOOCCHOHCHOHCOOH + H_2O_2 \xrightarrow{\text{Fenton reaction}}$$

$$2(HOOCCHO) + 2H_2O$$

These processes will eventually consume the available fSO_2 and when that has occurred, acetaldehyde (and glyoxylic acid), and the formation of yellow/brown, phenolic polymers begins. Hence, it is important to monitor and maintain SO_2 levels during the processing of wine postfermentation, even though this does not wholly prevent slow oxidation and polymerization of flavanols in red wine.

3.4 Condensation of Tannins and Anthocyanins

The benefits of judicious exposure of white wines to oxygen are that the formation of 'reduced' products such as hydrogen sulfide and other fault-associated thiols is prevented, or those formed are eliminated; and that fermentation is completed to dryness (but care is needed as this can also promote the growth of fault-causing bacteria). But in red wines, the benefits are mainly through the beneficial condensation of flavanols and related compounds.

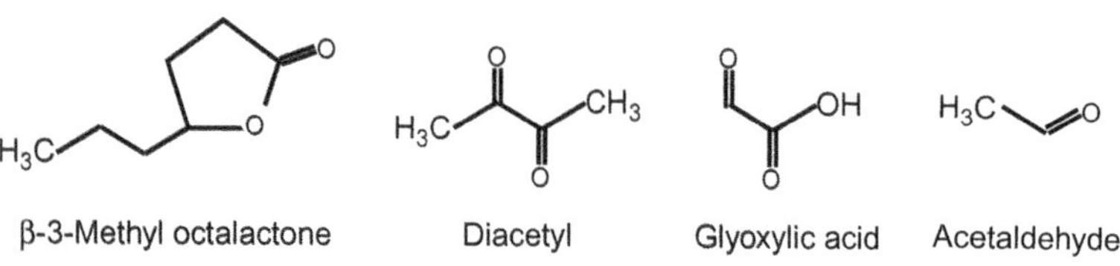

(2E)-2-Hexenal

Octanal

(2E)-2-Heptenal

Decanal

FIGURE 3.15 Structure of the main aldehydes and lactones found in wine.

Some of these products enhance apparent color and cause a shift in the absorption maximum, possibly as copigmentation cofactors but others presumably from condensation (Boulton, 2001; Timberlake and Bridle, 1977). Furthermore, these condensation products are color stable in the presence of SO_2.

The reactions are complex but the essential elements are set out in Figure 3.16. This figure shows how the aldehydes, arising as products of oxidation of diphenols, undergo a variety of condensation reactions, some of which lead to the formation of stable, colored end-products (Oliveira et al., 2011). The vinyl–flavanols can further react with other flavanols to produce orange- to blue-colored pigments. The polymeric procyanidins (tannins) are more sensitive to oxidation than the monomers but in wine are protected somewhat by other compounds including anthocyanins (de Freitas et al., 1998; Mirabel et al., 1999). Anthocyanin–tannin complexes are essential if tannins are to be held in solution and contribute to the quality of red wine. White wine made using a red wine process fails to retain these tannins in solution (Singleton and Trousdale, 1992).

4. AROMA: DIVERSITY AND DIVERSE ORIGINS

4.1 Introduction

Wine is not simple: flavors and aromas achieved in one particular ferment are virtually impossible to repeat in another, unless the winemaker goes to some length to limit the flavor and aroma options by using just the

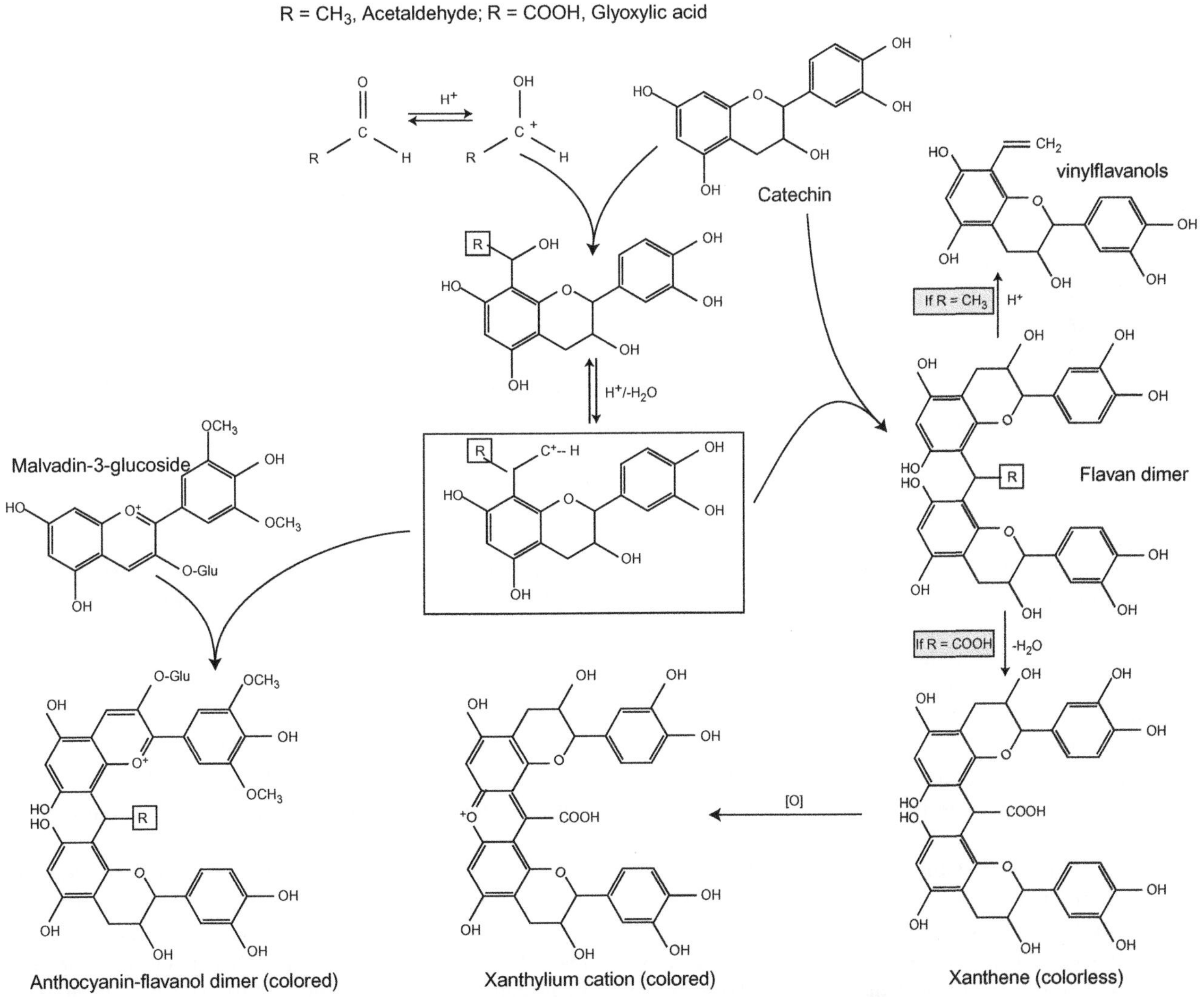

FIGURE 3.16 Outline of the process by which the oxidative products, acetaldehyde and glyoxylic acid, interact with catechins to produce reactive vinylflavanols and a range of colored and colorless dimers. *(Redrawn from Oliveira et al., 2011, with permission).*

free-run juice (omitting the complex products within skin and seed) and a neutral yeast, omitting a malolactic fermentation and conducting the whole under a blanket of nitrogen in an otherwise sterile, stainless steel environment. Even then, those brands that aim to provide a product that is constant from one year to the next must still blend batches to achieve a semblance of consistency.

That it is so difficult to achieve an 'industrial-standard' product in different years is evidence of the impact of not just variety, but also season, viticultural practice, microbial diversity and complex chemistry on the final product. This complexity is given expression especially through the volatiles that are, on one hand, typical of particular cultivars, but on the other, variable and enable the production of wines that are unique and probably never quite repeatable, even in the best hands. The numbers of volatile compounds probably exceed 1000 and about 700 have been identified (Swiegers et al., 2005). The prospect of surprise, joy and wonder is why this industry stands apart from most others.

4.2 Classes and Origins

While it may be said that all aromas and flavors originate in the fruit, it is the transformation of fruit products by microorganisms, yeasts and bacteria, in combination with direct chemical changes, acid catalyzed or oxidative, that imparts the flavors and aromas that comprise the character of a wine. This is so whether those are flavors or flavor precursors in their own right or simply basic metabolites such as sugars and amino acids. The only wines that have a character readily identified in the unfermented juice are those of the 'muscat' family in which a class of aroma compounds called terpenes is substantially evident in the berry and also in the wine. For the larger part, it is difficult to predict the aroma profile of a wine from that of the unfermented must or intact berry. It is the subtle transformation of minute quantities of chemicals that imparts the sensory experience of a wine. Rarely does one chemical dominate; more usually it is the combined sensory experience that provides the impact.

An example of the contribution of a range of volatiles that are characteristic of particular red wine cultivars is depicted in Figure 3.17. With the possible exception of linalool, all the other compounds are natural products processed by yeast and bacteria. Some of these will be predictive of cultivar but others will not be because they reflect the nutritional status of the vines rather than the cultivar, particularly those that are predominantly derived from the further processing of amino acids.

Table 3.4 and Figure 3.18 show a little of the diversity of aroma chemistry as evident in four wine styles: white, rosé, young and aged red wine (adapted from Francis and Newton, 2005). Dominating the profile of all wines are the ethyl esters, but also acetates, cinnamic esters and organic acids, all of which are products of microbial, mainly yeast, metabolism (Swiegers et al., 2005). Monoterpenes and sulfur compounds are characteristic of 'fruit-driven' white wine styles. The range of levels evident varies markedly between wines and wine style and wine age. The changes apparent with aging of red and white wines can be remarkable and arise from purely chemical reactions in a low-pH, low-oxygen environment.

The importance of the role of microorganisms is such that winemakers select particular clones according to the cultivar. The rise in popularity of modern Sauvignon blanc and Semillon styles of white wine would not have occurred in the absence of the recognition and selection of particular yeast strains and species that were routinely able to evoke the aromatic potential of these cultivars (Anfang et al., 2009). Not depicted in Figure 3.18, because the levels were so high that they obscured the impact of other volatiles, is the rise in aged white wines of benzenemethanethiol to values as much as 30,000 times the threshold, enough to completely overwhelm the senses!

4.3 Fruit-Derived

4.3.1 Terpenes

Prominent among fruit-derived aromas are those of terpenes, which are a simple class within the isoprenoid family of substances. All terpenes share a common metabolic origin, although some are formed de novo while others are a product of senescence. Terpenes are common in nearly all aromatic plant products, otherwise known as essential oils. They are based on a repeating C5 skeleton (a C10 skeleton is understood to be a monoterpene, C15 a sesquiterpene, C20 a diterpene, etc.). Those present in grapes are found predominantly as β-glycosides in which a glucose or disaccharide is bound through one of the hydroxyl groups (Figure 3.19). As such, they are not volatile nor can the enzymes in saliva (α-amylase) or many yeasts (zero or very low levels of β-glycosidase) release the free, volatile, aroma compound. The natural hydrolases present within the grape have a low activity at wine pH, are inhibited by sugars in the juice and are not active with all sugar glycosides found naturally (reviewed by Maicas and Mateo, 2005). The conditions in wine also inhibit many microbial glycosidases, although some are effective in some circumstances. Glucoside bonds are, however, susceptible to hydrolysis under acidic conditions and this is the reason for the rise in volatiles as wines ferment and age (Francis et al., 1996). Muscats are atypical and as much as 50% of the volatile terpenes are present as the free, volatile form (Figure 3.20).

4.3.2 Nor-*Isoprenoids*

Another group that have their origin in terpene biosynthesis are the *nor*-isoprenoids (Figure 3.19). These are degradation products of carototenes and xanthophylls (neoxanthins). Two that are important in fragrances and flavorings are β-ionone and β-damascenone. A characteristic of these chemicals is not only that they may be important in their own right as potent aromas, but also that they enhance or mask the aroma of other fragrances, even if present at subliminal levels. Thus, 2-methoxy-3-isobutylpyrazine (IBMP), a compound that imparts a strong vegetative character, is difficult to detect in the presence of β-damascenone while the threshold for the fruity attributes of ethyl cinnimate and ethyl caproate are lowered. It seems that the *nor*-isoprenoids are more important with respect to their impact on the perception of other aromatic compounds rather than in their own right (Pineau et al., 2007). Thus, while a range of red wines from the premium regions of France had levels of β-damascenone that are well in excess of the odor threshold in isolation, these levels were undetectable when presented with the other aromas present in red wines (Figure 3.21). However, that is not to say that those levels were unimportant in the sensory experience of those wines.

A character important especially in aged Riesling is TDN, a derivative of carotenes, the intermediate of which is 'riesling acetal' (Daniel et al., 2009). As such, it originates in the skin of the berry or from contaminating leaves. It is rarely detectable in young wines but increases with age depending on sun exposure history of the fruit, region, storage temperature and closure; levels are lower under cork than under synthetic closures, possibly due to scalping by the natural product (Black et al., 2012).

4.3.3 *Thiols and Mercaptans*

The human senses are attuned to detect sulfur compounds at levels that are far lower in concentration than any other

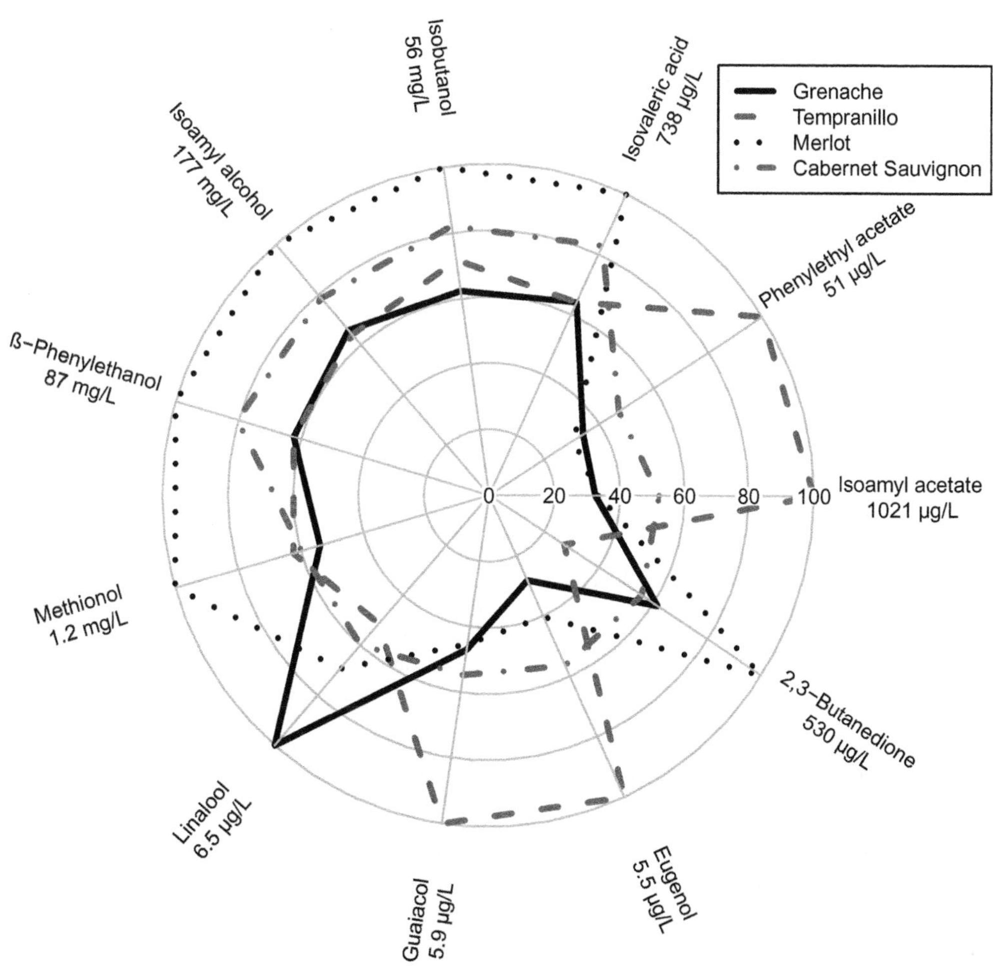

FIGURE 3.17 Radial plot of volatile aromas found to be diagnostic for wines made from four red wine cultivars growing in a number of regions of Spain and for two seasons. *(Data from Ferreira et al., 2000, with permission).*

TABLE 3.4 Categories of Common Volatiles Found in Wine, Their Descriptors and Sensory Threshold Level

Category	Chemical	Descriptors	Threshold (μg/L)
Ethyl esters	Ethyl isobutyrate	Sweet, rubber	15
	Ethyl 2-methylbutyrate	Apple	1
	Ethyl isovalerate	Fruit	3
	Ethyl butyrate	Apple	20
	Ethyl hexanoate	Apple peel, fruit	5
	Ethyl octanoate	Fruit, fat	2
	Ethyl decanoate	Grape	200
Acetates	Isoamyl acetate	Banana	30
	Phenylethyl acetate	Rose, honey, tobacco	250
	Ethyl acetate	Pineapple	12,000
Cinnamic esters	Ethyl dihydrocinnamate	Flower	1.6
	trans-Ethyl cinnamate	Honey, cinnamon	1.1
Organic acids	Isobutyric acid	Rancid, butter, cheese	2300
	Isovaleric acid	Sweat, acid, rancid	33
	Acetic acid	Sour, vinegar	200,000
	Butyric acid	Rancid, cheese, sweat	170
	Propanoic acid	Pungent, rancid, soy	8100
	Hexanoic acid	Sweat	420
	Octanoic acid	Sweat, cheese	500
	Decanoic acid	Rancid, fat	1000
Alcohols	Isobutanol	Wine, solvent, bitter	40,000
	Isoamyl alcohol	Whiskey, malt, burnt	30,000
	2-Phenylethyl alcohol	Honey, spice, rose, lilac	10,000
	Methionol	Sweet, potato	1000
C6 alcohols	1-Hexanol	Resin, flower, green, cut grass	8000
	(*Z*)-3-Hexenol	Green, cut grass	400
Monoterpenes	Linalool	Flower, lavender	25
	Geraniol	Rose, geranium	30
	cis-Rose oxide	Green, flower, lychee, rose	0.2
	Wine lactone	Coconut, spice, lime	0.01
Phenols	Guaiacol	Smoke, sweet, medicine	10
	4-Ethylguaiacol	Spice, clove	33
	Eugenol	Clove, honey	5
	4-Vinylguaiacol	Clove, curry	10
	4-Ethyl phenol	Phenol, spice	440
	Vanillin	Vanilla	200
Lactones	*cis*-Oak lactone	Coconut, flower	70
	γ-Nonalactone	Coconut, peach	25
	γ-Decalactone	Peach, fat	0.7
	γ-Dodecalactone	Sweet, fruit, flower	7
	4-Hydroxy-2,5-dimethyl-3-(2H)-furanone	Caramel	5
	Z-6-Dodecenoic acid gamma lactone	Soap	0.1
	Sotolon	Cotton candy, spice, maple	5
Nor-isoprenoids	β-Damascenone	Apple, rose, honey	0.05
	β-Ionone	Seaweed, violet, flower, raspberry	0.09
	1,1,6-Trimethyl-1,2-dihydronaphthalene	Kerosene, toast, marmalade	2
Sulfur compounds	3-Mercaptohexyl acetate	Box tree, grapefruit, passionfruit, cat urine	0.004
	4-Mercaptomethyl pentan-2-one	Box tree, passionfruit, cat urine	0.0006
	3-Mercaptohexanol	Sulfur, passionfruit, cat urine	0.06
	2-Methyl-3-furan-thiol	Meat	0.005
	3-Methyl thio-1-propanol	Sweet potato	1000
	Benzenemethane-thiol	Struck match, struck flint	0.0003
	Dimethyl sulfide	Cabbage, sulfur, gasoline, cooked asparagus	10
Miscellaneous	2,3-Butanedione	Butter	100
	Acetoin	Butter, cream	15,000

(*Continued*)

TABLE 3.4 (Continued)

Category	Chemical	Descriptors	Threshold (μg/L)
	3-Isobutyl-2-methoxypyrazine	Earth, spice, green pepper	0.002
	Acetaldehyde	Pungent, ether, bruised apple	500
	Phenyl-acetaldehyde	Hawthorne, honey, sweet	5
	1,1-Dioxyethane	Fruit, cream	50

Note: See also Figure 3.18 and http://www.flavornet.org.
(Data from Francis and Newton, 2005).

known series, and only through application of relatively recent advances in mass spectrometry and stable isotope-based techniques has the importance of this group been unveiled. Human sensitivity to this group is akin to that expected of pheromones (Tables 3.4 and 3.5, Figures 3.22 and 3.23). Descriptors for this group vary from the exquisite to the disgusting, presumably depending on concentration and matrix (Francis and Newton, 2005). Many of the unpleasant-smelling sulfur compounds originate from the degradation of the sulfur-containing amino acids, cysteine and cystine, which are important protein-forming amino acids. However, the important, usually pleasant, trace thiols are derived from the degradation of the tripeptide glutathione (γ-L-glutamyl-L-cysteinylglycine). Glutathione and ascorbic acid jointly interact within living cells to regulate oxidation status. The aroma products of interest to winemakers are degradation products. These chemicals are thought to be essential components for white wines, especially Sauvignon blanc and Semillon (Figure 3.23).

Choice of yeast clone is particularly important in releasing the free 3-mercaptohexan-1-ol (4-MMP) from its cysteine-linked precursor. In a comparison of eight strains selected at the Australian Wine Research Institute (AWRI 125-130), differences of more than 100-fold were observed in the release of the active aroma compound (Howell et al., 2004). 4-MMP is present both as the alcohol and as the acetate, and both are highly active aroma compounds, with levels of 60 and 4 ng/L, respectively (Swiegers et al., 2005).

4.3.4 Methoxypyrazines

Vegetative notes are an important element of the aroma profile of many red and some white wine cultivars, but, in excess, are a fault. One substance now well established as a primary contributor to the vegetative range of aromas is IBMP (Figure 3.24). Values for this are now reported routinely for Bordeaux (www.bordeauxraisins.fr). This substance is widely distributed in vegetative parts of plants; it increases in fruit in the vegetative, preveraison phase of growth (Figure 3.25) and then declines as the berry ripens. Levels of IBMP are negatively correlated with temperature, and positively correlated with high water status, shading and high levels of vegetative growth (Roujou de Boubée et al., 2000; Ryona et al., 2008; Sala et al., 2005). Studies have shown that there is little that can be done in the winery to modulate this character in the wine and therefore attention must be paid to viticultural practice, especially in wet harvests (Roujou de Boubée et al., 2000, 2002; Ryona et al., 2008; Sala et al., 2005).

4.4 Yeast-Derived

The vast majority of aromas shown in Table 3.4 and Figure 3.18 are microbial in origin, either through direct biosynthesis or produced during senescence and autolysis: esters, alcohols, organic acids, lactones and the majority of the sulfur compounds. The metabolic origin of these has been thoroughly reviewed (Swiegers et al., 2005) and extensive accounts are provided in more specialized texts (Boulton et al., 1996; Ribéreau-Gayon et al., 2006a).

4.4.1 Esters and Acetates

Esters are important across wine styles. They generally impart fruit-related aromas (Table 3.4) (Swiegers et al. (2005) and are usually the dominant feature (Figures 3.18 and 3.26). They are formed by the condensation of an organic acid with the hydroxyl group of an alcohol or phenol.

Thus:

$$\text{Acetic acid} + \text{Ethanol} \rightarrow \text{Ethyl acetate} + H_2O$$

In general, they are produced in parallel with ethanol during fermentation. Acetates (also butyrates, octanoates, etc.) include esters but also salts, e.g. sodium acetate, but these are generally not volatile and thus for sensory purposes acetates and esters are synonymous terms, with the acetate referring to an ester formed with acetic acid rather than one of the many higher aliphatic and aromatic acids that are present in wine (aliphatic: linear chain of carbon; aromatic: ring of carbon, usually 6 and unsaturated). The

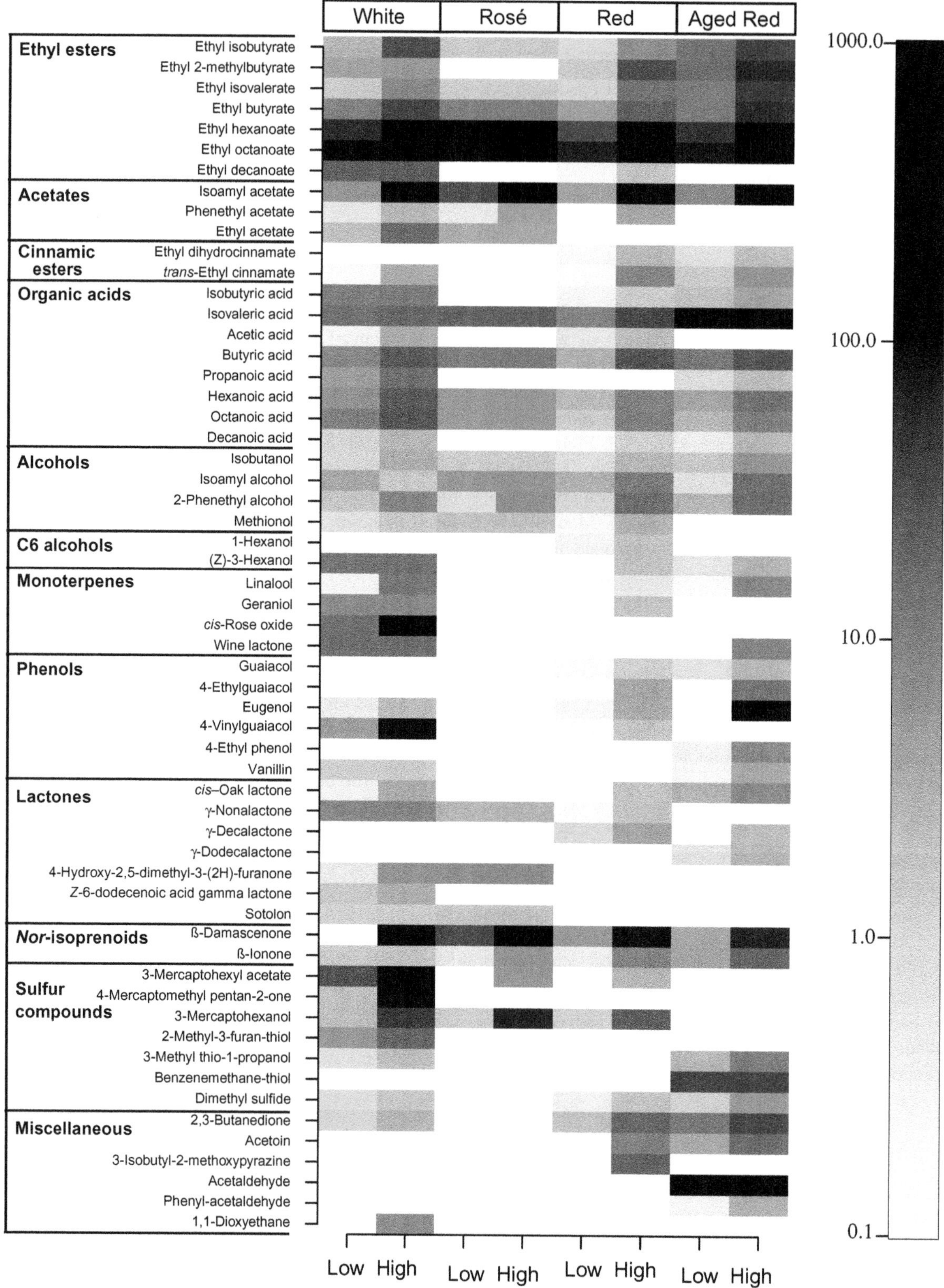

FIGURE 3.18 Map of relative intensity of a selection of wine aroma compounds classed as per Francis and Newton (2005). The shades are the $\log_{10}$ of the ratio of the observed level divided by the aroma threshold (dose–response trends are generally logarithmic in physiology). For each wine style, two columns are presented; one represents the low end of the observed range and the other the high. The aroma threshold used in these calculations was the lower of the reported values.

FIGURE 3.19 **Structure of the main terpenes and their derivatives as found in wine (occur usually as glycosides).**

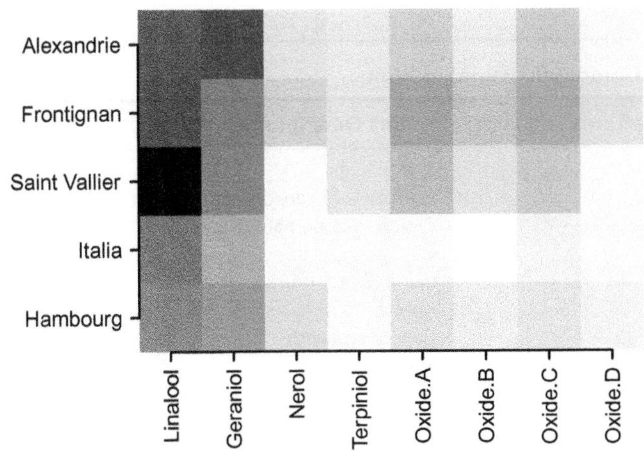

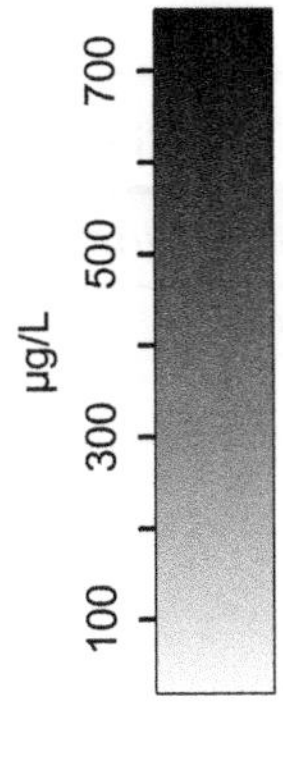

FIGURE 3.20 **Map of levels of free (volatile) monoterpenes in juice extracted from fruit of related Muscat cultivars.** *(Data from Ribéreau-Gayon et al., 1975).*

occurrence and level of particular esters vary with fermentation conditions, cultivar and yeast strain/species (Figure 3.20). The ester blueprint characteristic of particular cultivars arises as a result of a synergy between micro-organism and cultivar (Swiegers et al., 2005), although the organism tends to be the dominant factor (Nykänen, 1986). Thus, ethyl anthanilate in particular, as well as ethyl cinnamate, ethyl 2,3-dihydrocinnamate and methyl anthanilate, is a potent odorant that characterizes the aroma of Pinot noir wines from Burgundy (Moio and Etievant, 1995).

4.4.2 *Organic Acids*

Organic acid formation is central to respiration and particular members of the tricarboxylic acid (TCA) cycle, and organic acids such as succinic acid may accumulate under anaerobic conditions. Other organic acids are formed either directly, as is acetic acid (primarily bacterial), or indirectly through metabolism of amino acids or fatty acids, although not all of these are volatile. Under nitrogen-deficient conditions pyruvic and α-ketoglutaric acids may accumulate as degradation products of certain

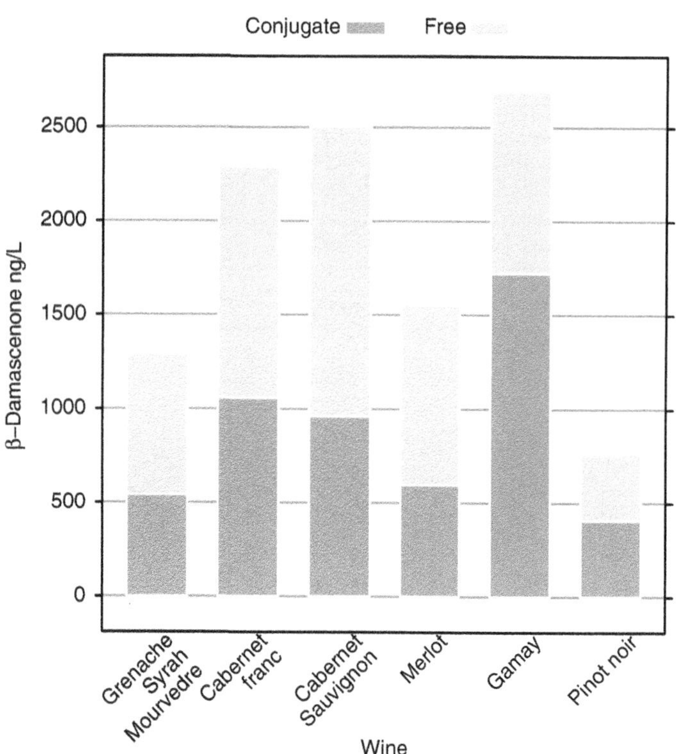

FIGURE 3.21 Average concentration of the *nor*-isoprenoid β-damascenone as measured by gas chromatography–mass spectrometry in red wines of France. The odor threshold in red wine was estimated as 7000 ng/L, cf. 50 and 140 ng/L in water/ethanol and a model white wine, respectively. *(Data simplified from Pineau et al., 2007).*

TABLE 3.5 Additional Sulfur Compounds, Including Thiols, Commonly Found in Wine

Compound	Concentration (μg/L)	Aroma Threshold (μg/L)	Aroma Descriptor in Wine
Hydrogen sulfide	Trace to >80	10–80	Rotten egg
Methanethiol (methyl mercaptan)	5.1, 2.1	0.3	Cooked cabbage, onion, putrefaction, rubber
Ethanethiol (ethyl mercaptan)	1.9–18.7	1.1	Onion, rubber, natural gas
Diethyl sulfide	4.1–31.8	0.93	Cooked vegetables, onion, garlic
Diethyl disulfide	Trace to 85	4.3	Garlic, burnt rubber
Benzothiazole	11	50	Rubber
Thiazole	0–34	38	Popcorn, peanut
4-Methylthiazole	0–11	55	Green hazelnut
2-Furanmethanethiol	0–350 ng/L	1 ng/L	Roasted coffee, burnt rubber
Thiophene-2-thiol	0–11	0.8	Burned, burned rubber, roasted coffee

Note: Reviewed by Swiegers et al. (2005).

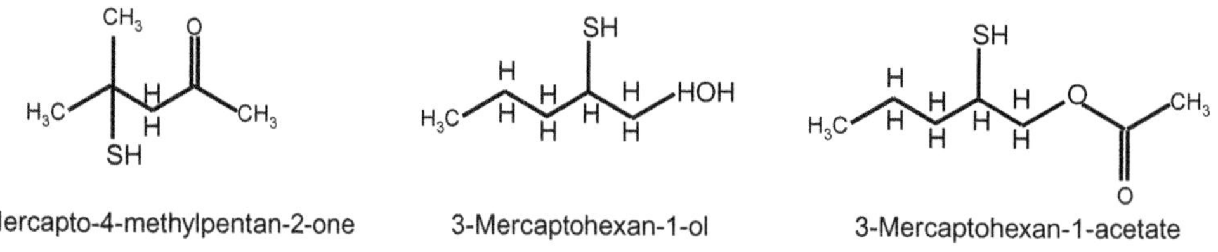

4-Mercapto-4-methylpentan-2-one

3-Mercaptohexan-1-ol

3-Mercaptohexan-1-acetate

FIGURE 3.22 Structures of key thiol aroma compounds found in wine (usually as precursors).

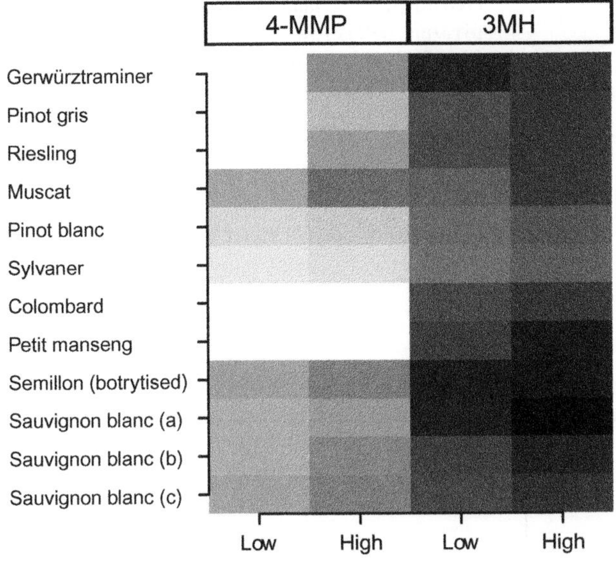

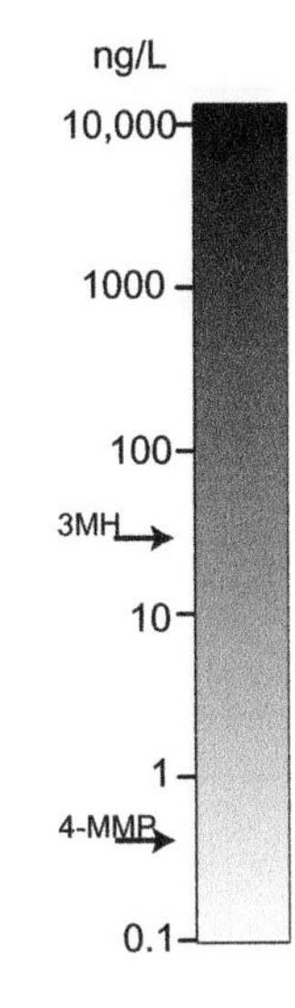

FIGURE 3.23 Range of observed values for two thiol aroma compounds, 4-mercapto-4-methylpentan-2-one (4-MMP) and 3-mercaptohexan-1-ol (3MH), for a number of white wines. Three Sauvignon blanc comparisons are: (a) four Bordeaux wines, 1996 vintage; (b) three Sancerre wines, 1996 vintage; and (c) four vintages of a single Bordeaux wine, 1992–1995. The arrows indicate the approximate threshold values for sensory assessment of the two compounds. *(Data from Tominga et al., 1998, 2000).*

amino acids and cause stability problems through reaction with SO_2 and phenols (reviewed by Swiegers et al., 2005). The levels produced during fermentation are dependent on the cultivar and the yeast clone/species and while the levels are correlated, differences in relative concentration of the two acids do occur (Figure 3.27). These acids are not volatile, whereas those with a fatty acid origin usually are.

Acetic acid is the principal volatile organic acid and levels above 0.7 mg/L constitute a fault, imparting a pungent 'vinegar' character. High levels also contribute to the production of ethyl acetate, another important component of the volatile acidity (VA) fault. The level produced by yeast depends on the yeast clone/species and temperature: *S. cerevisiae* > *S bayanus*; high temperature > low temperature (reviewed by Swiegers et al., 2005); but usually the levels are tolerable.

The aromas associated with the longer chain length (fatty) acids may contribute to complexity at low levels but, if appreciable, constitute a fault—what may be desirable in a cheese is not acceptable in a wine!

4.4.3 Alcohols

The higher alcohols, like the long-chain acids, can exert a positive influence in small quantities, adding to complexity, but in higher quantities are likely to be regarded as a fault (2-phenylethanol and hexanol aside) (Table 3.4). As for the acids, these may be alphatic (isobutanol) or aromatic (2-phenylethanol). The formation of these may be through deamination of amino acids (reviewed by Nykänen, 1986). Several studies have been conducted on the relationship between the amino acid content of fruit and the aroma profile of the wine. In general, the findings are that, while amino acid profiles tend to be cultivar specific, uptake and processing by the yeast during early fermentation mean that the aroma profile ultimately is more of a reflection of the particular yeast (Bell and Henschke, 2005; Ferrari, 2002; Hernández-Orte et al., 2002, 2006; Lohnertz et al., 2000).

CH_3 CH_3 N N CH_3

2-Methoxy-3-isobutyl pyrazine

FIGURE 3.24 Structure of the compound responsible for the vegetative character of many wines. Note the N in the pyrazine ring.

4.4.4 Phenols

Phenols have been discussed in the context of anthocyanins, tannins and oxidation, but the low molecular weight phenols are also often volatile, especially if esterified, and contribute to wine aroma, sometimes positively but more often negatively, imparting a disinfectant/medical off-aroma ('Band-Aid'). These phenols have diverse origins. They may originate in the fruit naturally or from smoke from a wildfire, or diffuse from oak (Kennison et al., 2008; Ribéreau-Gayon et al., 2006b). Experimental treatment of Merlot vines with smoke shows that extraction and release of the active phenolics continues throughout vinification and is stable once bottled and at levels well in excess of the sensory threshold (Figure 3.28).

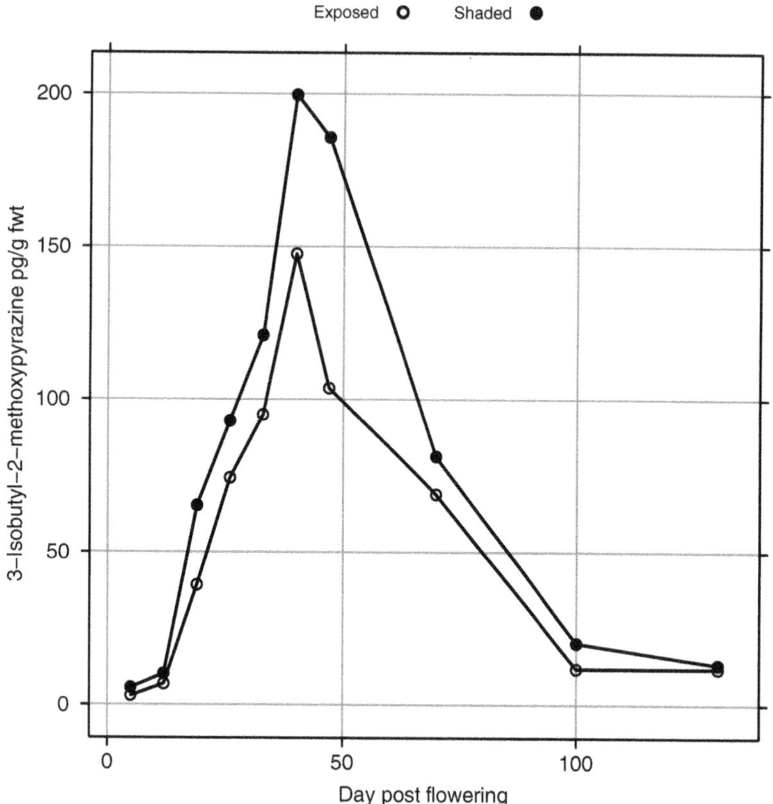

FIGURE 3.25 Change in levels of isobutylmethoxypyrazine (IBMP) in fruit of Cabernet franc clone CL327 growing at Geneva, New York State. *(Data from Ryona et al., 2008, with permission).*

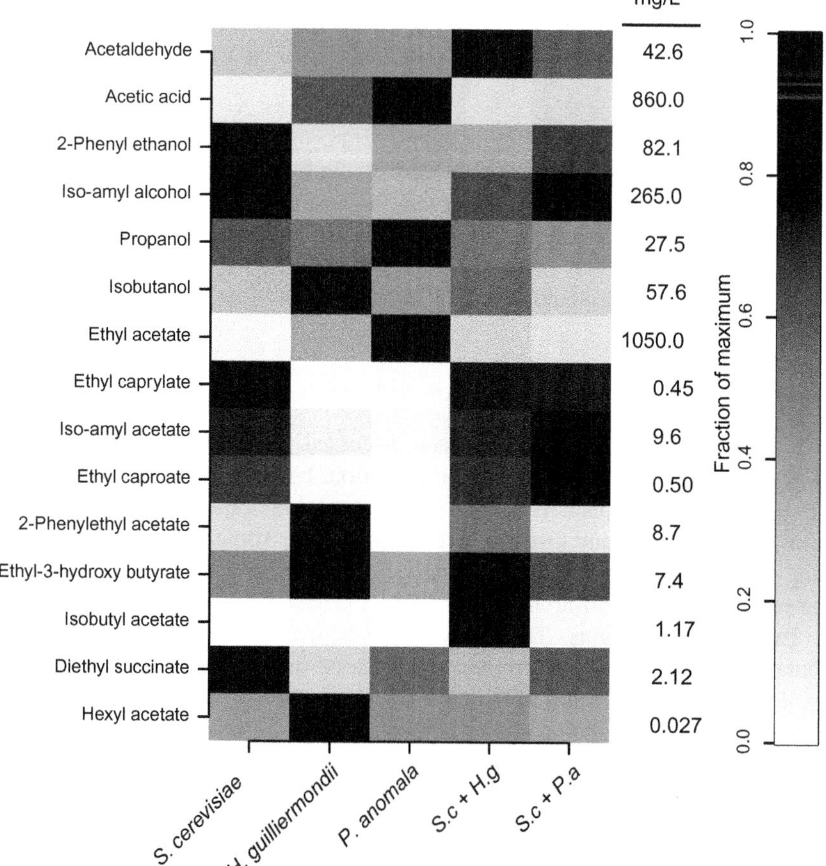

FIGURE 3.26 Volatile products produced during fermentation of cv. Bobal by pure or mixed cultures of *Saccharomyces cerevisiae*, *Hanseniospora guilliermondii* and *Pichia anomala*. *(Data from Rojas et al., 2003).*

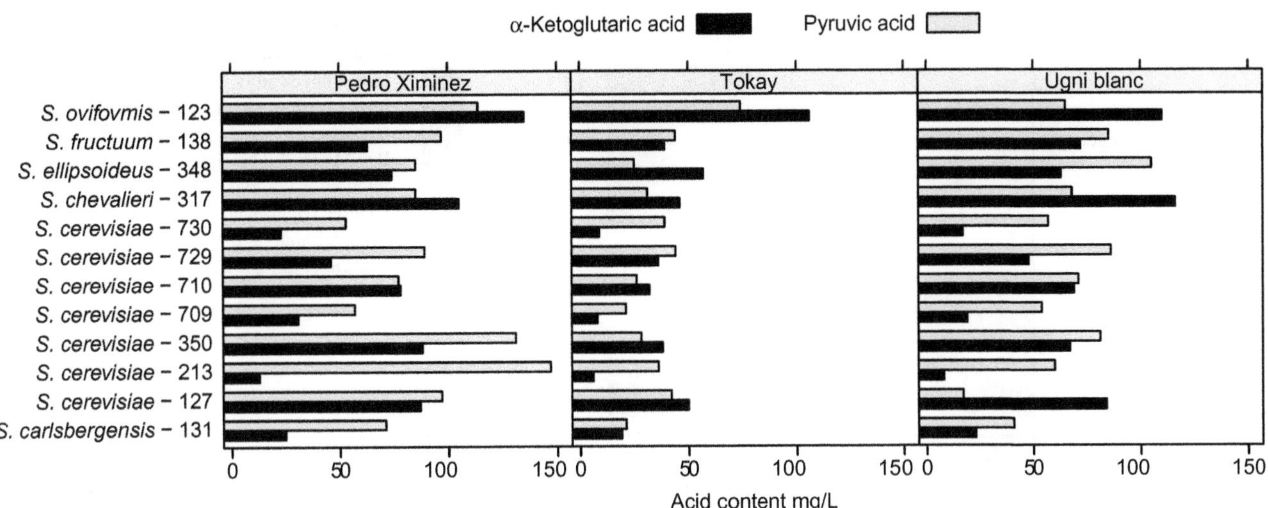

FIGURE 3.27 Impact of yeast clone and species on the production of two key organic acids, α-ketoglutaric acid and pyruvic acid. *(Data from Rankine, 1968).*

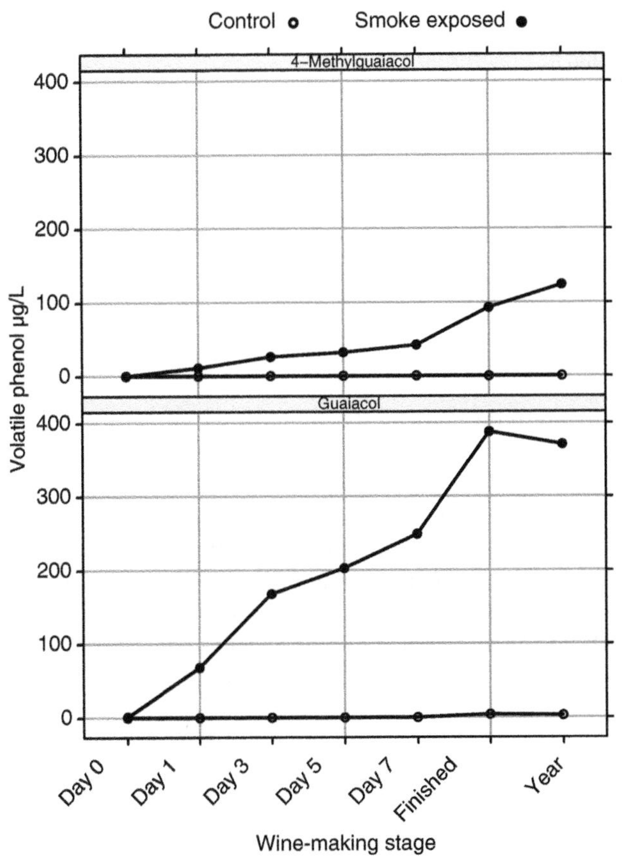

FIGURE 3.28 Extraction of volatile phenolics into wine from fruit of cv. Merlot exposed to smoke for eight successive, 30-min periods during days 0–24 postveraison. The wine was pressed at 3.6° Bé, finished to near 0° Bé, inoculated with *Leuconostoc oenos*, settled, stabilized and then bottled. The sensory threshold for guaiacol is ca 10–75 μg/L and that for 4-methylguaiacol is reported to be 65 μg/L. *(Data from Kennison et al., 2008).*

Otherwise, the primary source of these and related phenolics is oak (Chatonnet and Dubourdieu, 1998). Eugenol and vanillin (Figure 3.12) are the two most prominent in this category, although the lactone, methyloctalactone (oak, whiskey lactone), may be more important as a source of aroma. It differs greatly among the three oak species studied, with *Quercus robur* > *Q. petraea* > *Q. alba* (pedunculate, sessile and white oak, respectively).

4.4.5 Lactones

Lactones are ring structures derived usually from a carboxylic acid; a lactone can be thought of as being an ester formed with itself rather than a second chemical (Figure 3.15). Ellagic acid and its derived hydrolyzable tannins contain these groups (Figure 3.7). They are named after the organic acid they are formed from, usually an aliphatic but sometimes branched. Thus, nonalactone has nine carbons, dodecalactone has 12 carbons, and so on. These are important aroma compounds and while not present in high quantities in wines (Figure 3.18) they impart important, usually pleasant aromas to a wine. Sotolon is associated with fruit that has been infected with *Botrytis.* They have their origin in fatty acid metabolism.

4.5 Bacterial

Bacteria can have a major impact on wine sensory attributes, from the strongly negative due to infection with an *Acetobacter* producing volatile acidity, to the added complexity associated with a malolactic fermentation. Both classes of organism affect the organic acid content of the wine through either the fermentation of residual sugar or

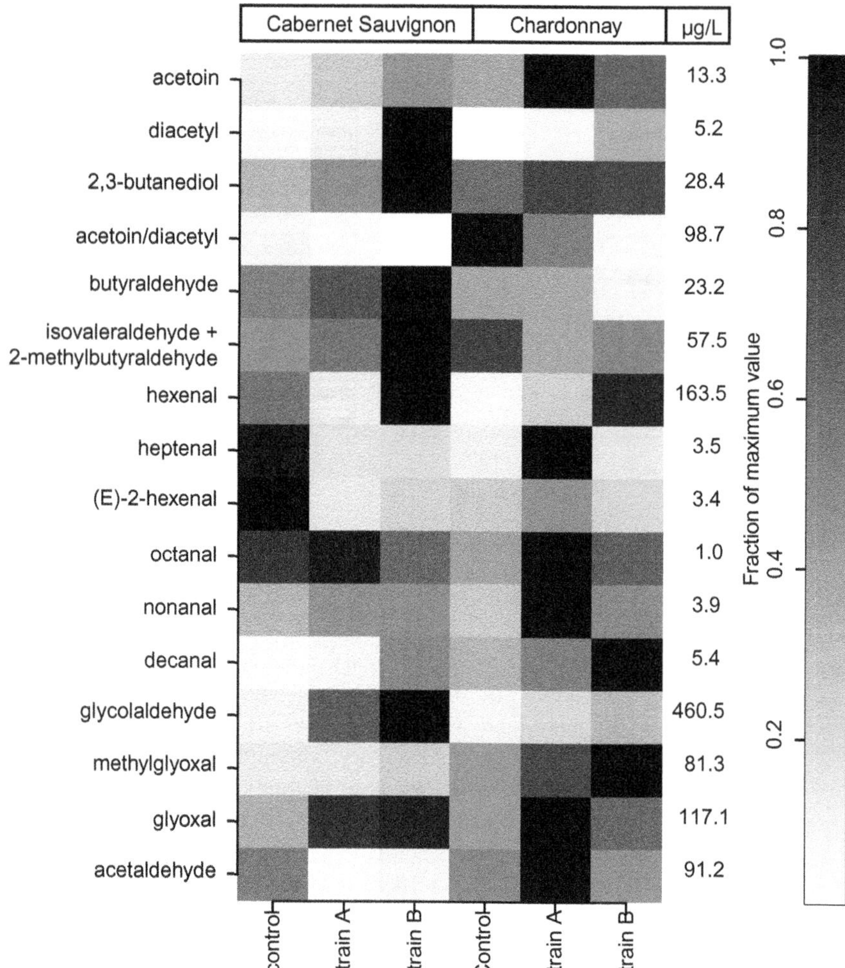

FIGURE 3.29 Change in the profile of volatiles in two wines, Cabernet Sauvignon and Chardonnay, in response to malolactic fermentation with particular strains of *Oenococcus oeni* (A = Uvaferm ALPHA; B = MLD). The figure shows the value as a fraction of the maximum value found for each volatile (right-hand side). The control is the wine without a malolactic step. *(Data from Flamini et al., 2002).*

transformation of other organic acids, notably malic acid to lactic acid. While stability is usually the primary outcome of a malolactic fermentation, the impacts are much broader depending on the strain of organism used and the vessel (if oak).

An example of the impact of a malolactic fermentation (no oak) on the suite of volatiles is depicted in Figure 3.29. Diacetyl imparts a 'buttery' character to a wine with a sensory threshold of <1 μg/L. The production of this is strongly influenced by the strain of the bacterium and by the wine. Acetoin and 2,3-butanediol (butanedione) also impart a 'dairy' character but their sensory threshold is above the levels detected in this example (Table 3.4). Acetaldehyde, while responsive to the microbiology, is also probably undetectable. Its decline in Cabernet Sauvignon is probably unrelated to the microbiology but is related to the presence of anthocyanins and the formation of adducts.

Chapter 4

Safety

Chapter Outline

1. INTRODUCTION

Your safety is your responsibility.

Be aware that the outcome of an accident may not be able to be cured or repaired!

It is your employer's or your educational institution's responsibility to provide a safe working and learning environment, and to provide appropriate training. If you find fault then it is your responsibility to raise it with a supervisor.

This chapter highlights issues of personal safety but does not claim to be complete or to absolve the responsible party, organization or manager from developing and maintaining a safe working environment. Each circumstance will have its peculiar safety characteristics, and recommendations may have changed since printing. So check current regulations and safety advice, including material safety data sheets (MSDSs), e.g. those available from a manufacturer or from government authorities. An audit of safety issues should be conducted for every new situation and the audit should be reviewed and revised at least annually. A person with first aid training should be available at all times.

But, remember, your safety is your responsibility; suing someone after an accident will not restore your eyesight if, for instance, that accident resulted in the loss of an eye!

1.1 Generic Safety Issues

1.1.1 Environment

- **Electrical**: All electrical services should be fitted with an earth leakage detector and these should be tested regularly (monthly); this is especially important in wet areas. This is a problem area in many small wineries, causing regulators to conduct audits.
- **Ventilation**: Gases used, or generated, during fermentation are dangerous and good ventilation and monitoring, especially of carbon dioxide (CO_2), are essential to personal safety. Solvents and corrosive agents must be stored separately and in appropriate containment cupboards.
- **Pressure**: Bottled gases are under very high pressure and frequently have side-delivery ports. Inadvertent opening of a valve, at head height, could lead to death or at least serious injury. An open bottle, if inadvertently dislodged, may become a missile. Secure on site and in store and supervise use.

- **Heat**: Hot water (80° C) and high-pressure steam (ca 130° C) can cause severe burns. Appropriate protective gloves, full facial protection and clothing must be worn when working or cleaning with such hot liquids and vapors.
- **Safety equipment**: This must include fire extinguishers; breathing apparatus if enclosed fermenters or fermentation rooms are used, or if toxic gases such as ammonia or sulfur dioxide (SO_2) are used; CO_2 concentration meter; safety showers and eye wash stations; a fume hood if preparing solutions containing noxious vapors, aerosols or particles. Professional advice and training may be required.

1.1.2 Personal Safety

Particular clothing and protective garments and equipment are required in specific areas:

- **Winery and related work areas**: Hard hats if people or equipment are located above head height; safety glasses; steel cap boots, preferably rubber boots for work in wet areas; gloves for handling dry ice.
- **Laboratory and related work areas** (e.g. stores): Protective gown or coat, full cover boots or shoes; safety glasses; suitable gloves for handling corrosive substances. Note that the minimal safety level must be maintained when working between environments, e.g. a hard hat when entering a 'hard hat' area such as a winery.
- **Training** in safe operation of equipment such as destemmer–crushers, must pumps and presses BEFORE these items are used; no unlicensed use of equipment such as forklifts or tractors.
- **Food and drink**: Do not store, handle or consume food or drink in laboratories, cool rooms, etc.
- **Never run** in a laboratory or winery environment, except in an emergency.
- **Be aware** of the actions of others around you. You have a duty of care toward all other users—including non-technical staff: cleaners, office staff and emergency workers—as they have to you.
- **Smoking**: Do not smoke in a laboratory or any enclosed work environment.
- **Lifting**: If you are required to lift heavy items, ensure that you learn how to do so safely and what the limits are.

1.2 Laboratory-Specific Issues

1.2.1 General

Know the location of:

- **Exits**
- **Isolation switches and taps** for power, gas and water
- **Fire extinguishers**: Know how to use the extinguisher and which extinguisher to use when; know when to run!
- **Spill kits**
- **Safety shower** and eyewash station
- **First aid kit**
- **Emergency contacts**
- **Emergency evacuation point**.

1.2.2 Chemical, Solution and Solvent Handling

Always:

- **Check MSDS statements** for all chemicals and solvents prior to handling; keep up-to-date copies in a convenient location.
- **Adopt safe pipetting practices**: Never mouth pipette; instead, use a pipette bulb or aid or, preferably, use an autopipette and disposable tips.
- **Add acids and bases to water**; NEVER the other way around.
- **Direct** the opening of a flask away from your body when making an addition.
- **Adopt safe disposal practices** (see MSDS); check local authority regulations.
- **Dispense volatile solvents** in a certified fume hood.
- **Transport and handle dangerous substances** in an appropriate containment vessel (e.g. Winchesters for strong acids).
- **Maintain a high state of cleanliness**; this will minimize the risk of inadvertent, adverse chemical reactions.
- **Label containers** fully and durably; labels can be downloaded for many substances, e.g. from http://www.chemalert.com/ (subscription required):

  Handling warning if appropriate (e.g. Corrosive, Toxic, . . . ; Dangerous good label (http://www.chemalert.com/).
  Product name, chemical name(s), United Nations number(s), ingredients, concentration.
  Date prepared, dispose-by date.
  Name or person who prepared solution (if appropriate).
  Name of supplier/importer.
  Risk and safety phrases, e.g. http://europa.eu.int/comm/environment/dansub/main67_548/index_en.htm or http://hsis.ascc.gov.au/risk_phrases.htm.
  MSDS reference.

  Note: Labeling extends not only to the original container, but also to stock and working solutions, and to wastes for disposal.

The purpose of labeling is not only to minimize errors and mistakes by competent and experienced staff members but also to protect the inexpert, the cleaner, emergency services staff, etc.

Never:

- **Return excess chemical** to its container.
- **Store volatile solvents** in domestic refrigerators – these are not explosion proof!
- **Store acids and bases together**.
- **Store oxidizing and reducing agents together**.
- **Store dangerous substances** above head height.
- **Store heavy items above** waist height unless a lifting aid is available.
- **Sniff a chemical** or solvent directly.

1.2.3 Microbiology

A winery is a microbiological factory. The load of microbes and their spores in the air will be exceptionally high. Fortunately, those present in the main part are not a health risk (regarded as risk level 1, the lowest health risk). Be aware, though, that the environment from which the fruit came will contain microorganisms from risk level 2, mainly soil-borne microorganisms. Thus, it is advisable to treat all cultures and samples as being potential pathogens and to handle them accordingly.

The greater risk is of unintended contamination of samples leading to results that may be misleading. Standard practice includes sterilization of vessels and sampling–dispensing apparatus, disinfection of benches, use of gloves, face masks, protective clothing (laboratory coat and full cover shoes) and ensuring that wounds and abrasions are covered and protected. Sterilization may be by heat, alcohol (80% ethanol v/v), chemicals (chlorine or peroxides) or medical grade detergents. If using a commercial laboratory for analysis, ensure that you adhere strictly to its protocols, which will be designed to minimize the risk of cross-contamination.

Inadvertent accidents such as needle-stick injuries must be reported and treated medically.

Legal requirements with respect to laboratories that handle microbiological specimens are usually defined by local regulations (e.g. AS 2243.3: 2010 in Australia and New Zealand). These usually include specific advice on cleaning, cleaning agents and personal protection. (Note: some listed agents are highly inappropriate for wine laboratories: formaldehyde, glutaraldehyde, peracetic acid.) Otherwise, cleaning procedures in the cleaning protocols will suffice.

1.2.4 Spills

If a spill is minor then follow the directions set out below; if a spill is major or toxic, evacuate and call the emergency services. Be aware of those substances that produce highly toxic fumes or gases (e.g. ammonia, SO_2, sodium hydroxide). In those cases, evacuate all people and call the fire brigade. If the spill is volatile and flammable, switch the power supply off at the main switchboard.

- **Acids**: Evacuate non-essential people; check that you are appropriately clothed and protected (if volatile agent, e.g. hydrochloric or formic acid, wear a respirator with the appropriate cartridge); contain the spill with vermiculate absorbent and then absorb the remainder with additional vermiculite: pour an excess of sodium bicarbonate neutralizing agent over the top, mix thoroughly, wait until the reaction is complete (5 min) and then scoop it up and dispose of it in a waste bag.
- **Bases**: As above, but use citric acid as the neutralizing agent; take care (if ammonia or other volatile base, wear a respirator with the appropriate cartridge).
- **Solvents**: As above, but you may need to wear a respirator in addition to appropriate clothing. Contain the spill with absorbent, then use activated charcoal or a proprietary absorbent.

1.2.5 Waste Disposal

Neutralize acids and bases before diluting and disposal in sink (beware of possible heat of neutralization). This is appropriate for small quantities only; otherwise, use a commercial, licensed, waste disposal company. Label and stockpile solvents and dispose through a registered agent for incineration—do not blend.

1.2.6 Sources of Information

- Chemical supplier
- Industry association
- Government agency
- Supervisor
- CRC *Handbook of Chemistry and Physics*
- Merck Index
- Material Safety Data Sheets (MSDS)
- Internet.

1.2.7 Emergency Contacts

Office/Supervisor/Safety Officer: (________________).
Hospital/Ambulance/Emergency Services: (_________) (________________) (________________).
Doctor: (________________).
Poisons Information Center (24 hours): (________________) 13 11 26 (Australia).

2. SELECTED MATERIAL SAFETY DATA SHEET DOCUMENTS

2.1 Potassium/Sodium Metabisulfite

This is classified by WorkSafe Australia as a Hazard:

- **Slightly corrosive**: When added to water it may release a toxic gas which may present a problem in poorly ventilated areas. Long-term exposure may lead to respiratory and skin sensitization. Avoid inhalation of dust, and skin and eye contact.

- **Eye contact**: Corrosive and severe irritant. May result in pain, redness, corneal burns and ulceration with possible permanent damage.
- **Inhalation**: Irritant, slightly corrosive. Exposure may lead to irritation of the nose and throat, coughing and inflammation–ulceration with breathing difficulties. Upon addition to water toxic and corrosive gases are evolved.
- **Skin**: Slightly corrosive to severe irritant. Contact may result in itching, pain, redness, rash and dermatitis. Prolonged contact may result in burns and possible sensitization.
- **Ingestion**: Slightly corrosive. Ingestion may result in irritation of the gastrointestinal tract, nausea and vomiting. Owing to the oxidation in the body of sulfites to sulfates, sulfites are well tolerated. However, in large doses, sulfurous acid is formed. Some individuals may have an allergic reaction.

2.1.1 *First Aid*

- **Eye**: Hold eyelids open and wash with running water for 20 min. Seek immediate medical treatment.
- **Inhalation**: If overexposure occurs, leave immediate area immediately. If problems arise seek medical attention.
- **Skin**: Gently flush affected areas with soap and water. If irritation develops seek medical attention.
- **Ingestion**: If poisoning occurs contact a doctor or the Poisons Center (Australia-wide 13 11 26). Do not induce vomiting. Give the patient a glass of water to drink. Seek urgent medical attention.
- **Non-flammable**: May release toxic sulfur oxides when heated to decomposition.
- **Incompatible with oxidants** (hypochlorites and peroxides) and acids (e.g. sulfuric acid). May release toxic SO_2 gas.

2.2 Carbon Dioxide

Note: Risks associated with confinement in high CO_2 areas should be assessed professionally and strict action taken to avoid such risks by use of appropriate design criteria to prevent inadvertent entry and to ensure ventilation prior to entry to enclosed areas.

CO_2 is classified by WorkSafe Australia as a Hazard (Under Review):

- **Asphyxiant gas**: Non-irritant. Exposure to concentrations greater than 10% leads to death. Use safe work practices to prevent gas accumulation. Symptoms are directly related to the replacement of oxygen in the lungs. Chronic exposure, even at low levels, may result in calcium accumulation in body tissues and kidneys.
- **Eye contact**: Non-irritant but exposure to liquid or solid may result in corneal burns and severe frostbite with possible permanent damage.
- **Inhalation**: Non-irritant, asphyxiant. Overexposure may result in rapid breathing, elevated heart rate, drowsiness with loss of mental alertness, lack of coordination, emotional instability, vomiting, shaking, unconsciousness, coma and death.
- **Skin**: Non-irritating vapor. Contact with liquid (e.g. cold vessels or pipes containing liquid or vapor) may result in frostbite with severe tissue damage. Dry ice has a temperature of $-78.5°$ C and even momentary contact can cause severe frostbite with permanent tissue damage.

2.2.1 *First Aid*

- **Eye**: Flush gently with running water, holding the eye lids open for 20 min. Seek immediate medical attention.
- **Inhalation**: Leave the area of exposure immediately. Medical emergency—call for medical assistance. If assisting a victim, avoid becoming a casualty; wear an air-line respirator or self-contained breathing apparatus (SCBA). *Note that the use of such apparatus requires specific training*. If the victim is conscious, move him or her to an uncontaminated area to breathe fresh air. Keep him or her warm and quiet. If the victim is unconscious, move him or her to an uncontaminated area and give assisted respiration. Continued treatment should be symptomatic and supportive.
- **Skin**: For cold burns, remove contaminating clothing and gently flush affected areas with warm water (30° C) for 15 min. For large burns, immerse in warm water for 15 min. Seek immediate medical attention.

2.2.2 *Precautions*

- **Flammability**: Non-flammable gas. Do not allow storage temperature to exceed 45° C.
- **Reactivity**: Moist CO_2 is corrosive, hence acid resistant materials are required (stainless steel). Certain properties of some plastics and rubbers may be affected by gas or liquid, e.g. embrittlement, leaching of plasticizers.
- **Ventilation**: Use with adequate ventilation. Open windows and doors where possible. In poorly ventilated areas, mechanical extraction may be necessary. Note that CO_2 is a heavy gas ($d = 1.55$, compared with air and nitrogen $d = 1.0$), settles in low areas and tends to fill containers such as fermenters.

2.2.3 Personal Protective Equipment

Wear safety glasses and leather gloves. Where inhalation risk exists, wear SCBA or an air-line respirator. *Note that this requires specific training*—reduce risk by the use of appropriate design and protocols.

3. EQUIPMENT

3.1 Crusher–Destemmer

3.1.1 Safety Issues

- **Auger**: NEVER attempt to clear a blockage or move fruit around while the auger is operating and the machine is connected to power—DISCONNECT before cleaning or clearing blockages.
- **Electrical**: Smaller instruments may rely on an extension lead connection and as wine making is always a 'wet' area activity there is a constant risk of electrocution (it is best to wire a lead directly into a waterproof fitting). NEVER use unless connected to an earth leakage device that has been recently tested and certified. Take sensible precautions to keep exposed connections away from water and conducting liquids, e.g. 'juice'. Ensure that electrical leads are placed such that they do not become a trip hazard.

3.2 Press

This relates to a small 80-L water press with a recirculating pump to conserve water. If using other equipment refer to its standard operating procedure (SOP) and safety recommendations.

3.2.1 Safety Issues

- **Electrical**: The instrument may rely on an extension lead connection and as wine making is always a 'wet' area activity there is a constant risk of electrocution (it is best to rewire with an appropriate lead directly into a waterproof fitting). NEVER use unless connected to an earth leakage device that has been recently tested and certified. Take sensible precautions to keep exposed connections away from water and conducting liquids, e.g. 'must'. Ensure that electrical leads are placed such that they do not become a trip hazard.
- **Lifting**: In some circumstances it is necessary to locate the equipment above a larger receival vessel. Take care in setting this up to ensure a stable and safe working environment. Fruit boxes weigh as much as 18 kg (40 lb); ensure that safe lifting practices are adopted.
- **Slip hazard**: Crushed grapes and wet surfaces make for a slip hazard. Therefore, never run and always take care to avoid falling.
- **Cuts and abrasions**: The edges of metal surfaces are often sharp. Therefore, gloves should be worn; also, gloves are required to avoid damage to hands when mixing the fruit between presses.

3.3 Cold Rooms

3.3.1 Safety Issues

- **Containment**: Small-scale research and teaching fermentations are usually carried out in cold rooms rather than in ventilated spaces as is standard practice in commercial-scale wineries. Such rooms must provide an emergency internal door release.
- **Gaseous**: We have measured CO_2 levels within a fermentation room of greater than 10%, a lethal level. A calibrated CO_2 meter MUST be used to check that the level is safe before entry. ALWAYS ventilate any room containing an active fermentation before entering. Design criteria should provide for forced air ventilation and an interlock to prevent entry while CO_2 concentrations exceed safety limits.

Chapter 5

Table Wine Production

Chapter Outline

1. PLANNING

Successful wine making involves planning. The development of a checklist and timeline (including critical path and risk analyses) is the essence of success. This is the role of the intake manager in a commercial winery. A guide to a list of items of equipment that may be required depending on scale and style is provided in the Appendix.

The styles of wine making adopted here are limited in their scope in order to restrain complexity and to provide a basis on which to build with experience. If you, the winemaker, have a particular style or research objective in mind, then additional reading and research will be required. However, it is worth appreciating that most wineries develop their own particular practices and guard the details closely: in fine wine production, it is the detail that counts. Other texts and treatises elaborate these processes and should be consulted where further information is required (Boulton et al., 1996; Jackson, 2008; Ribéreau-Gayon et al., 2006a, 2006b). As a young winemaker, the best way to develop fine skills is to work as an apprentice in a range of respected wineries.

Legal requirements governing wine composition vary from one jurisdiction to another and should always be consulted (e.g. Standard 4.5.1, Wine Production Requirements; search http://www.comlaw.gov.au as the actual reference changes with updates).

2. QUANTITY

In general, the weight of fruit harvested will be about one-and-a-half to two times the volume of juice desired (Rankine, 1989). The proportion of a bunch that is flesh varies from about 75 to 90%, with the remainder being stems, seeds and skin. The yield of juice is commonly lower in white wine cultivars because standard practice is to press before fermentation and thus the juice present in the skins is discarded or used for secondary products (distillation). Yield is also lower in research and teaching than in commercial-scale production, where it is normal to recover juice and wine from pressings and lees. However, only hard pressing would yield the high commercial values cited by Rankine (1989), and then only for low phenolic cultivars such as Sultana (syn. Thompson seedless), Pedro Ximinez or Doradillo.

The actual yield will depend not only on the degree of pressing but also on berry size, cultivar, maturity and texture of the pulp at harvest (degree of cell wall breakdown), all factors that vary from one season to the next (Blouin and Guimberteau, 2000). The degree of cell wall breakdown at harvest can vary widely between cultivars and even between sites for a single cultivar and sometimes a pectinase enzyme may be added to assist in the extraction, although this means significantly more extraction of skin and seed components (see Figure 5.4). Finally, high-quality white and *méthode champenoise* wines may often be produced from the free-run juice obtained by whole bunch pressing. In these instances the yield may be between only 45 and 55%.

In each of the three scenarios considered in this text, teaching, research and semi-commercial, we assume that the purpose is to produce sufficient wine for a formal sensory tasting of the bottled product. We further assume a capacity for statistical rigor in the assessment. Twenty liters per replicate is a minimal value to enable sensory and analytical analysis, postbottling (thus 60 L+ if you have three fermentation replicates). If this is not your intention then the quantities may need to be adjusted accordingly.

3. CHOICE OF MAJOR EQUIPMENT ITEMS

This section provides a guide to equipment and facilities for teaching and research and small-scale wine making only. Figure 5.1 shows an example of an experimental wine/teaching facility with racks of small 5-L and 20-L fermentations and stocks of 100-L fermentations on the floor. It is therefore limited in its scope. Mechanization of grape and wine production at the commercial scale is subject to ongoing innovation, usually by manufacturers rather than by academic institutions. Therefore, manufacturers'

FIGURE 5.1 Example of an experimental small-scale wine fermentation facility with racks of 5-L and 20-L fermentations being monitored by a researcher. *(Photograph with permission from photographer, Ms J. Wisdom and researcher shown).*

web sites and their agencies may be the best sources of current information. For example, a recent innovation has been the development of vibrating conveyors and electronic berry sorting and grading machines to replace hand sorting and which reduce damage during transfers (e.g. Pellenc, 2011).

3.1 Construction Materials

Wine is a beverage for human consumption and all materials used to hold or process grapes, juice and wine are required to meet food regulation standards. For details of these standards refer to appropriate, local, regulatory authorities, such as the Food and Drug Administration in the USA, Food Safety Australia and New Zealand [e.g. Australian (Food) Standard 4.5.1] and the Organisation Internationale de la Vigne et du Vin (OIV).

Excessive exposure to metals, although perhaps meeting those standards, can also pose problems with the formation of faults and hazes (e.g. copper and iron). Durability is often of concern and modern industrial machinery and containers are therefore constructed of corrosion-resistant stainless steel (austenitic stainless steel, e.g. 304, 316—even these require care, protection and passivation; this process is outside the scope of this text as it requires specific training). Acidity and chelating properties of fruit and wine acids corrode lesser grades of stainless steel, metals or alloys and therefore contact with such should be strictly limited (see Chapter 8 for cleaning procedures).

Alcohol in wine leaches low molecular weight plasticizers (e.g. polyvinyl chloride) from inappropriate grades of plastic, even if classified as food grade (ASA, 1999). Furthermore, tolerance to sanitizing treatments is essential (sodium hydroxide, steam, caustic detergents). Do not use containers purchased from general merchandise stores, even for the storage of chemicals or preparation of solutions and additives.

Purchase items from a reputable food or wine equipment retailer and state on the order that the purpose for the items is for the production of wine. Non-compliant materials may serve as a container but must be lined with a compliant material, a plastic or a natural product such as beeswax or tartaric acid (e.g. concrete open fermenters).

Food Standards Australia and New Zealand recommend against the use of plastic materials when preparing foods for analysis and suggest borosilicate glass (Pyrex®) or stainless steel. Commonly available carboys or demijohns are formed from soda glass and may contain air bubbles. Such glass is brittle and subject to failure on rapid heating or cooling (e.g. filling with juice from cold fruit; if juice is chilled, deliver into chilled vessels). Borosilicate glass, while more expensive, is considerably more resilient to thermal and physical stress. The presence of fragments of glass in a wine is a serious contaminant and handling large fermenters, even when enclosed in raffia or plastic mesh harness, presents a safety issue. Otherwise glass meets the requirements as being chemically resilient and non-contaminating.

High-density plastics, while suitable for primary ferments, are not usually acceptable for storage or secondary ferment purposes because they may be permeable to oxygen (Boulton et al., 1996) and are difficult to maintain in a hygienic state, being subject to deep scratching. However, new polymers are claimed to overcome many of these limitations and to suit the construction of high-performance wine storage tanks (Flecknoe-Brown, 2004). Glass and stainless steel, in contrast, are not easily scratched and therefore hygiene is more readily maintained in these vessels.

Rubber or neoprene is not suitable as a bung, cork or seal in small fermenters. Silicone rubber should be used or the bung enclosed in an inert, food-grade stretch film such as Parafilm®.

Wood is a traditional material and continues to be used for secondary fermentation, occasionally for primary fermentation when prolonged contact with oak and the flavors it imparts is desired, along with extended lees contact (as in traditional Chardonnay production) and for aging fine red wines. Wood presents a challenge in that it is porous and hygroscopic (absorbs water and swells when wet), and wooden vessels are constructed of abutted staves providing opportunities for microbial contamination. Wooden barrels, once used, must be treated with care and remain hydrated and in a hygienic state (see Chapter 8 for cleaning techniques).

3.2 Destemmer–Crusher

A wide range of machines is available from the major wine equipment suppliers. Design and construction are important considerations. The machine should be corrosion resistant and easy to clean (and disassemble/reassemble), and any bearings should be external to the processing area. All electrical items must be fully waterproofed. Small, semi-commercial machines are generally fitted with single-phase electrical motors and a short electrical lead. A safety switch should be fitted on the machine itself and the short lead replaced with an industrial-quality lead that will reach the power source without an intermediary connection. For examples of recent developments in technology, see Bucher Vaslin (www.buchervaslin.com); but these machines are designed for high-volume, commercial production wineries.

Two types are generally available: crusher–destemmers and destemmer–crushers. The latter machines are generally considered to be more appropriate because they reduce contact between juice and stem/leaf material, reducing microbial contamination and extraction of vegetative flavors and aromas. In general, the berries are caught in a hole in a rotating cage and separated from the pedicel, giving a product not unlike that produced when fruit is harvested mechanically. Destemmer–crushers also enable the preparation of fruit for whole-berry processing such as carbonic maceration. Ideally, the speed of rotation of the screw conveyor will be variable, a range of screen sizes will be available and the gap between the crusher rollers will be adjustable. The rollers must made be of food-grade material. Bearings should be sealed and, if requiring lubrication, should be lubricated only with food-grade lubricants. The auger must be guarded during use.

3.3 Presses

A commercial winery may have access to a wide range of presses: air-bag (membrane), basket, continuous and gravity, with air-bag (membrane) presses generally being the mechanism of choice. These are gentle, causing little damage to the skins and, consequently, juice of low solids content and, importantly, can be automated. While one manufacturer once produced a small-scale membrane press suitable for research and teaching, this is no longer the case and small-scale winemakers must accept the constraints associated with other mechanisms. Their options are water-bag presses and small basket presses. The current authors use a modified water-bag press. For further information on commercial machines and commercial suppliers, see Boulton et al. (1996) and Ribéreau-Gayon et al. (2006a).

Small basket presses are fitted with a screw mechanism and, usually, wooden staves only. A hydraulic mechanism is much preferred to a screw because it enables reliable control of the process (cf. large commercial presses). If installing a hydraulic mechanism, it should include overpressure safety switches and a pressure gauge. Only in this way can the process be controlled and repeated. Even in this case, if the press is of a wooden stave construction, it should be modified by the addition of an internal stainless steel screen. An external barrier or plastic bag to retain a carbon dioxide (CO_2) cover may also be a useful addition. A problem common to many basket presses is uneven pressure distribution across the radius of the machine, leading to overpressing and underpressing of berries. However, Ribéreau-Gayon et al. (2006a) regard these as eminently suited to whole bunch pressing and the extraction of juice for fine wines, although laborious. They are laborious because the fruit must be redistributed by hand between presses. The alternative of a screw mechanism is not suited to research or quality wine production except perhaps for small-scale commercial red wines, and may be found in many boutique wineries.

A water-bag press modified with a recycling pump, pressure relief valve and a removable, stainless steel external shield (see Appendix) is suitable and meets environmental issues of minimizing water use. While not ideal and labor intensive, this type of press enables controlled pressing of whole bunch, whole berry, crushed berry white wine and red wine pressing. These presses are available in 20, 40, 80 and 160 L capacities. A limitation is that when the bag fails, water contaminates the pressings. A competent workshop may be able to modify the press to enable it to be air operated, eliminating that problem. Vertical orientation of the three smaller presses also means that there is an uneven distribution of material. This is not ideal and places excessive stress on the upper area of the bag, contributing ultimately to its failure. It is important to possess a spare bag in case of such a failure.

3.4 Fermenters

Fermenters are classified as open or closed, and if closed, then of fixed or variable volume, static, rotary or autofermenters. Closed fermenters are appropriate to white wine production, completing red wine fermentation, storage and such processing activities as fining, settling and clarifying wines and juices. For small-scale fermentations, sophisticated design features found on industrial-scale fermenters are rarely available, although they can be custom built to meet the requirements of scientists and educators.

Stainless steel kitchen stock pots of 25–50 L volume make excellent miniscale open fermenters suited to red wine production for teaching and research purposes. The availability of a close-fitting lid enables extended extraction on skins and lees with a CO_2 cover. For larger volume ferments, small, variable-volume stainless steel fermenters are suitable but are considerably more expensive if replication is required or large numbers of students are involved. Industrial food-grade plastic bins, which may be leased, are an economic solution for semi-commercial red wine fermentations of 300–400 L (sufficient to fill a barrel). Once again, these are usually fitted with a lid and may be filled directly with fruit from a mechanical harvester, eliminating the need for a destemmer–crusher (but note that such fruit may contain contaminants; 5% is a common value). Otherwise, a commercial open fermenter with a door and sloping floor may be purchased or constructed to custom requirements.

Temperature is an issue that needs careful management: as the volume of a ferment vessel increases, the heat-exchange problem increases (Boulton, 1979). Larger volume fermenters will require an immersion or an external heat exchanger to maintain effective temperature control. Generally speaking, wine making for teaching and research purposes is conducted in large, cool or cold rooms, enabling many comparisons and replicates to be fermented under similar conditions.

White wine and the finishing stage of red wine fermentations require closed, anaerobic fermenters as both are subject to oxidative faults if left open to oxygen/air. In some regions of the world, and for particular white wine styles, oxidative prefermentation/fermentation practices are adopted and the brown products removed at clarification, postfermentation (e.g. Boulton et al., 1996; Ribéreau-Gayon et al., 2006a). This style of wine making is not included here. Open access to remove skins and lees is not usually a problem at this scale and for this purpose and therefore bottles, demijohns and carboys are highly suited to small ferments when fitted with an airlock and a bung. For safety reasons, vessels larger than about 5 L should be fitted with a mesh holder to minimize the risk of an accident in the case of breakage. For larger scale ferments, stainless steel kegs (beer kegs) are highly suited as these also may be fitted with a bung and an airlock and are not subject to breakage. Alternatively, food-grade plastic barrels may be used by fitting a bung and an airlock to the lid. These are readily available from brewing and wine-making suppliers.

3.5 Cooperage

Historically, oak barrels were the vessel of choice for fermentation and for storage of red and white wines. In modern times, oak is used sparely for fermentation but widely for storage of red table wines and for aging fortified wines. Oak is not a neutral material and imparts characteristic flavors including vanillin, oak lactone isomers and oak tannins (Spillman et al., 2004), although this diminishes to near zero after 3–4 years of use (the barrel is then said to be neutral).

Oak is chosen not only for the happenstance of the complementarity of the flavors and aromas but especially because of its suitability for tight cooperage (Singleton, 1974). Oak species used include *Quercus robur*, *Q. petraea* and *Q. alba*, respectively, French common and white oaks and American oak, but grades exist for all and producers of quality wines pay great attention to selection. The common volumes are 225 L (barriques), 300 L (hogsheads) and 480 L (puncheons), but barrels are available up to 700 L (e.g. Philippe Morin, Seguin Moreau, Tonnellerie Quintessence). The use of barrels with a variety of sources, use history and volume gives a winemaker many options for blending.

Selecting barrels is an art as the characteristics of the barrel need to be matched, sensorily, to those of the wine. A great diversity of sources occur that differ in quality and character along with diverse methods of manufacture. Factors to consider are species, origin (locale, forest), method of drying and toasting (heating) and degree of toasting, which imparts a tar character to the wine, and volume of barrel: the smaller the volume, the greater the oak character imparted.

Time in oak is important because oak is permeable, resulting in evaporative loss of water (the so-called ‘angel’s share’) and ingress of oxygen (microoxygenation). Volume can be restored by topping up with wine set aside and stored for the purpose. However, prolonged storage can lead to overoxidation—there is an optimal period of storage for particular wine styles and regular tasting should be conducted to assess the integration of flavors and aromas and the acquisition of oak and oxidative characters. Commonly, this is about 9 months, but may extend to 24 months or be as short as 3 months depending on wine style and barrel volume and age.

There exists also a large range of oak barrel alternatives: staves, sticks, chips, blocks, beads, rice and powders. Research and development (R&D) for these is mainly commercial and in-house. They are generally intended for lower price-point wines but can deliver good sensory outcomes (discuss with a supplier representative, e.g. Seguin Moreau, Philippe Morin). Oak barrel alternatives are a good option for teaching purposes.

3.6 Transfer Pumps, Lines and Fittings

A wide range of pumping mechanisms is used in wine making (e.g. Boulton et al., 1996). Most of these are not relevant for the purpose of small-scale wine making except perhaps for demonstration in a teaching context. Transfer of juice is usually done by hand with a container rather than by pumping. A general-purpose, positive-displacement pump is desirable for small-scale wine making, one that can operate with both liquid and some solids and with small-diameter lines, e.g. 25 mm. Diaphragm and peristaltic types fall into this category, as do certain types of piston pump, although these are usually fitted with larger lines owing to the slurry of skins and seeds with which they are designed to work. Flexible impeller pumps (e.g. Jabsco®) may also be used, as may centrifugal pumps, but neither of these is good at handling even small quantities of particulate matter, skins and seeds. Both should have a screen fitted to the intake end of the tubing. Centrifugal pumps are not self-priming and will need to be primed or placed such that there is a positive head on the intake side. Nor do they cope with entrained air. Ideally, pumps are fitted with a variable speed controller and must be constructed so that they can be used safely in a wet area.

All lines must be of food-grade materials: stainless steel for fixed lines and fittings, and a food-grade plastic for flexible lines and hoses. Exposure to copper and iron should be kept to a minimum. Food-grade polypropylene, polyethylene and silicones are suitable.

All fittings should be designed for ready cleaning. British Standard Milk (BSM) fittings and unions are widely used in the wine industry (stainless steel). Other fittings that are suitable are clean-in-place (CIP), clean-in-place flat-face (CIPFF) gasket fittings and Tri-Clover® joints. Ball valves are generally used as they have an open center and allow access and the passage of solids.

3.7 Filters and Bottling Equipment

The equipment consists generally of a stainless steel keg into which the wine is transferred and may be pressurized, a coarse, in-line filter, a final filter, food-grade tubing, filling apparatus, a closure machine, a labeler and a capsule or foiling machine. The keg is connected to a pressurized cylinder of food-grade nitrogen fitted with an appropriate food-grade gas regulator. Again, ideally, filtering and bottling should be conducted in a dust-free, food-grade room that can be cleaned daily.

Filtering is an important aspect of wine making, even for the home winemaker. Both red and white wine should be filtered through a coarse pad (plate and frame), in-line filter of about 5 μm exclusion pore diameter. Laboratory-scale filtration equipment is suitable for teaching and possibly for research, although the small capacity of the filters generally means that only a few bottles at a time may be filtered, unless the wine is very thoroughly settled and carefully racked. Larger capacity, pad filtering equipment is required for semi-commercial-scale wine making.

Final filtering of wines requires a membrane filter with an exclusion diameter of less than 0.45 μm to ensure brightness and stability of white wines and microbial stability for both red and white wines.

3.8 Controlled-Temperature Rooms

Production of microscale to semi-commercial-scale wines for research and teaching can be carried out successfully in controlled-temperature rooms. In many cool-climate regions, these may not be required for small-scale wine making because of the low air temperature. However, while controlled-temperature rooms enable the production of many wines under similar conditions, they are a safety risk, unless continuously ventilated, because otherwise the room may contain toxic levels of CO_2. Therefore, such rooms must be able to be rapidly ventilated to provide safe working conditions for those who enter to monitor the wines, as well as an internal release mechanism for the exit and an external CO_2 monitor panel.

To minimize movement of ferments, the refrigeration capacity should be such that the temperature range can be maintained between −5 and +25°C or have separate cold and freezer rooms with an efficient means of moving ferments between the two. A fermenting wine, if thermally insulated, will rise in temperature by about 0.65°C per° Brix (Boulton et al., 1996) and will normally be several degrees warmer than atmospheric temperature during active fermentation. A cold room or freezer capable of holding wine at −4 to −2°C is required for settling and holding white juice prior to fermentation and for settling and stabilizing white red wines postfermentation. White wine ferments will be carried out at 14–20°C and prefermentation cold-soaking or maceration may be carried out at 10°C. Thus, a system that enables considerable flexibility is required (and will usually be designed by a professional refrigeration engineer).

Refrigerated sea containers are an item that may be leased and are generally suited to the purpose and have ample refrigeration capacity. However, the standard container is not fitted with a door that can be opened from inside, although such a door should be fitted for reasons of safety. Alternatively, custom-made jacketed fermentation vessels can be constructed, giving the student or researcher full control, at least comparable with a commercial facility. Their design does not, however, fall within the ambit of this text.

All internal surfaces, including shelving, must be able to be washed down and cleaned. The floor should be fitted with a large-capacity, grated drain.

Barrel room and bottled wine storage rooms have an additional requirement of humidity control (relative humidity >75%) to reduce loss of volume due to evaporation, drying of corks and leakage.

4. BACKGROUND TO WINE-MAKING PRACTICE

4.1 Record-Keeping

As in all other aspects of wine production, complete and accurate records are essential to quality assurance and for problem solving. An example of a simple set of records is provided as an Excel spreadsheet on the companion website to this book http://booksite.elsevier.com/9780124080812 but we advise the use of written records or Microsoft Access, or another relational database, as such software limits the risk of inadvertent mis-sorting and corruption of data. Hand-written records are safest.

4.2 General Issues

Fine wine production is, in good part, the technology of extracting desirable aroma and flavor precursors while limiting undesirable compounds, and the art of being able to make fine judgments. Along with choice of harvesting date, prefermentation treatments play a vital role in achieving that goal. Prefermentation treatments determine the scope you have to work with in the latter phases of the wine-making process.

In red wine production, the key is to ensure maturity and minimal contamination with leaves and stems—not only from vines but perhaps also from windbreak and ornamental species, small quantities of which may have a profound influence on a wine's sensory attributes (Capone, 2012).

The task is more demanding for white wine production because the pressing technique interacts strongly with fruit maturity to determine the balance of extraction and the final quality of the wine. In general, handling and extraction should be gentle and produce a juice of high clarity, ideally about 200 NTU (nephalometric turbidity units), but 500 NTU is regarded as satisfactory (Ribéreau-Gayon et al., 2006a). Aspects of wine chemistry that may usefully be monitored are pH (rises with extraction), optical density at 280 nm [OD_{280}; rises with phenol extraction, compare with polyvinylpolypyrrolidone (PVPP) blank], OD_{420} (rises with oxidative browning) and sensory attributes (e.g. bitterness and herbaceousness). Protocols for these analyses are provided in Chapters 10 and 11.

The general process for red wine is outlined in Figure 5.2 and that for white wine in Figure 5.3. These figures show the stages (alphabetical values) and generalized trends in the key parameters: ° Baumé, pH, anthocyanins, tannins, free sulfur dioxide (fSO_2) and temperature, while the process is outlined more explicitly in flowchart form in Chapter 8.

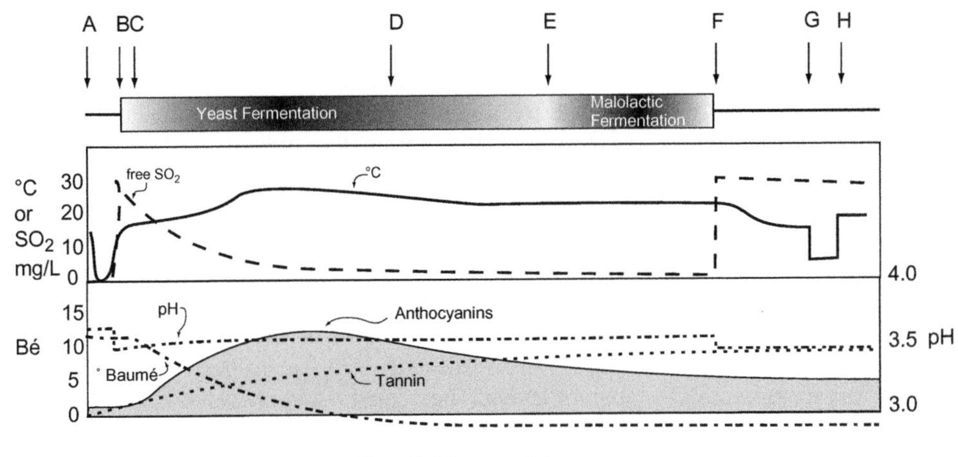

FIGURE 5.2 General course of a red wine fermentation. Note that the scale for tannins and anthocyanins is arbitrary. A: Harvest; B: crush; C: initiate yeast fermentation; D: press off skins and seeds; E: initiate malolactic fermentation; F: aging, G: fine, cold-stabilize; H: final adjust–filter–bottle.

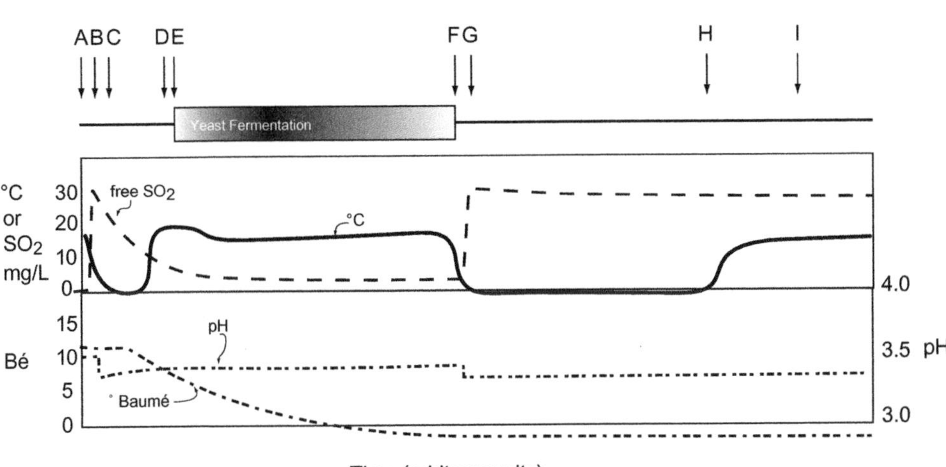

FIGURE 5.3 General course of a white wine fermentation. A: Harvest; B: destem–crush–press; C: adjust and settle; D: rack, bring to temperature; E: initiate and monitor yeast fermentation; F: rack–fine–adjust; G: cold–stabilize; H: rack, final adjust; I: filter–bottle.

4.3 Protocol Development

This section describes a minimalist protocol, one suitable for novices. However, the individual may design their own protocol based on style, cultivars, locations and season, or discuss with local winemakers, and some additional guidance is provided (but see also texts by Boulton et al., 1996; Jackson, 2008; Ribéreau-Gayon et al., 2006a, 2006b).

4.4 Preparation for Ferment and Choosing When to Harvest

Check all equipment and supplies well before the date of anticipated harvest. Ensure that everything is clean and in good working order, and that essential spare parts are on hand. It is a good idea to develop a checklist rather than rely on memory.

One of the most important preparations is following maturity and determining an approximate harvest date by monitoring fruit in the field. Regular maturity assessments should be carried out from veraison to harvest (see Chapter 7), on a weekly basis at first, and then twice weekly as harvest approaches. A commercial winery will plan its harvesting on the basis of capacity optimization and with respect to climatic events, adopting a risk minimization strategy. Small-scale/lot winemakers can adopt more of a purist approach, harvesting selectively, even by berry in extreme instances (e.g. some Sauternes).

The duration of ripening is approximately 6 weeks, with sugar accumulating at about 1° Brix per 3 days or 1° Bé every 5 days (Blouin and Guimberteau, 2000; Coombe and Iland, 2004). Fruity, aromatic wines are usually made from fruit harvested between 23.5 and 24.5° Brix. Vegetative, crisp and acidic table wines are made from fruit harvested 1–2° Brix earlier or from blends of fruit harvested at a variety of maturities, perhaps even as low as 17° Brix, through to fully mature fruit. In cool climate regions, lack of thermal ripening conditions at the end of harvest may set the maturity at a less than ideal level. Each region will, however, have its own 'ideal' maturity and this too will vary seasonally as climate is a principal determinant. Discuss your options with a local winemaker.

Brix is not the sole determinant of ripeness for wine making: titratable acidity (TA), pH and flavor maturity are as or even more important (see also Chapter 7, Berry Maturity by Sensory Analysis). TA, pH and, in some jurisdictions, even sugar can be adjusted but flavor cannot be, so this aspect is the most important element once the fruit has a sufficient level of sugar to become stable following fermentation (ca 9–10%). In cool areas, rachis (stem) maturity may also be used as an index because immature stems will impart a bitter 'stemmy' character to the subsequent wine if in contact with the juice for any significant period. This is an issue for hand-picked fruit only, and especially for those styles (e.g. Pinot noir) where some winemakers use extended cold soaking and the addition of a proportion of stems in an attempt to increase color extraction and color stability (through copigmentation with tannins and other phenolics, principally procyanidins). Making and comparing wines made from fruit at different states of maturity and with or without some stems is a good teaching activity.

4.5 Harvest, Transport and Storage

Harvesting seems like a simple and uncomplicated, if at times laborious, activity. However, the modern viewpoint is that 'wine making begins in the vineyard'. Harvesting is the end of a long and thoughtful process designed to deliver fruit of a particular quality to the winemaker. At least as much thought should be put into producing fruit to this point, and ensuring that it is delivered in an optimal state to the winemaker, as is put into the wine-making process itself. Therefore, picking should be approached intelligently and be guided by the advice of the winemaker: damaged and diseased fruit should be avoided. The optimum window for harvest depends on cultivar and style; thus, cultivars such as Semillon or Riesling may be picked for a dry, table wine style or a sweet, late-pick style. Some such as Verdelho may have a very short optimum harvest period, as short as 2 days, while others such as Shiraz (Syrah) are more forgiving. This too depends on location and climate, so experience counts. Quality, quantity and legal (food-safe) issues each requires deliberate consideration. Protocols for sampling and defining maturity are presented in Chapters 7 and 10.

The task of choosing when to harvest is often assigned to the winemaker rather than the viticulturist but the winemaker will rely on the viticulturist for evidence to inform his or her decision making. Most countries have legislation governing legal limits for pesticides and fungicides termed maximum residue limits (MRLs). These are set by the individual nation. Those of the intended destination are those that apply to your fruit, regardless of where the fruit is produced (e.g. AWRI, 2012b). Commercial wineries will require vineyard managers to keep and provide at a harvest a Spray Diary that records the date of application and the rate of all chemicals applied to the vine. In Australia and New Zealand a guide is available online from the Australian Wine Research Institute (AWRI, 2012a).

It is important that the winemaker and grower reach an understanding regarding appropriate use of chemicals before flowering because some have very long withholding periods. However, there are other legal limits that may be beyond the control of the viticulturist that also need to be measured, such as salinity and especially sodium levels. Thus, good communication between winemaker and viticulturist is essential.

Even small vineyard blocks or patches may be quite variable. These variations have important consequences for the winemaker. Traditionally, sampling techniques have assumed that variation within the vineyard is random. However, this is only a part of the story and stable, spatial variation generally contributes about 50% of the overall yield and maturity variance. Measure this if you can and adjust your harvest procedures accordingly (see Chapter 7).

Harvesting should be conducted in cool conditions. This is a simple matter with machine-harvested fruit which is usually harvested in the evening, after heat accumulated during the day has dissipated. A few vineyard managers even hand harvest at night, but usually, hand harvesting is carried out from first light until mid-morning and ceases before the fruit temperature rises above about 20°C. Fruit, once harvested, should be processed immediately or within 24 h, otherwise serious off-flavors may develop. If long-distance transport is unavoidable, then a higher than normal level of SO_2 should be applied (as much as 200 mg/L; note that this will slow the start of fermentation) (Makhotkina et al., 2013). If feasible, the fruit should be chilled before transport. Postharvest defects are mainly oxidative faults and thus maintaining cold temperature and sulfuring at harvest will minimize the risk of damage.

Modern mechanical harvesters are capable of providing fruit in excellent condition to a winery, sometimes in a better state than hand-harvested fruit (Allen et al., 2011). This provides the simplest and most convenient way of obtaining fruit for semi-commercial-scale wine making. The one disadvantage of mechanical harvesting is that it does not discriminate between intact and damaged fruit or fruit infested with fungi such as *Botrytis*, unless a sorting table or machine is available. The usual way to manage this is to rogue damaged fruit before harvesting (i.e. hand pick damaged fruit and drop on the ground).

Hand harvesting can be selective and provide superior fruit for super-premium grades of wine. As a supplementary means of roguing damaged or infected fruit, harvested fruit can be passed over a sorting table or conveyor prior to destemming and crushing/pressing. With hand harvesting allow 1–2 h per 100 kg per person. Hand-harvested fruit has been shown to require less bentonite for fining owing to a lower content of unstable proteins (Pocock and Waters, 1998), an issue that is more important in the production of white wines than red (Waters et al., 2005).

It is usual to add 70 mg/kg total sulfur dioxide (tSO_2) as $K_2S_2O_5$ in liquid form to bins prior to harvest if mechanically harvesting. This is important for machine-harvested fruit because most machines harvest single berries and leave the stems attached to the vines. Vertical impact-style harvesters will harvest whole bunches but require specially designed trellis systems (Intrieri and Poni, 2000) and may not suit all cultivars. For hand-harvested fruit in perforated containers, it is sufficient to sprinkle a little solid SO_2 over the fruit and to add the balance at crushing. Pointed picking secateurs are the most common tool used, for occupational health and safety reasons. Some cultivars with thin, unlignified peduncles may be picked without the aid of a tool. Hand harvesting is the usual process for teaching, research and small-scale wine making.

4.6 Destemming and Crushing/Pressing

4.6.1 *In Brief*

Red Wine

Fruit should be cool; below 20°C.

With hand-harvested fruit, the emphasis is on destemming with minimal but complete crush; free-run juice and berries go directly to the fermentation vessel. Mechanically harvested berries may be gently crushed with emphasis on separating material other than grapes (MOG) as far as possible before passing to the fermenter. Some stems may be returned to cultivars such as Pinot noir and Grenache (possibly only during a cold-soak treatment) but only if the stems are fully mature and free of obvious fungal infection.

If $K_2S_2O_5$ has not already been added, add this in liquid form progressively during the crush.

White Wine

Fruit should be cold; below 10°C.

Destem hand-harvested fruit and crush and collect free-run juice (generally the first two pressing cycles) into a vessel filled with CO_2. Alternatively, use a whole bunch press. Mechanically harvested fruit may go directly to a drainer, possibly after a sorting process to remove MOG (in commercial circumstances and for particular wine styles, pectinase enzymes may be added to help with extraction, e.g. Sauvignon blanc). When all free-run juice has been collected, transfer the remaining material, gently and directly, into a press. Press through two or three successive press cycles and collect the pressings into a vessel under a cover of CO_2.

If $K_2S_2O_5$ has not already been added, add progressively during the crush.

Commercial wineries may retain some or even all stems if they have the option of using a drainer (a very tall vessel) and the stems are fully mature. This is not an option for the small-scale winemaker. However, some skin contact may be experimented with, especially if using pectinase enzymes to improve flavor-precursor extraction and juice from skins (Figure 5.4). Be aware that this process will also lead to an increase in juice pH as K^+ is extracted from the skin.

4.6.2 *Detail*

Hot water or steam clean the destemmer–crusher just prior to the operation. Calculate the weight of $K_2S_2O_5$ required according to the weight of fruit and the anticipated yield of juice. This is usually prepared as a 1% m/v solution and added progressively during the crush if not already added at harvest. Note that the concentration of $K_2S_2O_5$ required may be varied according to the state of the fruit, or, if wishing to delay the onset of fermentation to increase color or aroma extraction, then increase, and vice versa (e.g. Makhotkina et al., 2013).

It is feasible for the small-scale winemaker to ferment berries from red wine cultivars directly, uncrushed, from a mechanical harvester or destemmer into a small, open fermenter (cf. carbonic maceration, although that would be into a closed fermenter under CO_2 and controlled temperature). This is good for building muscle fitness (i.e. the ferment requires a lot of physical input to ensure complete mixing and extraction) during the early stages of fermentation and maceration.

Run a small part of the fruit through the destemmer–crusher and check the adjustment of the rollers and the speed of the auger. All berries should be removed; if they are not, you may need to change the size of the screen provided your machine has a variety of screen sizes available. Check that the berries are crushed but that no seeds are damaged by the rollers. Thorough crushing will speed the onset of the ferment and make cap management easier during the early stages, but may reduce color and tannin extraction. It is common to run the juice into a receiving vessel with a cover of CO_2 (chips of dry ice).

Clean up immediately to minimize the risk of creating a source of contaminating microorganisms for future crushes.

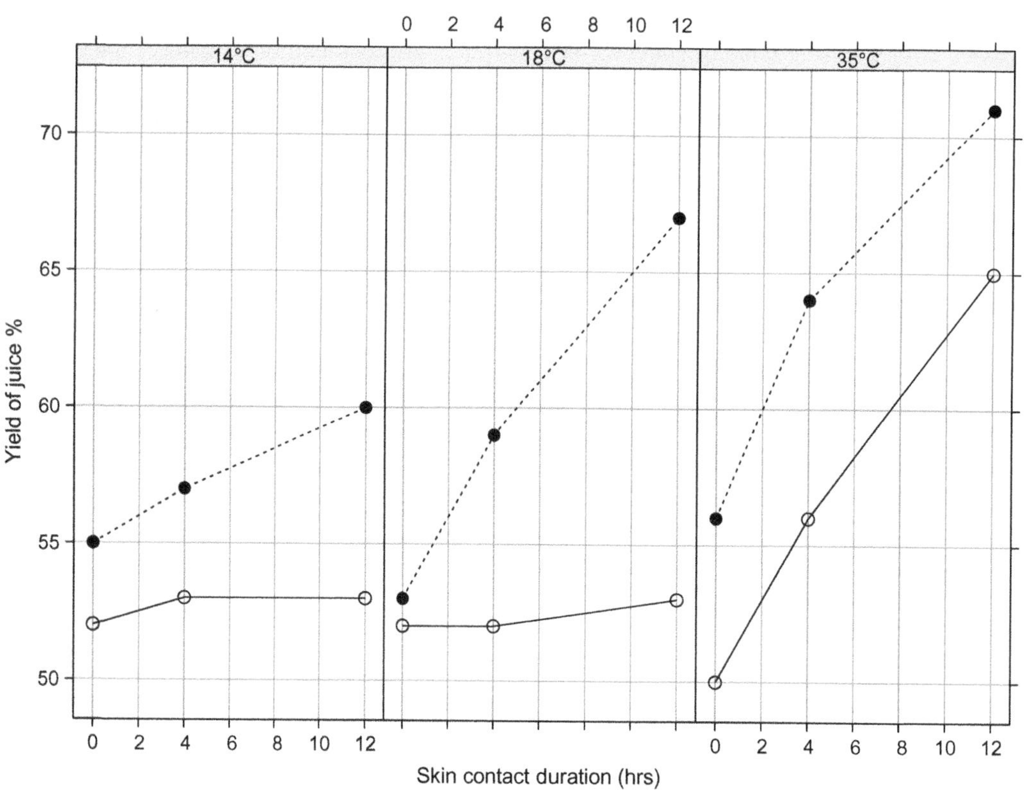

FIGURE 5.4 Effect of temperature, skin contact duration and addition of pectinase at crushing on yield of Chenin blanc juice. Note how important pectinase addition is in promoting juice release and thus recovery at moderate temperatures. The crushed berries would need to be raised to an unrealistically high temperature to achieve a similar result. Open circles: control; closed circles: plus pectinase. The labels in the upper rectangle indicate the temperature at which the juices were held while on skins. *(Data from Ough and Crowell, 1979, Figure 1, redrawn with permission).*

4.7 Clarifying and Making Preliminary Adjustments (White Wines)

Depending on prefermentation treatments, cultivar and seasonal issues, white wine juice and pressings will contain appreciable levels of pectins (from cell walls) and proteins. Retaining these will lead to excessive frothing and possibly stable froths that will be problematic. These are removed by cold settling or possibly centrifugation. To aid this process a pectinase enzyme is added to reduce the gel-forming behavior of the pectins and thus to allow them to settle under gravity with other cell wall and cell materials. Pectinases are available from a wide range of sources but only those from fungi are allowable in Europe (OIV regulation, *Trichoderma hazarium* and *Aspergillus niger*). The temperature optimum for these enzymes is about 27°C but such a high temperature is inconsistent with modern wine making and more commonly a temperature of 16–18°C is used initially and then lowered to complete the settling process (e.g. Figure 5.4).

It is also possible at this stage to remove excess soluble proteins that may contribute to frothing and subsequently to haze formation postbottling by the addition of a small quantity of bentonite (e.g. 0.5 g/L) as a settling aid. Be aware, however, that this may reduce the aroma potential of the wine and will preclude *sur lie* treatment postfermentation as may be practiced with Chardonnay or with sparkling wine production.

Other adjustments at this stage include pH (see Chapter 8) and a check on SO_2 levels.

This process is usually carried out close to fermentation temperature and under a cover of CO_2. Note that low temperature will slow the activity of the pectinase enzyme, but check the manufacturer's advice (Ough and Crowell, 1979; Ribéreau-Gayon et al., 2006a). In commercial wineries the settling process may be avoided by passing the juice through a continuous centrifuge with considerable savings in energy and time and of juice. A heat exchanger is then used to bring the juice temperature down to the level required at the beginning of the fermentation. In the small-scale operation, the settled (cellar bright) juice is racked from the pressing lees into the fermentation vessel and, as usual, a blanket of CO_2 is provided to protect the clarified juice from oxidation.

While the juice may appear clear, it will retain some colloidal solids that assist in reducing the problem of yeast settling out toward the end of the fermentation and impairing the completion of the fermentation to dryness ($<-1°$ Bé).

4.8 Analyses, Adjustments and Additions

4.8.1 *In Brief*

Red Wine

Fruit weight (kg).
Chemistry:
Essential: pH, TA, Bé, fSO_2, tSO_2
Ideal: ammonium (NH_4^+), fermentable amino nitrogen (FAN)
Optional: anthocyanins (content and extractability).
Benchmarks: pH 3.3–3.4, TA 5.5–7.0 g/L tartaric acid equiv., Bé 12.5–14.0, tSO_2 50 mg/L, NH_4^+ 50 mg/L, FAN 110 mg/L.
Sensory: bitterness, astringency, sugar/acid, hydrogen sulfide (H_2S) and other off-flavors and aromas.

White Wine

Fruit weight (juice volume/wt): L/kg.
Chemistry:
Essential: pH, TA, Bé, fSO_2, tSO_2
Ideal: NH_4^+, FAN
Optional: turbidity.
Benchmarks: pH 3.2–3.4, TA 5.5–7.0 g/L tartaric acid equiv., Bé 12.5–14.0, tSO_2 50 mg/L, NH_4^+ 140 mg/L, FAN 165 mg/L.
Sensory: bitterness, astringency, sugar/acid, H_2S and other off-flavors and aromas.

4.8.2 *Detail in Common*

Getting the adjustments right at the outset provides for better integration of the palate than making adjustments later in the process. More importantly, pH adjustments assist in avoiding contamination from fault-causing microbes (pH <3.5) and may enhance palate and color intensity. Prior to this innovation, as much as 30% of commercial wines in Australia were spoiled due to high pH (low acidity at harvest) (Speedy, 2012). Very low pH (2.9–3.2) will slow the growth of both yeast and *Oenococcus*, cause an overly acidic palate and should in most instances be adjusted upward to a range that can be modified by lactic acid catabolism. A good compromise is a pH of about 3.3–3.4.

It is exceedingly difficult to predict the final pH because several processes modify it during wine making. pH will generally rise during the fermentation of red grapes as a consequence of extraction of K^+ ions from the skin, especially in grapes from warm regions and alkaline soils. Malolactic fermentation will also raise the pH by between 0.1 and 0.2 units depending on the content of malic acid and the initial pH (change is greatest at pH 3.2). However, pH may also fall as a consequence of precipitation of potassium bitartrate during the cold-stabilization step (provided the pH is below 3.56 initially). The amount of change will depend on the overall buffering capacity of the juice and this varies with season, cultivar and vineyard owing to differences in final organic acid content, content of amino acid, ammonium and metal ions, particularly K^+ but also Na^+.

For white wines in particular, protein content may be a problem, causing excess foaming and instability. A minimal level of bentonite may be added at pressing to reduce foaming (ca 0.4–0.5 g/L), although this will mean that the wine should not be left on lees at the end of the ferment. Bentonite addition may also lead to loss of aroma. The whole-bunch pressing technique can minimize the need for bentonite addition, because stems tend to remove certain protein fractions; as can crushing/pressing under an inert gas blanket and delaying the addition of SO_2, because the presence of SO_2 in the juice tends to increase protein extraction. Thus, time on skins should also be minimized for this reason, but possibly for other reasons as well (Ribéreau-Gayon et al., 2006a).

4.9 Nutrition and Hydrogen Sulfide

Yeasts are heterotrophic organisms. That is, they depend wholly on their environment for their primary nutrients, sugars for energy, nitrogen for the production of proteins and nucleic acid during growth, vitamins for normal cellular functioning and essential minerals. This is the basis for the guidelines given in the Analyses, Adjustments and Additions section, above (Section 4.8). Sources are those contained within the yeast cells at the commencement of the fermentation and that are present in the juice. The sources within the yeast are influenced by the rehydration medium and the composition of the production culture (reviewed by Ribéreau-Gayon et al., 2006a).

Nitrogen is commonly deficient in juices and deficiency is usually signaled by the production of H_2S and a slowing of the progress of the ferment; it may become sluggish and 'stinky' (Park et al., 2000).

The production of H_2S occurs as a by-product of the catabolism of sulfur-containing amino acids (cystine, cysteine and methionine) found principally in proteins that are being degraded as a source of amino acids rather than free NH_4^+ or fermentable free amino acids. Any deficiency in FAN is normally compensated for by the addition of diammonium phosphate [DAP, $(NH_4)_2PO_4$]; although this may not be ideal, it is economical because amino acids are usually expensive and the addition of crude sources such as hydrolysates of yeast may induce off-flavors and aromas.

It is important not to add excess DAP, especially in wines that are planned to have a malolactic secondary fermentation, because lactic acid bacteria cannot use ammonium as a nutrient and the presence of ammonium may encourage the growth of spoilage microorganisms. In addition, legal limits are set in most jurisdictions and these vary widely so it is best to aim for the lowest level that is consistent with completing the fermentation and, when doing the additions, by taking into account the natural levels of ammonium and FAN in the fruit.

Note that excessive levels of primary amino acids, particularly arginine, one of the most common in grape juice, may lead to the formation of suspected carcinogens during wine storage/aging. The principal compound is ethyl carbamate, formed as products of the metabolism of urea and ethanol (reviewed by Butzke and Bisson, 1997). Many countries now define limits for ethyl carbamate ($<10\ \mu g/L$). This is a good reason to minimize the content of arginine by maintaining a moderate nitrogen status in the vineyard and using DAP as the primary source of nitrogen for yeast metabolism (Henschke and Ough, 1991). However, as noted earlier, amino acids are an important source of volatile esters and higher alcohol aromas, so serious nitrogen deficiency in the vineyard should be avoided. Alternatively, if arginine is excessive then a bacterial urease may be added at the end of the ferment to wines that are to be aged, but this does not guarantee a successful reduction in the problem (Butzke and Bisson, 1997).

4.10 Yeast Choice and Preparation

One of the most reliable yeast strains used in wine making is that known as 'EC1118', 'Prise de Mousse' or 'Champagne'. This is *Saccharomyces bayanus* (syn. *S. cerevisiae* race *bayanus*), a champagne yeast that tolerates stress (high alcohol, high or low temperature, nitrogen limitation) to a greater degree than *Saccharomyces cerevisiae* (syn. *S. cerevisiae* race *cerevisiae*). As a champagne yeast, it is generally considered an aromatically neutral yeast. It is the standard yeast for experimental wine making because of its robust nature. This race is also widely used to restart stuck or sluggish ferments where another yeast strain has been used and has experienced problems (although new, fructose-tolerant yeast strains are becoming available; check with suppliers as the choice is subject to constant change).

A wide range of yeasts is available; with careful management, and when applied to wines with appropriate juice chemistry and fermentation conditions, they will ferment to dryness. These are usually selected on the basis of trials within a winery and are designed to promote the development of particular attributes (especially aroma) for given wine styles and cultivars.

Some of the finest wines are made from blends produced from fermentations of batches of wine, each fermented with a particular strain of yeast to achieve balance and length of flavor and aroma throughout the palate. Blends are also used to enable the production of wines with a consistent style from year to year, despite varying environmental and viticultural circumstances. Thus, it may be important to conduct small-scale fermentations of particular varieties with a range of yeast strains in order to select appropriate strains for particular wines, wine blends and styles.

We have now entered the 'new biologies' era, and advances in knowledge in flavor and aroma chemistry/biochemistry are leading to the identification and development of strains of yeast specifically adapted to meet particular requirements in wine making. For example, the mercaptans 4-methyl-4-mercapto-2-pentan-2-one (4-MMP), 3-mercaptohexan-1-ol and its acetate (3MH and 3MHA) are vital components of the aroma characteristics of Sauvignon blanc. Certain yeast strains produce enzymes that encourage the release of these aromas, enabling a winemaker to manage the aroma profile of a particular wine (Anfang et al., 2009; Howell et al., 2004).

Small-volume fermentations are almost always started with the addition of rehydrated yeast prepared from freeze-dried yeast. This has the advantage that the strain of the yeast is known and was produced under strict quality assurance protocols by the manufacturer (if you have a problem, the manufacturer will have records provided you have written down the batch details). The yeast cells are produced under nutrient-rich, oxidative, non-fermentative conditions that equip the yeast to tolerate the stress of fermentation; the yeast cells accumulate stress tolerance factors, especially fatty acids, sterols and nutrients (reviewed by Ribéreau-Gayon et al., 2006a). This may be further boosted by rehydration in a vitamin-rich nutrient medium (several proprietary media are available from commercial suppliers). A further advantage is that the initial cell density is well defined and may be standardized (ca 5×10^5 to 10^6 cells/mL, 200 mg/L of freeze-dried cells).

An alternative is to produce a liquid culture of a selected local race or strain, but the challenge is to maintain culture purity (axenic), and considerable resources are required to select the strain in the first place and to maintain its purity. Sometimes, a proportion of a mid-ferment wine may be added to start the next culture, but this carries the disadvantages that yeast vigor is reduced with successive numbers of cell division and the content of stress-tolerance factors declines with succeeding generations. Small and microscale fermentation protocols are usually used in the testing and development of these strains, but during research rather than in a teaching or hobby environment.

Wild or natural ferments are still practiced in some regions but this is a high-risk strategy unless the dominant yeast present in the winery is capable of producing quality wine. Usually such ferments will begin with non-*Saccharomyces* yeasts as these dominate in the field and, subsequently, the rarely occurring *Saccharomyces* strain may take over and complete the ferment. Few wild yeasts are capable of fermenting a wine to dryness.

4.11 Temperature Management

4.11.1 *In Brief*

Red Wine

The ideal temperature for red wine production is between 23 and 28°C; 30°C is the maximum. It is possible to conduct small fermentations without temperature control if mean ambient temperature is below 20°C, but research wines will require control.

White Wine

The ideal temperature for white wine production is much lower than for red wine, being about 15–18°C. Small fermentations will require an ambient temperature of about 2°C lower than these values. In cold climate regions, it may be possible to conduct white wine fermentations without temperature control but this is not realistic for most regions of the world. Again, research wines will require temperature control to avoid the vagaries of the weather.

4.11.2 *Detail in Common*

Fermentation is a strongly exothermic process and it has been estimated that it may generate a temperature rise of about 140 kJ/mole of sugar fermented or a theoretical rise of about 26°C for a juice of 24° Brix (Williams, 1982). This heat needs to be dissipated and studies have shown that the maximum temperature achieved varies by the cube root of the volume ($\sqrt[3]{\text{volume}}$) of the ferment (Boulton, 1979). For small fermentations, up to about 500 L, temperature control is not critically important for red wine fermentation as heat loss to the environment will normally be sufficient to keep the maximum temperature rise to 5–0°C above ambient. In hot climates and for larger volume fermentations, some arrangement for cooling will be required; for example, a temperature-controlled room, a jacketed fermenter or an immersion heat exchanger.

The temperature of a ferment has both direct and indirect effects on subsequent wine quality, but these are poorly understood. In general, the higher the temperature, the greater the rate of CO_2 generation and the greater the loss by entrainment of volatile aroma compounds. Thus, high-temperature fermentations are characterized by a lack of fruit character and may also develop peculiar flavors with a caramelized attribute. For these reasons winemakers generally attempt to restrict the temperature to a maximum of 30°C for red wines, but for white, aim to keep it below 20°C.

A temperature above 32°C in a red wine may also lead to a stuck fermentation as yeast cells become progressively sensitive to high-temperature stress as alcohol levels rise during the fermentation (Boulton et al., 1996). Styles requiring a higher level of fruity aroma (e.g. Rosé styles and Pinot noir) may be fermented at lower temperatures. Higher temperature, however, promotes the extraction of tannins and procyanidins, so maintaining a temperature that is too low may also have detrimental impacts on wine quality and color stability. It is for this latter reason that some winemakers allow ferment temperature to exceed 30°C in the mid-phase of a ferment before alcohol levels are sufficient to cause yeast death.

4.12 Cap Management in Red Wines

Cap management is vitally important in red wine making. It minimizes the risk of faults arising from contamination by surface-growing yeasts, maximizes (ideally optimizes) the extraction of anthocyanins and tannins, brings the cap to the temperature of the liquid and facilitates fermentation through mild oxidation.

At the commencement of a fermentation, the pressed berries, skins and seeds will be dispersed through the juice, and then the seeds and any stems will tend to sink to the bottom. As the ferment begins, CO_2 bubbles develop inside the crushed skins or attach themselves to the skins particles, giving them buoyancy and causing them to rise.

4.12.1 *Open Fermenters*

A good rule of thumb is to plunge the cap and thoroughly press the skins to the base of the fermenter twice daily. In microferments, a stainless steel flat plate (paint) mixing plunger will suffice. Excessive plunging may lead to unacceptable levels of bitterness and astringency, whereas inadequate plunging will lead to low color in the wine with inadequate tannin levels to provide good mouth-feel attributes. Occasionally, winemakers will use a submerged barrier to prevent exposure of the cap to air, but this method generally gives poor extraction unless combined with pumping-over operations.

4.12.2 *Closed Fermenters*

Pumping-over is regarded as providing the best method of cap management although this facility is usually only available in commercial-scale wineries or national R&D

research facilities (e.g. Hickinbotham Winery, University of Adelaide). In operation, juice is taken off through a racking valve sited about 10% of the height from the base, through a screen and into a sump, and then pumped to the top and distributed either manually or mechanically over the surface of the cap. Usually about 1 volume of the ferment is pumped over twice daily. Horizontal, rotary fermenters are rotated intermittently to achieve a similar outcome.

4.13 Aroma Management

Sluggish and stuck ferments are less common in red wines than white, because the skins are a good source of nutrients; however, a stinky ferment indicates the need for immediate addition of DAP (e.g. 50 mg/L) to overcome the problem. A problem should not emerge if sufficient DAP is added as a matter of practice. This is best done in split additions to ensure a smooth fermentation (e.g. 30 mg/L on day 1, at the end of the lag phase, 20 mg/L once the ferment drops by 3–4° Bé and a further 20 mg/L when at about 4° Bé. This is a rule of thumb only, and additional DAP may be added as required provided the level does not exceed regulatory standards. Ideally the additions will be estimated based on analyses of ammonium and FAN. Addition of DAP is not recommended once the ferment nears completion (<3° Bé).

4.14 Optional Additions: Red Wines

Optional additions for red wines are oak chips (ca 50 g of oak per 10 L) or staves and/or tannin extract (Gutierrez, 2003; Pérez-Coello et al., 2000).

4.15 Press-off and Complete Red Wine Fermentation

The final one-third of a red wine fermentation is generally completed off skins and seeds to avoid a tannic and possibly bitter wine that might arise as a consequence of excessive extraction of tannins and procyanidins, especially from seeds (de Freitas et al., 2000; Ribéreau-Gayon et al., 2006a). It is very difficult to define, analytically, the optimum stage to terminate skin and seed extraction (Boulton et al., 1996). Experience and palate are probably the best guides. Anthocyanin extraction peaks on day 2 or 3, and skin tannin and procyanidin extraction by the end of alcoholic fermentation, but seed tannin and procyanidin extraction is extended (de Freitas et al., 2000). Procyanidins and anthocyanins form complexes through chemical condensation reactions which at first stabilize color but ultimately lead to the formation of precipitates in the bottle (Ribéreau-Gayon et al., 2006b).

Extended maceration, on skins, will increase the level of seed-derived procyanidins and may enhance the keeping quality and mouth feel of the wine, but only if the fruit and seeds are fully mature. In cool seasons, cool climates and high biomass (vigor) vineyards, it is worth noting that seed maturity may be uncoupled from sugar accumulation, and delayed. In this case, extended maceration or cold-soaking prior to fermentation will almost certainly diminish the palatability of the wine.

Extended maceration is allowed to proceed by simply sealing the fermenter (using an airlock to enable gas pressures to equilibrate). Palate evaluation is used to determine when to terminate the step.

Normally, the juice is first run off through a racking valve, passed through a grill and pumped or poured into a fresh container. In experimental and teaching-scale wine making, a racking tube (siphon) or plate is used to achieve a similar end. The remaining solids are then dumped into a press and run through three pressing cycles until the 'marc' is dry. These pressings may be kept separate but usually they are added to the rackings, because they are highest in color and tannins. The ferment is then allowed to run to dryness. Red wine will tolerate some exposure to air in this step but it is wise to maintain an inert gas cover in any receiving vessel.

It is essential to maintain an inert gas cover if the wine has been allowed to go to dryness before pressing.

4.16 Malolactic Fermentation

The purpose of the malolactic fermentation is to deacidify excessively acid wines, provide a degree of in-bottle microbial stability, and enhance palate and aroma. There is, however, no guarantee that each of the goals will be attained unless care is taken to manage the microbial ecology of the ferment. That is, sound hygiene practices must be used to avoid contamination by fault-causing microorganisms (see Chapter 6).

Malolactic fermentation is a contentious aspect of red wine making (and for some white wines). Three genera of bacteria are associated with malic acid metabolism: *Oenococcus* (syn. *Leuconostoc*), *Lactobacillus* and *Pediococcus*. Australian and French practice is usually to separate the two fermentations in the belief that this will minimize the risk of development of off-flavors and aromas. American practice is to run coferments, accepting that there may be a risk of a sluggish or stuck ferment while achieving more uniform and rapid malic catabolism due to the lower alcohol levels and higher nutritional status of unfermented juice (reviewed by Boulton et al., 1996). The relatively high levels of SO_2 in the early phases of wine making will prevent undue growth of lactobacilli.

Wild strains of *Oenococcus* are rare in the vineyard and therefore must be added, either through infection in the winery or by deliberate inoculation; the latter is the better option. Other lactobacilli are present and, while susceptible to low levels of SO_2 in the juice, are heterotrophic and particular strains may give rise to faults such as ropiness and mousiness (reviewed by Boulton et al., 1996). It is worth noting that the sensitivities to SO_2 reported in most literature refer to wild strains and those for commercial strains may differ. While some strains of alternative lactobacilli are commercially available, *Oenococcus* is probably the organism of choice because it tolerates pH levels below 3.5, whereas *Lactobacillus* and *Pediococcus* generally do not. Furthermore, *O. oeni* is homofermentative and thus unlikely to cause faults. Studies show that it is much more reliable and effective than wild lactobacilli, which are slow and may not complete the process (Puppazoni, 2007).

A temperature of about 18°C is generally recommended to optimize the rate of progress and minimize the risk of contamination. Nevertheless, this process proceeds much less vigorously than the primary ferment and can be monitored by observing the release of CO_2 through an airlock. Formal assessment of progress is done by paper or thin-layer cellulose chromatography (or gas–liquid chromatography if available) and confirmed at the end by enzyme analysis.

4.17 Clarifying, Aging, Fining and Stabilizing

These processes should be considered together, as they interact. Aging a wine before bottling inevitably leads to improved clarity through settling of particles. Aggregation of colloids to produce particles large enough to settle also occurs through their interaction with tannins (Ribéreau-Gayon et al., 2006b). It follows that aging in-bottle should not be carried out unless these colloids have first been removed. Otherwise, unsightly sediments and hazes may form in the bottle, limiting sales potential, especially in white wines.

Achieving resistance to the formation of hazes and sediments postbottling is central to the process of fining and stabilizing. Aging, while not appropriate for most white table wines, may nonetheless improve heat and protein stability by removing larger, heat-unstable proteins and replacing them with smaller stable proteins and peptides (Ribéreau-Gayon et al., 2006b). Therefore, protein fining in such wines should not be carried out until after aging *sur lie*. Aging is not a panacea and may, if not carried out correctly, lead to the development of flavor and aroma faults that require additional fining or moderation by blending immediately before bottling.

A wide range of fining and stabilizing agents is available for red and white wines (Table 5.1). Normally, wines are fined before they are stabilized, although sometimes the processes are combined.

Fining or re-fining wine means treating a wine with agents that will enhance its appeal by removing unattractive aromas and tastes.

Fining should start with a careful sensory assessment of the wine. Define the sensory attributes of the wine: is it fruity or floral or are those characters masked by reduced or sulfurous aromas? Does the palate finish cleanly or do bitter or aggressive tannins detract from the taste and the palate? It is usual to conduct sensory trials, beginning with a coarse range of selected fining agents and then with a narrow range. Generally speaking, the less agent added the better, not merely economically but also because most fining agents remove more than the target compound and may diminish the sensory attributes of the wine or even add their own malattributes.

Methods for optimizing the additions of copper (Cu^{2+}) and bentonite are set out in Chapter 8. The remaining fining agents may be required if one or more aspects of the process was flawed or, in the case of Cu^{2+}, copper-based fungicides have been unduly relied upon in the vineyard, as may be the case in organic vineyards (e.g. Bordeaux mix).

Excess copper in a wine (>0.5 mg/L) is likely to lead to a brown casse in the reducing conditions found in white wines (note that this may be higher than the legal limit in some jurisdictions). The presence of copper can be deduced by adding one-quarter volume of 10% HCl or a few drops of hydrogen peroxide (H_2O_2). In both cases the haze should dissipate. Its presence can be confirmed by the addition of a few drops of 5% m/v potassium ferrocyanide (Margalit, 2004). Copper can be measured using specialized equipment (atomic absorption spectrometer) although a wet-chemistry method is available (Ough and Amerine, 1988).

4.17.1 Clarifying

Achieving clarity is a continuous process and a part of all postfermentation procedures. Clarity is achieved by:

- settling and racking
- adding enzymes to fragment polysaccharides, especially *Botrytis* glucans, if present (Ribéreau-Gayon et al., 2006b), and pectins
- settling colloidal tannins with protein agents (egg albumen, fish protein, etc.)
- addition of clay minerals and silicates to remove temperature-unstable proteins
- chilling in the presence of crystalline tartaric acid powder to remove super-saturated organic salts
- centrifugation and/or filtering.

Filtering is usually carried out at bottling and will be discussed in Section 4.19.1. Settling is generally best done in small or low vessels to minimize the distance that

TABLE 5.1 Target and Timing of Fining and Stabilizing Additions

Target		White Wines		Red Wines	
		When	Agent	When	Agent
Polysaccharides	Pectins	Preferment	Pectinase	Nil	
	β-glucanes	Preferment or prefiltration	β1 → 3 glucanase	Prefiltration	β1 → 3 glucanase
Tannins		Prebottling	Proteins	Prebottling	Proteins
Procyanidins		Prebottling	PVPP	Prebottling	PVPP
Proteins		Prebottling (or preferment)	Bentonite	Prebottling	Bentonite
Sulfur compounds	H_2S	Prebottling	$CuSO_4$	Prebottling	$CuSO_4$
Urea				Prebottling	Urease
Metal ions	Cu^{2+}, Fe^{2+}, Fe^{3+}	Prebentonite	K-ferrocyanide[1], ascorbic acid		
KHT		Postfining	Cold	Postfining	Cold
Color	Anthocyanins	Prebottling	Charcoal and/or PVPP		

Notes: [1]Use of ferrocyanide requires authorization in Europe and strict limits apply to levels of residual cyanide (< 0.1 mg/L in Australia). KHT = potassium hydrogen tartrate; H_2S = hydrogen sulfide; PVPP = polyvinylpolypyrrolidone; $CuSO_4$ = copper sulfate.

particles must fall in order to accumulate at the bottom or, in casks, on the sides. However, even wines that have gone through a number of settling and racking processes and are 'cellar clear' may still contain a significant content of suspended fine particles and colloids that can only be detected spectrophotometrically and which may ultimately impair wine quality if not removed.

Racking of barrel-fermented or aged wines is a repeated process, usually of transferring the wine that is located above the coarse lees into a fresh vessel (Figure 5.5). This may be a direct transfer, in which case barrel differences are maintained or even enhanced, or via an intermediary container, in which case the barrel differences are blended. The fresh vessel will have been sulfured and the wine should achieve about 25 mg/L fSO_2. This transfer should be done at about 3-month intervals and at no time should this value drop below 15 mg/L.

Racking is achieved by taking the wine off by gravity or pump at a point just above the coarse lees. Storage vats will contain a racking valve; some barrels may have a racking spigot on one end but more commonly a racking plate is used to set the intake point just above the lees to minimize disturbance of the surface of the lees. Racking should occur slowly and may or may not involve aeration depending on the needs of the wine. Be cautious if aerating (i.e. do not deliver to the bottom of the new vessel and ensure adequate fSO_2) as oxidation is a difficult fault to correct. In the case of white wine, it is probably wise to add a pellet of dry ice to avoid oxidation.

4.17.2 Aging

Aging of wine comprises several distinct processes:

- chemical reactions that release flavor and aroma precursors
- reactions that produce new flavor and aroma compounds; condensation and complexing reactions that lead to aggregation of low molecular weight procyanidins and anthocyanins (in red wines) moderating the tannin mouth feel, stabilizing early color but also ultimately reducing color
- extraction of flavor components from oaken storage vessels or staves.

These processes begin at crushing and continue throughout the life of the wine. The reactions are temperature dependent and thus the drinking-life of a wine depends strongly on storage temperature. One could say the colder the longer the life, but a temperature of between 10 and 15°C is usually regarded as optimal. In general, wines should be aged following stabilization, although some aging on lees is a characteristic of the production of champagne and some Chardonnay and Sauvignon blanc style wines.

Commercially, fine red wines are aged in a mixture of new and old oak, held at an fSO_2 level of about 30 mg/L, and the casks are kept full (topped up) to minimize oxidation. The wine is transferred to a fresh barrel (or to a blending tank under CO_2 cover and then transferred), and the used barrel is washed before being reused (see

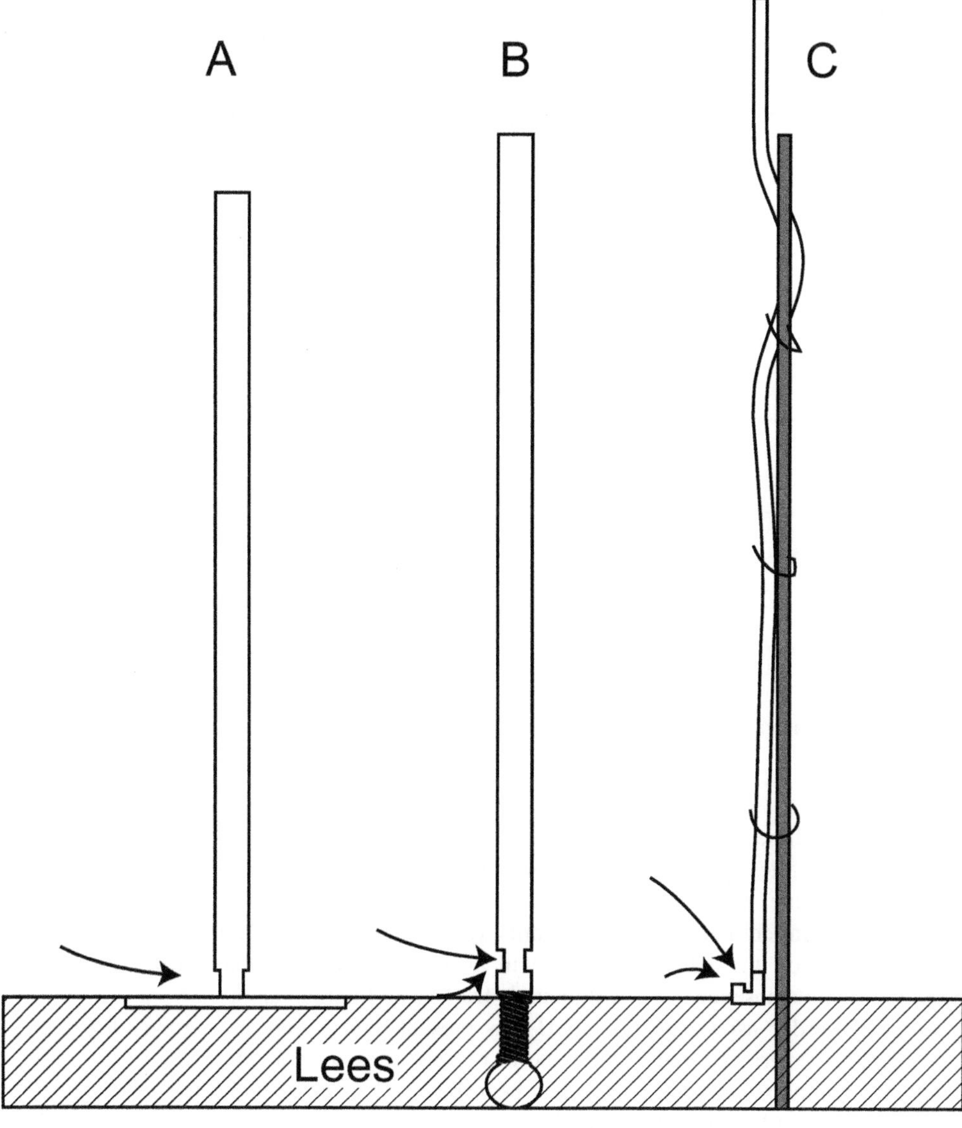

FIGURE 5.5 Drawings of three racking arrangements for vessels without a racking spigot (tap). A: Fixed racking plate; B: adjustable racking tube; C: a simple siphon.

Chapter 8). This is usually done at 3 month intervals and may be repeated for 6–24 months depending on the wine. Standard and teaching wines are not usually held in barrels but may be exposed to oak staves in steel tanks. Whatever approach is adopted, frequent sensory assessment is mandatory. Final maturation is done in vitro and the wine is held in the cellar for 2–5 years.

Some aging in oak is usually adopted for red wines but be aware that extended aging in oak may damage the wine. Oak choice is very important depending on style: French oak is usually the choice for fine Cabernet Sauvignon and Chardonnay but American oak for Shiraz (Syrah). Commonly, a batch of wine is aged in a mix of new and old barrels (and/or large and small) to enable the best balance to be achieved by blending before bottling.

4.17.3 Fining and Stabilizing

Note that under Australian legislation the use of some products (e.g. fish, egg, milk) may need to be declared on the label (http://www.foodstandards.gov.au) and almost certainly legal limits exist in other jurisdictions. The principal components targeted are presented in Table 5.1:

- proteins (problematic for high-temperature stability): see Chapter 8, Bentonite Protein Stability Test
- low molecular weight phenolics (bitterness and astringency)
- colloids (clarity and stability)
- flavor and aroma faults (sensory)
- supersaturated organic acids (cold stability, especially in the presence of nucleating agents such as colloids)
- color faults (appearance; oxidation).

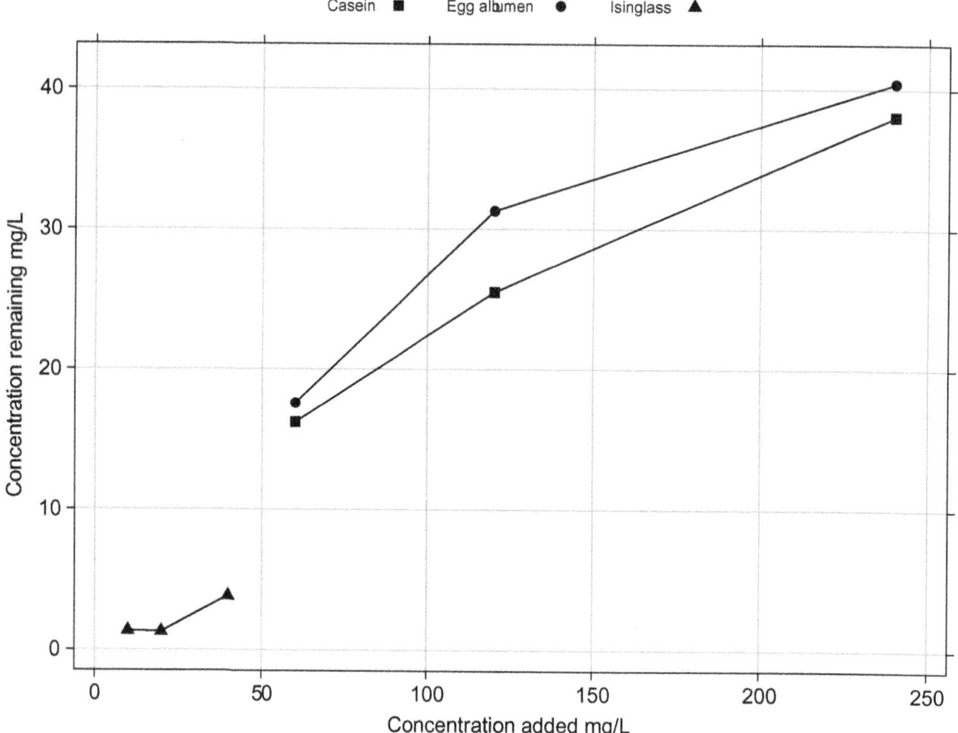

FIGURE 5.6 Plot of the effect of fining agent and concentration on amount remaining dissolved in wine, postsettling (mean of three wines). *(Data from Watts et al., 1981, with permission).*

It is important to realize that some of these reagents may remain in the treated wine, even after protein removal with bentonite or silica gel (these wines are over-fined) (Figure 5.6). Increasingly, countries are demanding that the agent be included in a warning on the label, unless the product has been tested to prove its absence.

Fining is a complex and very high-order aspect of wine making, one requiring considerable experience and knowledge. We will not deal with the complexities here, but those wishing to investigate this matter further should seek the advice of an experienced winemaker or read reviews on the topic such as those presented in Ribéreau-Gayon et al. (2006b). But this should not deter the aspiring winemaker from the learning process. The most common issues to be dealt with in both red and white wines are temperature stability, bitterness and H_2S. Cold stability is more important in white wine than in red wine, as is clarity. We will consider those here.

4.17.4 Flavor and Aroma Faults

These are normally dealt with first as the products of the fining process are removed during the settling or filtration processes that are associated with achieving temperature stability. The most common aroma fault is that of H_2S, which at low levels may not be particularly obvious but may still mask other, desirable fruity aromas. This is readily removed with cupric salts that form an insoluble copper sulfide precipitate. A protocol for determining the optimum concentration is provided in Chapter 8. Care needs to taken as copper is a heavy metal and, as such, strict limits apply to its concentration in wine (0.2 mg/L); copper will also strip other desirable sulfur-based aroma compounds from wine (mercaptans).

If H_2S is present then it may be a good idea to treat the wine to remove it as soon as the cause of its production has been rectified to prevent the formation of more complex sulfur compounds that may be more difficult to remove. Cadmium may be used to test for the presence of these because it removes only H_2S, leaving other, reduced (rubbery) sulfur compounds, the treatment of which may demand other actions (see discussions in Jackson, 2008; Margalit, 2004; Ribéreau-Gayon et al., 2006a). Cadmium and silver, while effective, are forbidden additions in wine making for consumption owing to their toxicity.

Excessive bitterness and astringency are other common flavor faults than should be treated before stabilizing the wine. Somewhat ironically, these may be treated by the addition of proteins, the very substances that cause heat instability. However, the proteins used in treating wine are of animal origin and quite different from those that occur naturally in grapes. The enological grade proteins are collagens from animal cartilage (gelatins) or fish swim bladders (isinglass), albumens from blood or egg white, and casein from milk. These proteins absorb polyphenols and tannins on their surface with weak, reversible bonds (van der Waals') and coalesce to form colloids which aggregate and form insoluble precipitates. Sometimes this process does not proceed to completion, giving an overfined wine, i.e. one that retains unstable proteins from the fining agent. Silicates or bentonite are frequently used to ensure complete removal of the fining protein. The interactions are not simple, vary from agent to agent, and are affected by pH and the concentration of salts in the wine (reviewed by Ribéreau-Gayon et al., 2006b).

4.18 Blending

Cultivar blends are traditional in the Old World, especially for red wines, and are becoming increasingly important in the New World, just as diversity of yeasts in the primary ferment is gaining popularity as a means of increasing the diversity and range of aromas and flavors (Anfang et al., 2009; Ferreira et al., 2001). The traditional means of blending is to coferment the different cultivars; as many as eight to 12 may be used, although more commonly the number is limited to four or fewer. The French industry provides many good examples of cofermented, blended wines, especially from southern and south-eastern regions. The practice is less common for white wine, where single cultivars tend to be the rule and blending involves ferments of the same cultivar but from separate ferments or blocks or maturities. However, many excellent wines are made by blending (e.g. Semillon with Verdelho and/or Chenin blanc; and Cabernet Sauvignon with Merlot and Petit Verdôt).

Cofermentation, however, while delivering great integration, allows for little control and success depends strongly on the selection of the grapes, the blend and the availability of mature fruit of each cultivar. In traditional wine making, this was driven largely by availability and it is more economical. However, with economic pressures driving the industry toward consistent, quality outcomes, such an approach is rarely adopted in the New World, or even now in the upper echelons of the Old World. A more refined approach is to keep the fruit as separate parcels and fermentations until the end of the primary and secondary ferments, and then blend before aging or bottling. Occasionally, red and white cultivars are blended, as in the traditional Champagne blend of Pinot noir, Chardonnay and Pinot Meunier. Some Rhône-style red wines include a little Viognier to add floral aromas and in the New world Orange Muscat may be used. Labeling laws rarely demand identification of minor component cultivars. Often only about 1% of total volume is sufficient to impart a 'lifted' aroma.

Most red wines and many white wines are indeed blends, if not of different cultivars then of different ferments or barrels. Wines may also be blended to achieve particular alcohol, acid or pH outcomes, rather than by making additions or adjustments. Blending is the highest order of wine making and usually involves duo–trio evaluations of fine mixtures at a range of ratios, observed both immediately and following 24 h or more of standing to allow the chemistry to come to equilibrium. Making wines at an experimental scale provides an economical way of learning the craft of blending to achieve quality, consistency and, if necessary, technical or legal outcomes (alcohol content, pH, TA, residual sugar, color).

If you wish to make blends, the first step is to write down the objectives and then assess whether the wines you have at hand are appropriate for the purpose. It may be necessary to buy in wine or at least ensure that your ferments are managed to provide the requisite material, not simply in terms of cultivars but in technical terms as well.

4.19 Prebottling Adjustments and Bottling

The technical aspects of the wine, including heat and cold stability and its sensory attributes, should be checked before bottling. The pH should first be adjusted to the target, either up or down using the selected organic acid or alkali; then organoleptic attributes, fining as necessary, and then the fSO_2 should be adjusted. Clarification and tests to ensure cold and heat stability are required prebottling. It is not enough to assess clarity by eye for white wines; they should be assessed spectroscopically.

4.19.1 Bottling

Bottling involves first the choice of an appropriate closure and vessel. Cork is the cheapest for small-scale wine making but is far from ideal for experimental wines. For such wines, artificial closures, even a 'crown' seal, are preferable. Crown seals (as used for beer) are increasingly used for sparkling wines (they have traditionally been used for the secondary ferment) and replaced by a cork for maturation and sale, but may be used for any wine.

For a review of bottle closures see Godden et al. (2001, 2005). The central problem with cork has been tainting associated with 'cork' taint (TCA, 2,4,6-trichloroanisole), which is formed by a reaction between chlorine, phenol and the presence of fungi (Sefton and Simpson, 2005). It is not always due to cork contamination but may be due to contamination of barrels or timber in the winery. This is an emotive topic and the Portuguese industry in particular is working to develop solutions.

The critical issue in bottling is hygiene. New bottles are usually sterile but may quickly become contaminated. All closures, lines and filters must be sterilized with either a propriety cleaning agent or heat (water of $\geq 80°C$).

Both red wine and white wines are passed through a 5 μm prefilter and a 0.45 μm final filter. Ideally, this is done under an inert gas such as nitrogen. Care must be taken not to overfill and normally a headspace of a gap of about 30 mm is required between the surface of the wine and the base of the closure. Quality assurance procedures would apply in a bottling run and samples are taken at intervals and tested for microbial sterility (see Chapter 6).

4.20 Waste Disposal

Wastes are an integral part of the production of wine (Chapman et al., 2001). Local authorities will refuse an application to construct or operate a winery unless a waste management plan is provided. In general, local industries agree on a Code of Practice or Best Practice Guide (e.g. FSA, 2006; WFA, 2008). Regulatory authorities (environment protection, conservation or planning approval organizations) provide guidelines that apply to individual localities.

Winery wastes include:

- stalks and marc
- lees and bitartrate
- waste water and detergents from washing down processing areas and cleaning of items
- cleaning chemicals
- spills: juice, chemicals and wine
- spent filter aids (diatomaceous earth)
- chemical and laboratory wastes
- fermentation emissions.

Small operations can normally be accommodated within existing organic and sewage systems but only following consultation with local authorities. The design of larger wineries usually includes settling and oxygenation ponds with some provision for the disposal on land of the treated waste waters. Marc can be processed by composting (usually mixed with other 'green' materials or simply distributed in the vineyard).

5. PROBLEMATIC FERMENTS

5.1 Sluggish and Stuck Primary Ferments

Stuck or sluggish ferments are those that cease fermentation before all sugar is utilized or which display a much reduced rate of fermentation (Bisson, 1999; Bisson and Butzke, 2000). The symptoms are usually readily apparent if one is graphing the progress of a ferment on a daily or more regular basis: the trend should be to a negative ° Bé; if not, then the condition should be addressed as soon as it becomes apparent. While usually this problem becomes obvious early in the fermentation, this is not always the case (Bisson and Butzke, 2000). Recovering a problematic ferment usually requires access to consultants or winemakers of considerable experience and expertise.

The causes are many and include inappropriate choice of yeast, inadequate nutrition, temperature outside range (<15°C for white or >32°C for red wines), competition from other yeasts and microorganisms (Edwards et al., 1999a) and excessive SO_2, ethanol and/or fructose as fermentation progresses. Careful diagnosis is necessary if further occurrences are to be prevented. This will normally require both counts and plating for contaminating microorganisms and determination of the levels of fructose and glucose (see Chapters 6 and 11).

Under stress, many yeasts, and some strains in particular, express flocculation proteins that cause the yeast cells to aggregate and settle, leading to further stress and death. Such situations may be managed by the addition of limited amounts of oxygen (microoxidation) to relieve the oxygen stress and/or the addition of yeast hulls to provide nutrition and to limit flocculation (Jin and Speers, 1998). In small ferments this may be achieved by hand mixing (swirling) the ferment. Oxidative damage is rare during an active fermentation.

Toward the end of a ferment, the fructose to glucose ratio increases substantially and fructose is metabolized much less efficiently than glucose (Berthels et al., 2004). Recovery of such ferments demands the addition of a fructose-tolerant strain of yeast. This is often associated with high-sugar fruit and therefore a 'bayanus' strain, tolerant to higher alcohol levels, should be used (and indeed should have been selected initially). Restarting such ferments is particularly problematic and is usually achieved by adding as much as 1 part to 2 of a fresh, active ferment using an appropriate strain. With white wines, the temperature may need to be raised somewhat to re-establish rapid fermentation.

Failure may also have a biological basis. For example, a 'killer-yeast' susceptible yeast strain may become infected (see Chapter 6), or other wild yeasts (e.g. *Kloeckera* spp.), which are very common on grape skins, may outcompete the chosen yeast. Initially these may dominate, especially if cold soaking is practiced (e.g. *Kloeckera* spp. are reported to tolerate low temperature and high SO_2). Also, *Lactobacillus kunkeii* has been shown to be highly inhibitory, but is quite sensitive to SO_2 so should not be a problem (Edwards et al., 1999b). If the cause is microbiological then hygiene and sulfuring practices should be reviewed.

Ecological problems arising from competition or toxins cannot be resolved by the usual means. A valuable wine may possibly be recovered by resulfating or pasteurization before reinoculation, otherwise it is waste! Organic-style wines and wine-making practices generally that are based on indigenous yeast strains are particularly at risk of these problems and good microbiological skills are demanded if the risks are to be minimized, i.e. low-tech wines demand high-tech skills!

Almost inevitably, the wine resulting from a problematic ferment will have organoleptic faults that will need to be resolved by careful fining and blending.

5.2 Sluggish and Stuck Malolactic Ferments

The malolactic fermentation is also prone to problems. These may arise if a wild *Lactobacillus* is used in the place of a wine-adapted commercial strain of *O. oeni* or if the starter is in poor condition, aged or inappropriately stored. *Oenococcus* is generally intolerant of SO_2, even the bound form, and the total concentration should be less than 5 mg/L at inoculation (<0.8 mg/L fSO_2) (Kunkee, 1967). It is also intolerant of high levels of alcohol ($>14\%$), and sensitive to low pH (<3.1) and to excessive levels of malic acid and, indeed, the end-product, lactic acid. Like yeasts, malolactic bacteria require nutrients, and commercial preparations of these are available which may be necessary in white wines undergoing a malolactic fermentation. As with yeast fermentation, the best approach is to minimize risk by ensuring that all parameters are within the range of the strain of *Oenococcus* selected (Miller, 2009). As with failing yeast fermentations, assays for bacterial contamination and for *Lactobacillus* counts will be required as a part of the problem-solving approach.

Chapter 6

Microbiology and Methods

Chapter Outline

1. INTRODUCTION

Microorganisms are ubiquitous in nature, holding niches from volcanic cores to arctic tundra. Their small size provides for rapid growth and hence a rate of genetic change suited to adaptation to a great diversity of ecosystems.

The majority of microorganisms are innocuous. A relatively small proportion is potentially harmful, causing illness (human, animal), disease (human, animal or plant) or spoilage of food and beverages. There are also microorganisms found to be beneficial to humans, assisting in the production of food and beverages. These organisms have the ability to ferment nutrients, producing stable and desirable end-products for our consumption.

Wine is the result of a microbiological process. The quality of the wine and its stability in the bottle are dependent on adequate control of the alcoholic and malolactic fermentation steps. This requires the use of the appropriate microorganisms and preventing the ingress of spoilage microorganisms.

The primary natural source of microorganisms in the winery is the grape. The surface of grapes is contaminated with a mixture of yeasts and bacteria, and to a lesser extent molds (e.g. Martini et al., 1996). Bacterial contaminant groups include lactic acid bacteria and acetic acid bacteria. The composition of the microbial flora changes as the berry develops, and is influenced by

A Complete Guide to Quality in Small-Scale Wine Making.

temperature, humidity, hygienic control of viticultural practices and the use of pesticides.

Hygiene control over harvesting equipment and transfer of grapes to the winery is important to prevent spoilage organisms dominating the winery environment. This is a critical control point (CCP) in the process flow of wine making.

Although over 500 species of yeast occur, only around 15 of these are significant in wine production and wine spoilage. The surface of the mature grape harbors up to 10^6 fungi (yeasts plus molds), but at this stage the yeast responsible for alcoholic fermentation, *Saccharomyces cerevisiae*, is virtually absent (reviewed by Ribéreau-Gayon et al., 2006a).

The yeasts present are generally oxidative and alcohol sensitive by nature and include mainly species of the genera *Pichia*, *Candida*, *Hanseniaspora*, *Dekkera*, *Kluyveromyces* and *Metschnikowia*. Once harvesting begins the microorganisms present on the grapes rapidly become established on equipment and in the winery through cross-contamination. Hygiene control in the winery environment is important to minimize establishment of spoilage organisms that may infect the wine at various stages (Fugelsang, 1997).

After harvest, transport and crushing, the grape juice contains a yeast flora similar in numbers and composition to the grapes preharvest. Under natural processes without the addition of starter cultures, *S. cerevisiae* starts to develop and coexist with the grape yeasts after about 20 h in the tanks. It is presumed that the *S. cerevisiae* originates from the winery environment (Martini et al., 1996). As spontaneous fermentation begins, *S. cerevisiae* quickly outgrows all other yeasts and is isolated almost exclusively as the dominant fermentation yeast after 2–3 days (Fleet and Heard, 1993; Ribéreau-Gayon et al., 2006a). As the fermentation develops, *S. cerevisiae* reaches around 10^8 cells/mL. The population progressively decreases during the final part of alcoholic fermentation, called the yeast decline phase, but still remains at a level of 10^5–10^6 cells/mL. Following the completion of alcoholic fermentation, *S. cerevisiae* numbers fall to less than 100 cells/mL.

Sound management conditions are required to prevent the introduction or development of other yeasts during alcoholic fermentation, which could result in spoilage of the wine. These include:

- must or grape sulfiting
- maintenance of anaerobic conditions, as many of the spoilage yeasts rely on oxidative metabolism
- rapid and complete exhaustion of sugars to prevent growth of spoilage yeasts such as *Brettanomyces intermedius*, which would result in olfactory flaws.

Support programs in the winery that include documented procedures for cleaning winery equipment and surrounds should be established (see Chapter 8).

As the ethanol level in wine increases those organisms adapted to growth and survival under high-ethanol conditions will dominate as the alcohol becomes toxic to other organisms. The use of a *S. cerevisiae* starter culture ensures a fermenting strain with properties that can rapidly outcompete the indigenous grape flora and produce appropriate flavor metabolites in the wine.

Lactic acid bacteria, which are part of the grape microbial population, are present in the must at levels of 10^2–10^4/mL. Three groups make up the lactic acid bacteria on grapes: *Lactobacillus*, *Leuconostoc* (*Oenococcus*) and *Pediococcus*. They are differentiated by their forms of fermentation and their cell morphology.

Oenococcus oeni is the principal malolactic fermentation organism. During the first few days of alcoholic fermentation its numbers increase slightly to around 10^5/mL. As fermentation progresses it reduces in number to 10^2–10^3/mL. When the yeast fermentation is almost complete the lactic acid bacteria start to increase again, and when their numbers reach around 10^7 or more per milliliter, malolactic fermentation begins (Ribéreau-Gayon et al., 2006a). This is generally about 2 weeks after alcoholic fermentation. Winemakers in the USA, however, commonly practice cofermentation as described in Chapter 5. If following this practice it is important to choose the appropriate strain as certain lactic acid bacteria are associated with fermentation faults.

Malolactic fermentation continues until after growth of the lactic acid bacteria has ceased and may still occur when the lactic acid bacteria are moving into their death phase. In summary, from the start to the end of fermentation, and even during aging and storage, the lactic acid bacteria alternate between successive growth and regression periods depending on the species and strain (Ribéreau-Gayon et al., 2006a). Detailed descriptions of the microbial dynamics of fermentation can be found in texts such as Fleet (1993), Boulton et al. (1996) and Ribéreau-Gayon et al. (2006a, 2006b).

Another group of bacteria present on the grapes which can significantly affect the wine-making process is the acetic acid bacteria. Sound ripe grapes contain around 10^2/g of these organisms but their numbers are around 10^5/g on damaged grapes. These bacteria are responsible for the production of acetic acid from ethanol, thereby increasing volatility of the wine and potentially causing spoilage. They can survive in microaerophilic conditions and their presence has been associated with sluggish and stuck fermentations (reviewed by du Toit et al., 2006).

It can be seen that the wine-making process is a fine balance between competing organisms which would be difficult to control without due consideration to:

- hygiene control
- the use of starter cultures of *S. cerevisiae* or *S. bayanus* and *O. oeni*

- sulfiting the wine at the end of malolactic fermentation to prevent further bacterial development and growth.

Even with best management practices in place many factors must be considered to ensure control over wine making to prevent stuck fermentations and/or spoilage.

2. STUCK FERMENTATIONS

A significant problem that may arise during wine making is the premature cessation of alcoholic fermentation, known as stuck fermentation. If the rate of fermentation is too slow it is regarded as being sluggish. The cause may be one or more of the following (Bisson, 1999; Bisson and Butzke, 2000; Henschke and Jiranek, 1993; Ribéreau-Gayon et al., 2006a; du Toit et al., 2006):

- nutrient deficiency: of nitrogen, vitamins or minerals
- inhibitory substances: ethanol, sugar, yeast-derived fatty acids, acetic acid, excessive sulfur dioxide (SO_2)
- killer toxins: toxins produced by yeasts against other yeasts
- microbiological: certain lactic acid bacteria, *Acetobacter* spp.
- pH too low (<3)
- glucose/fructose ratio: fructose tends to accumulate
- low oxygen: especially in presence of high ethanol, high fructose
- pesticide residues from the field (Calhelha et al., 2006; Čuš and Raspor, 2008)
- management practices such as overclarification
- temperature (too low at the beginning or too high during the ferment).

Microorganisms may directly or indirectly cause many of these problems. Excessive numbers of acetic acid bacteria in the ferment may result in high levels of acetic acid, for example. Likewise, wild yeasts that produce killer toxins against *S. cerevisiae* may survive in the ferment and inhibit growth of the desired fermenting yeast. With particularly ripe harvests that result in high sugar concentrations the juice is often low in nitrogen and the pH is high. Wild lactic acid bacteria may then grow rapidly, resulting in stuck fermentation and lactic disease. Control of malolactic fermentation at the end of alcoholic fermentation is then lost, resulting in loss of wine quality.

The microbial ecology in the wine—surviving organisms, their rate of growth, their impact on each others' growth and survival—is complex, and enhances the need for hygiene control throughout to effect adequate control over the microbial interactions.

Malolactic fermentation may also be inhibited or sluggish owing to the suppression of growth of *O. oeni*. Causes include:

- presence of SO_2 (<5 mg/L)
- low pH (<3.2)
- low temperature ($<15°C$)
- ethanol at 12–14% or higher
- high levels of CO_2
- bacteriophages: these are viruses active against specific bacteria; phage-resistant strains of *O. oeni* starter cultures must be used in wine making.

2.1 Killer Yeasts

Some yeast strains secrete toxins that kill other, sensitive yeast strains. The toxins are proteins or glycoproteins that disrupt the transport of amino acids in the cells, and cause acidification of cellular contents and adenosine triphosphate (ATP) leakage. Sensitive yeasts die within 2–3 h of contact with the toxin. The killer yeast phenomenon is common, although the incidence varies considerably around the world. Many wine strains of *S. cerevisiae* have been observed with the killer property, but strains of wild yeasts belonging to *Kloeckera*, *Hanseniaspora*, *Candida*, *Pichia*, *Kluyveromyces*, *Debaryomyces* and *Hansenula* have also been isolated with killer properties (Ribéreau-Gayon et al., 2006a).

The killer phenomenon may cause stuck fermentation and influence the organoleptic properties of the wine through the destruction of desirable yeast strains (Boulton et al., 1996; Ribéreau-Gayon et al., 2006a; Shimizu, 1993). The risk of killer yeasts to the wine can be minimized by:

- use of a starter culture *S. cerevisiae* strain with killer properties
- low contamination of grapes and juice with wild yeasts (use of SO_2).

3. MICROBIAL SPOILAGE OF WINE

'Vinification fermentations, good and bad, are a succession of microbial onslaughts' (Luthi, 1957).

3.1 Yeasts

Spoilage yeasts may be present in the wine in low numbers throughout the fermentation process. In good wine-making conditions these yeasts are unable to grow in significant numbers. However, in poor conditions they may be able to multiply to spoilage levels after alcoholic fermentation or indeed prior to strong growth of *S. cerevisiae* (Table 6.1). Control of sulfiting and the selected strain of *S. cerevisiae* are important features in preventing yeast spoilage. Lack of control at any stage of the process may also increase the risk of yeast spoilage during aging and bottle storage.

TABLE 6.1 Yeast-Related Faults, Causal Organisms and Their Management

Spoilage Effect	Causative Organisms	Control
Ester and aldehyde taints; increased volatile acidity	*Pichia, Hanseniaspora, Hansenula, Metschnikowia, Dekkera, Candida*	Maintain anaerobic conditions, cleaning program, use of SO_2
Formation of surface films	*Candida, Pichia*	Maintain anaerobic conditions, cleaning program, use of SO_2
Mousey, medicinal taints	*Brettanomyces*	Equipment cleaning, must sulfiting
Refermentation in the bottle	*Saccharomyces, Zygosaccharomyces bailii*	Cleaning program, sterile filtration, aseptic bottling
Oxidized taint from acetaldehydes	*Saccharomycodes ludwigii*	Cleaning program, complete alcoholic fermentation
Deacidification of wine	*Schizosaccharomyces*	Cleaning program, complete alcoholic fermentation

Note: SO_2 = sulfur dioxide.
(Source: Fleet and Heard, 1993; Ribéreau-Gayon et al., 2006a).

Yeasts capable of anaerobic growth, such as *Brettanomyces* and *Dekkera*, can utilize trace amounts of sugar left over and may reach levels of up to 10^5 colony-forming units (cfu)/mL in contaminated red wine (Ribéreau-Gayon et al., 2006a). *Brettanomyces* is commonly isolated from beer, wine and fruit juices and can grow in a pH as low as 1.8.

3.2 Bacteria

Wines are at greatest risk of bacterial spoilage toward the end of their fermentation and throughout storage. This is due to there being a higher pH at this stage, which allows for greater diversity in the bacterial flora and higher survival. Wine at this stage is more fragile. Keeping the wine below a pH of 3.5 is an important control. Spoilage may develop very slowly and cause problems many months into the storage of the wine.

The most common spoilage bacteria of wine are those belonging to the lactic and acetic acid bacteria groups (Table 6.2). Less commonly, spoilage may be caused by spore-forming bacteria of the *Bacillus* and *Clostridium* groups.

Lactic disease occurs when there is significant sugar available after the alcoholic fermentation for lactic acid bacteria to ferment, resulting in increased volatile acidity (acetic acid and D-lactic acid). This is a risk with particularly ripe harvests as there is high sugar concentration, high pH and often low nitrogen levels. A stuck fermentation might be expected, permitting rapid multiplication of lactic acid bacteria rather than the usual reduction in their numbers at this stage.

Acetic acid bacteria also have the potential to develop under these conditions and acidification in the wine may be due to the combined activity of lactic acid and acetic acid bacteria (reviewed by Sponholz, 1993). Keeping wine free of oxygen should control the growth of any surviving acetic acid bacteria (but see du Toit et al., 2006). However, exposure of the wine to air during pumping and transfer operations can stimulate their growth (Drysdale and Fleet, 1989; Joyeux et al., 1984). The use of undamaged grapes at optimum maturity will ensure that only low levels of these bacteria enter the winery, minimizing the potential for spoilage.

Bacterial degradation of glycerol with end-production of acrolein results in reaction with phenolic compounds in wine to produce a bitter taint. Referred to as 'amertume', this is, therefore, more commonly associated with high phenolic red wines than white wines (Sponholz, 1993). This spoilage is caused by various strains of lactic acid bacteria, including *O. oeni*, *Pediococcus parvulus* and *Lactobacillus* species. Wine is most vulnerable after completion of malolactic fermentation, although Australian wine has been less vulnerable than wine in Europe (Fugelsang, 1997).

Also after malolactic fermentation, the wine is most vulnerable to degradation of tartaric acid to acetic acid by *Lactobacillus*, known as 'tourne'. This results in increased volatile acidity and off-flavors. This fault is generally associated with other spoilage problems in the wine. Amertume and tourne are prevented by good winery hygiene and sulfiting.

Excessive production of diacetyl is detrimental to the sensory quality of wine, with levels over 4–5 mg/L (depending on the wine) causing a buttery character regarded as a fault (Rankine et al., 1969). The appropriate strain of *O. oeni* must be chosen to prevent the causative organisms from growing.

TABLE 6.2 Bacterial Faults, Causal Organisms and Their Management

Spoilage Effect	Causative Organisms	Control
Lactic disease—acidification	*Oenococcus oeni, Lactobacillus,* acetic acid bacteria	Equipment cleaning, must sulfiting
Amertume—bitterness	*Pediococcus parvalus, Lactobacillus cellobiosus, Leuconostoc mesenteroides*	Cleaning program, sulfiting
Tourne—increased volatile acidity	*Lactobacillus plantarum, Lactobacillus brevis*	Cleaning program, sulfiting
Ropiness	*Pediococcus damnosus, Leuconostoc mesenteroides, Streptococcus mucilaginosus* var. *vini*	Equipment cleaning, sulfiting
Buttery flavors from diacetyl production	*Lactobacillus, Pediococcus*	Good starter culture
Mannitol taint/mannite disease	*Lactobacillus brevis*	Equipment cleaning, must sulfiting
Mousey taints	*Lactobacillus* species	Equipment cleaning, must sulfiting
Geranium taint	*Lactobacillus, O. oeni*	Correct balance of sorbic acid and SO_2
Butyric acid taint	*Clostridium, Bacillus*	Cleaning program, SO_2, use of inert gases in headspace, sterile filtration, aseptic bottling
Histamine production	*Oenococcus oeni*	Correct starter culture strain

Note: SO_2 = sulfur dioxide.

Ropiness is characterized by a viscous, oily wine (Fugelsang, 1997; Ribéreau-Gayon et al., 2006a) and may be evident at the end of alcoholic fermentation. It may alert the winemaker to possible mannitol taint and high volatile acidity.

Various lactobacilli and the spoilage yeast *Brettanomyces* cause mousiness, an uncommon fault in wine resulting in a rodent urine odor, detected by rubbing the wine between the fingers. Low acid wines with insufficient sulfiting are at greater risk (Sponholz, 1993).

Some lactic acid bacteria can metabolize sorbic acid, resulting in a final reaction with ethanol to form a taint of crushed geranium leaves. If sorbic acid is used to control yeast growth in sweet wines then adequate sulfiting must also be used to prevent bacterial spoilage.

Particular strains of *O. oeni* may transform the amino acid histadine into histamine (Ribéreau-Gayon et al., 2006a). Apart from the possible toxic effects of histamine on the consumer, histamine will cause subtle but significant changes to the organoleptic properties during aging of the wine. Histamine production is an enzymatic reaction that can continue to occur after *O. oeni* cells are dead. Hence, the most effective control of this potential problem is the use of good inoculating strains that do not possess this property and then continuing good control over the wine microflora.

Spoilage of wine by *Clostridium* bacteria showing rancidity from butyric acid taint is rare, and will only occur after failed malolactic fermentation where the pH is greater than 4.0 (Sponholz, 1993).

3.3 Molds

Grapes are susceptible to mold attack in conditions of high humidity, and compact grape clusters enhance the transfer of molds between grapes. Molds causing grape spoilage are widespread in nature and easily transferred by insects and wind.

Gray rot is caused by infection of the grapes with *Botrytis cinerea* between fruit set and veraison. Although this is the same organism that causes noble rot, noble rot infection occurs on the mature grape. Gray rot produces fatty acids and terpenic compounds that produce characteristic moldy odors and cause aromatic flaws in wine (Ribéreau-Gayon et al., 2006a). *Botrytis cinerea* can also enhance the growth of acetic acid spoilage bacteria which utilize glycerol formed by the mold, leading to sour rot. It also contains laccase, a resilient oxidative enzyme that is resistant to SO_2 and which can only be removed by flash pasteurization or be minimized by treatment with bentonite, prefermentation, cooling to less than 12° C at pressing or, with red grapes, adding oenological tannin (AWRI, 2011b; Boulton et al., 1996; Margalit, 1996).

In dry white wines, gray rot promotes the development of rancid, camphorated and waxy odors that appear later during maturation and especially in the bottle. Other molds—*Aspergillus*, *Penicillium*, *Cladosporium*, *Mucor*, *Fusarium*, *Plasmopara*, *Uncinula*—metabolize amino acids and phenolic compounds and produce substances that are usually bitter. The resultant wine has phenol and iodine odors.

Molds are generally controlled on grapes with the use of fungicides, but the use of these poisons must be carefully managed to minimize residues in the juice and their potential inhibitory effect on the alcoholic and malolactic fermentations.

3.4 Cork Taint

Cork taint is generally recognized by a musty or moldy off-flavor. A low level of taint may simply result in a change in the aroma or taste of the wine. As cork is a natural product from the cork oak it is subject to microbial contamination and its quality is dependent on good agricultural practices and quality control during processing, transport and storage.

The microflora of corks is comprised largely of molds with lesser numbers of other soil-borne organisms including yeasts and bacteria. Molds are considered the most significant causative organisms of cork taint, with implicated genera including *Penicillium*, *Aspergillus*, *Cladosporium*, *Monilia*, *Paecilomyces* and *Trichoderma*. However, the chemistry of cork taint is complex and the biochemistry of the organisms involved is not fully understood. *Rhodotorula* and *Candida* yeasts have been implicated and *Streptomyces* bacteria are known to be associated with cork taint (Lee and Simpson, 1993). It is possible that *Bacillus* bacteria also have a role in cork taint (Fugelsang, 1997).

Exposure of corks to moisture can encourage the growth of these organisms to produce metabolites that have extremely low flavor thresholds and can be leached into the wine to give overpowering, deleterious taints. 2,4,6-Trichloranisole is the compound most responsible for cork taint but other chloranisoles and chlorophenols can also contribute to cork taint. In addition, the compounds geosmin, 2-methylisoborneol and guaiacol, all produced by *Streptomyces*, and 1-octen-3-ol and 1-octen-3-one cause cork taint (Lee and Simpson, 1993).

4. CRITICAL CONTROL POINTS IN SPOILAGE PREVENTION

While the details of the wine-making process may vary depending on the type of wine in production, common CCPs include the following.

4.1 Grape Production

Microbial contaminants at harvest may arise from infections at flowering or earlier; therefore, implementing a sound canopy, pest and disease management strategy is the first step in quality assurance.

4.2 Grape Harvest

The introduction of high numbers of potential spoilage yeasts and bacteria from poor-quality grapes may result in stuck fermentation and subsequent spoilage of the wine by lactic acid or acetic acid bacteria (Sponholz, 1993).

Mold contamination on grapes may lead to flavor taints in wine directly or via contamination of the winery. Production of ochratoxin in wine by the black aspergilli occurs (Hocking and Pitt, 2001); however, sound management practices through to the point of harvest minimize the risk of the presence of this toxin in the finished product.

Equipment used in harvesting, such as palletrons, bins and harvesters, must be hygienically controlled. Cleaning and sanitizing must be undertaken using water of appropriate quality.

4.3 Grape Processing

Cleaning and sanitizing of the winery are essential to minimize the development of spoilage microflora. Prevention of microbial buildup on equipment such as augers, crushers and presses must be managed through effective sanitation control. Potable water must be used in cleaning operations.

4.4 Alcoholic Fermentation

Alcoholic fermentation must be properly managed to prevent the uncontrolled growth of undesirable species and strains of yeasts that could dominate and result in stuck fermentation and/or flavor taints (Fleet, 2001).

4.5 Malolactic Fermentation

If this is undertaken by *Lactobacillus* or *Pediococcus* species rather than the desired *O. oeni*, off-flavors develop and the wine is spoiled (Fleet, 2001).

4.6 Bottling

The use of membrane filtration to minimize the presence of microorganisms in the finished product and preservatives to prevent the growth of those remaining are important controls to prevent spoilage of wine in the bottle. In addition, as the cork may be a source of microbial contamination producing flavor defects, the bottling of wine is the final CCP in its production (Lee and Simpson, 1993; Sponholz, 1993).

CCPs will generally be managed through hygiene control, support programs and management practices, but monitoring to verify compliance with the controls in place is essential. Implementation of an industry-recognized quality assurance program is advisable.

5. MONITORING THE PROCESS

To be confident that the process is under control, monitoring points should be chosen to test the microbial flora of the wine. Checking that the desired organisms are present in the right numbers and that the wine is free of spoilage organisms allows remedial action to be taken in the event a problem develops. You must decide on the appropriate way to monitor and either develop some level of internal testing capability or outsource testing to a suitably qualified laboratory.

In-house testing will require the setting up of a basic laboratory. Many materials needed for testing can be purchased in a ready-to-use form which minimizes the required equipment and technical expertise. This level of capability lends itself simply to monitoring fermentation progress. Where problems in wine making arise that require expert assistance, testing should be outsourced. Testing should only be outsourced to accredited laboratories that can demonstrate competence in the work (e.g. in Australia and New Zealand, NATA and Telarc are the accrediting agencies).

Careful consideration must be given to the benefits of setting up microbiology testing in the winery as there are risks associated with this activity. By its nature, microbiological testing is growing high numbers of organisms which can potentially contaminate the winery environment. If these include spoilage organisms there could be dire consequences. Sanitary control of the testing area, restriction of staff and equipment movement, and appropriate waste disposal are paramount.

5.1 Monitoring Points

5.1.1 Alcoholic Fermentation

- Sample the must after 2–3 days of fermentation to verify growth of *S. cerevisiae* to 10^7–10^8 cfu/mL and that numbers of wild yeasts are low.
- Sample again at the end of the fermentation to verify *S. cerevisiae* at $<10^2$ cfu/mL and an absence of spoilage yeasts.

5.1.2 Malolactic Fermentation

- Sample the wine about 2 weeks after alcoholic fermentation is complete to check adequate growth of *O. oeni* to 10^7 cfu/mL.

5.1.3 Bottling

A proportion of finished product should be sampled for analysis to verify aseptic filling.

6. SAMPLING

Sampling is conducted according to risk and sampling points chosen to provide information about the microbiological status of the wine. The purpose of sampling wine at the fermentation stages is to verify that the process is under adequate control. Sampling should be timely to allow a decision to be made to adjust or maintain process parameters. At bottling the purpose of sampling is to verify that the finished product is microbiologically stable and unlikely to spoil before consumption. It allows the winemaker to make a decision about the lot or batch. The sampling plan should be carefully worked out to ensure that it provides appropriate information.

Sampling is the largest cause of measurement uncertainty in any microbiological test procedure. Correct decisions based on the results of sampling can only be made if the samples are truly representative of the batch from which they are taken.

Wine is a homogeneous product but variations may exist between tanks or batches. Significant differences in test results may be observed between tanks of the same wine under identical control parameters. In clear, well-mixed wine, microorganisms are evenly distributed and it is easy to collect representative samples. Uneven distribution of microorganisms can be expected in musts and red wine ferments.

The costs of sampling and testing generally dominate the decision on how many samples to collect and often preclude the use of an ideal sampling plan. However, one should bear in mind that the cost of not sampling and testing sufficient samples can be significant and a balance must be struck. The number of samples to collect at fermentation will be determined by:

- the number and type of tanks in use
- the use of starter cultures
- the degree of control over wine-making parameters such as temperature.

Every tank in the ferment should be sampled and it is advisable to collect duplicate samples each time. In a well-controlled, hazard analysis and critical control point

(HACCP)-based process, one sample at each time-point may be adequate.

Contamination in the wine at bottling may be regular or irregular depending on the cause. No sampling plan can ensure the absence of spoilage organisms and hence the application of HACCP throughout wine making and particularly at bottling must be relied upon to reduce the risks of spoilage in-bottle. Some microbiological sampling plans are based on batches of five samples, with the multiples of five required, assigned by risk. However, a simple approach is to collect a bottle at the start, middle and end of the bottling run. This provides some basis on which to accept or reject portions of a batch as information is provided on the point at which contamination may have occurred during the bottling process.

For example, if the following pattern occurs:

- start bottle—sterile
- middle bottle—contaminated
- end bottle—contaminated

this would suggest that somewhere in the first third of the run the filling heads became contaminated. Either all bottles from the middle onward would be discarded or extra samples would be collected to pinpoint the time of contamination, therefore more reliably salvaging the sterile portion of the run.

6.1 Sampling Technique

Sampling should be conducted aseptically to prevent contamination of the sample and the introduction of contaminants to the container which is being sampled. Use sterile pipettes to withdraw samples into a suitably sized sterile container. Presterilized sample containers and pipettes are commercially available. Prior to opening bungs, stoppers or taps, sanitize the container opening by swabbing with cotton wool or tissue soaked in 70% ethanol (methylated spirits is a suitable alternative).

Only remove the lids of sample jars immediately before sampling and replace promptly after collection. Transfer the sample to the laboratory as quickly as possible and test within 5 h of sampling. Care must be taken while removing and replacing lids not to touch the opening or inside of the container with the hands or other items.

6.1.1 Sampling from an Open Tank

Mix the tank contents thoroughly immediately before sampling. Rapidly lower the sample container, with lid removed and opening face downward, into the middle of the tank, tilt to fill, withdraw and replace the lid promptly. For large tanks the sampling vessel should be attached to a pole that can be sterilized (e.g. stainless steel) to prevent contaminating the wine.

6.1.2 Sampling from a Tap or Valve

When taps are fitted to tanks, run the first flow to waste before collecting the sample directly into the sample container because any 'stagnant' wine in the line will not be representative of the body of wine in the tank.

The sample must be thoroughly mixed immediately before testing. Mixing is achieved by shaking the sample 25 times through an arc of 90–180° within approximately 10 s. The interval between mixing and removing the test portion should not exceed 3 min.

7. IN-HOUSE TESTING

Testing at the winery should be restricted to microscopic techniques and enumeration of the starter cultures to verify fermentation, using purchased, ready-to-use materials. In this instance, equipment and materials requirements are minimal.

- **Sampling equipment:**
  - Sterile plastic containers, min. 30 mL volume
  - Sterile pipettes
  - Alcohol 70%, or methylated spirits 70%
  - Cotton wool/tissues
- **Microscopic assessment:**
  - Compound microscope with 10×, 40× and 100× lenses
  - Immersion oil
  - Counting chambers, e.g. Neubauer with coverslips
  - Disposable Pasteur pipettes
  - Methylene blue stain
- **Enumeration by culture:**
  - Sterile pipettes
  - Disposable spreaders ('hockey sticks')
  - Alcohol 70%, or methylated spirits 70%
  - Pipettes
  - Sterile diluent, 9 mL volumes
  - Prepoured agar plates
  - Membrane filtration apparatus
  - Vacuum pump
  - Filter manifold
  - 0.45 μm pore size microbiological test filters
  - Incubation canisters or candle jars
  - Gas packs providing microaerophilic conditions
  - Incubator(s)
  - Contaminated waste disposal bags or bins.

If you choose to set up a fully equipped laboratory for extensive testing, more equipment is required and the laboratory may require microbiological media-making facilities. Additional equipment and material requirements would then include:

- Sterilizer, e.g. domestic pressure cooker or autoclave
- pH meter
- Balance (laboratory scale) to 2 decimal places or more
- Water bath for tempering molten agar
- Hot plate or microwave oven
- Glassware: measuring cylinders, Pyrex™ bottles
- Distilled or deionized water
- Petri dishes, 90 mm.

If you choose to manufacture microbiological media then a quality control program for the media needs to be implemented. This requires considerable knowledge and expertise and is not likely to be cost-effective for small wineries. If media are purchased as prepoured agar plates the manufacturer will provide quality control certificates with each batch.

Where access to ready-prepared microbiological media or suitable laboratory facilities in the winery is unavailable you may need to outsource testing to a suitably qualified commercial laboratory. This will also be required when troubleshooting studies are needed to investigate spoilage. The laboratory should be accredited under a national quality assurance program (e.g. NATA) but should also have the necessary knowledge in wine making and expertise in wine microbiology to be able to adequately advise on sample points, test the samples and interpret the results.

8. DETECTION AND DIFFERENTIATION OF WINE MICROORGANISMS

Microorganisms are generally detected by either culture in microbiological media or microscopic techniques, the former providing evidence of viability but the latter not reliably indicating viability.

8.1 Microscopic Examination

Conventional staining of yeasts and bacteria is used to determine cell morphology and, if from liquid suspensions, also allows a determination of the population using a counting chamber. The most commonly used chamber is the Improved Neubauer. Aseptically withdraw a few milliliters of sample into a sterile container and examine immediately. If multiple tanks are being assessed, label each container.

Set up the counting chamber with the coverslip in place. Mix the sample that has been collected. Using a Pasteur pipette place a drop of the sample on one edge of the coverslip. Capillary action will pull the liquid and fill the chamber.

If too much liquid is placed on the chamber the coverslip will be unstable, and the microbial cells will move around too much to count. If this occurs a tissue can be carefully placed on the edge of the liquid to mop up the excess.

Leave the liquid under the coverslip to settle for a few minutes and then place the chamber on the microscope stage.

Focus on the chamber grids using a low-power lens ($10\times$). You will see one large square that consists of 25 medium squares that are bounded by double lines. Within each of these are 16 small squares to give a total of 400 squares over the grid counting area. The large square measures 1×1 mm. Coverslips that are designed to match the counting chamber must be used as their thickness varies, which determines the depth and hence volume of the liquid being examined.

The microscope factor (MF) is used in calculations to determine the number of cells in the wine. The MF is the number of microscope fields in the slide area divided by the quantity of sample applied. The MF is obtained by dividing the slide sample area by the field area to obtain the number of microscope fields in the slide area, and then dividing this quantity by the sample volume. The final concentration of microorganisms in the wine is calculated by multiplying the average count per field by the MF.

Two kinds of counting chamber are used for this procedure:

- The Levy–Hausser is suited to yeast enumeration; a $40\times$ lens is used and the MF is 10^4 times the number of cells counted in the large square.
- The Petroff–Hausser is suited to bacterial enumeration; a $100\times$ oil-immersion lens is used and the MF is 2×10^7 times the number of cells counted in a small square.

It is not always practical to count all 400 small squares as this is time consuming if the number of cells is high. However, a minimum of six medium squares should be counted and the average of this count used in the final calculation. For highest precision, the chamber should be loaded several times and the counting repeated.

As cells will be observed resting on the lines of the grids you must establish a convention for counting. Cells that touch the borders on the top and right of the square being viewed are included in the count while those touching the bottom and left borders are excluded from the count.

When counting yeasts, count each individual cell and each budding cell as one yeast. For bacteria count a chain of cells as a single colony-forming unit (cfu).

For yeasts microscopic examination can also provide an indication of the viability of the culture. Before examination mix the sample approximately 1:1 with methylene blue stain. The mixing can be done directly on a microscope slide. Viable cells take up the stain and reduce it to its colorless form, giving an unstained cell, whereas the dead cells take up the stain without reduction, giving a blue cell.

Viability in this context may not exactly correlate with the ability of the cell to grow and reproduce on culture medium but will refer to a capacity to carry out some part of glycolysis (Boulton et al., 1996).

9. CULTURE OF WINE MICROORGANISMS

Culture techniques are designed for the growth of microorganisms with the ability to enumerate under specified incubation conditions. Not all viable organisms will grow but acceptable results are obtained under suitable quality control conditions. At various stages in the wine-making process it is useful to know the levels of fermenting flora and their viability. When spoilage of the wine is suspected, culture is the only means to confirm the type and numbers of organisms responsible at critical points in the process.

Numerous agar media are available for the isolation of particular groups of microorganisms. The medium of choice will depend on a variety of factors including cost, time of testing in the wine-making process, ease of use and availability as a prepared medium if required (see Tables 6.3–6.9).

Culture media are comprised of the following basic elements (Stephens, 2003):

- gelling agent, usually agar gel, which allows for the visualization of growing colonies
- amino-nitrogen compounds
- energy sources
- buffer salts
- mineral salts and metals
- growth-promoting factors

and may also include:

- selective agents such as antibiotics that allow the growth of the target organisms and inhibit the growth of non-target organisms or indicator dyes that aid in differentiation of microbial colonies.

TABLE 6.3 Yeast Extract, Peptone, Dextrose (YEPD) Agar Medium: Use for the Growth of Fermentation Yeasts

Material	Quantity
Yeast extract	10.0 g
Peptone	20.0 g
Glucose	20.0 g
Chloramphenicol	100 mg
Agar	20.0 g

Mix all the above ingredients in 1000 mL deionized water.
Autoclave for 15 min at 121°C.
Chloramphenicol (100 mg) may be added before autoclaving to suppress bacterial growth.

9.1 Yeasts

When the total number of viable yeasts needs to be determined, yeasts are isolated on yeast extract, peptone, dextrose (YEPD) agar or WL nutrient agar (WLNA) media (see Tables 6.3 and 6.4). These agars will isolate wine

TABLE 6.4 WL Nutrient Agar: Use for the Growth of Fermentation Yeasts

Material	Quantity
Yeast extract	4.0 g
Tryptone	5.0 g
Glucose	50.0 g
Potassium dihydrogen phosphate	0.55 g
Potassium chloride	0.425 g
Calcium chloride	0.125 g
Magnesium sulfate	0.125 g
Ferric chloride	0.0025 g
Manganese sulfate	0.0025 g
Bromocresol green	0.022 g
Agar	15.0 g

Mix all of the above in 1000 mL deionized water.
Autoclave for 15 min at 121°C. The pH of this medium will be 5.5.
If required, the pH may be adjusted to 6.5 by the addition of 1% sodium bicarbonate solution.
Chloramphenicol (100 mg) may be added before autoclaving to suppress bacterial growth.

TABLE 6.5 Malt Extract, Yeast Extract, Glucose, Peptone (MYGP) Plus Copper Sulfate Medium: Use for the Growth of Spoilage Yeasts

Material	Quantity
Peptone	3.0 g
Malt extract	3.0 g
Yeast extract	5.0 g
Glucose	10.0 g
Agar	10.0 g
$CuSO_4.5H_2O$	0.625 g

Mix all ingredients in 1000 L of deionized water.
Adjust to pH 4.7.
Autoclave for 15 min at 121°C.

TABLE 6.6 De Man, Rogosa, Sharpe (MRS) Agar: Use for the Growth of Lactic Acid Bacteria

Material	Quantity
Peptone	10.0 g
Lab-Lemco powder	8.0 g
Yeast extract	4.0 g
Glucose	20.0 g
Sorbitan mono-oleate	1 mL
Dipotassium hydrogen phosphate	2.0 g
Sodium acetate $3H_2O$	5.0 g
Triammonium citrate	2.0 g
Magnesium sulfate $7H_2O$	0.2 g
Manganese sulfate $4H_2O$	0.05 g
Agar	10.0 g

Mix all ingredients in 1000 L of deionized water.
Adjust pH to 6.2.
Autoclave for 15 min at 121°C.

TABLE 6.7 Lafon–Lafourcade Agar Medium Plus Actidione: Use for the Growth of Lactic Acid Bacteria When Yeasts Are Also Present in High Numbers

Material	Quantity
Yeast extract	5.0 g
Meat extract	10.0 g
Trypsic peptone	15.0 g
Sodium acetate	5.0 g
Ammonium citrate	2.0 g
Manganese sulfate	0.05 g
Magnesium sulfate	0.2 g
Glucose	20.0 g
Actidione	50 mg

Dissolve above ingredients in 400 mL deionized water.
Adjust the pH to 5.4 using 1N sodium hydroxide or 1N hydrochloric acid.
Add 1 mL Tween 80.
Weigh 20 g agar in 500 mL deionized water and boil to dissolve.
Cool and mix the two solutions.
Make to 1000 mL with deionized water.
Autoclave for 20 min at 121°C.
Warning: Actidione (syn. cycloheximide) is toxic. Avoid skin contact, aerosol formation and inhalation.

yeasts, *S. cerevisiae* and wild yeasts such as *Pichia*, *Candida* and *Brettanomyces*. However, once isolated, the different yeasts will look similar and it is not possible to differentiate wine from wild yeasts by colony morphology. Without added chloramphenicol these media will also isolate bacteria, which may limit their applicability.

When information about the type of yeasts is desirable, a different agar is used for detection. One example, MYGP (malt extract, yeast extract, glucose, peptone) plus copper sulfate agar, inhibits the growth of *S. cerevisiae*, allowing only wild yeasts to grow (see Table 6.5). This agar is therefore useful when monitoring a wine for spoilage organisms. Copper sulfate in the medium inhibits the growth of culture yeasts and hence any yeast growth is assumed to be that of wild, contaminating strains. However, when the level of culture yeast in the wine is high, such as toward the end of alcoholic fermentation, the medium may allow *S. cerevisiae* to grow. Differentiation of wild yeasts may need to be made through the use of additional selective media and further diagnostic tests including microscopic examination. When this is required it is advisable to outsource testing to a laboratory with the necessary expertise.

9.2 Lactic Acid Bacteria

Lactic acid bacteria are isolated from wine using a non-selective medium such as De Man, Rogosa, Sharpe (MRS) agar or a selective medium such as Lafon–Lafourcade agar medium plus actidione (see Tables 6.6 and 6.7). The choice of medium will be determined by the stage in wine making at which testing is being conducted. When viable yeast numbers in the wine are expected to be high the use of Lafon–Lafourcade agar medium will be required to detect the lactic acid bacteria.

TABLE 6.8 Malt Extract Agar (MEA) Plus Acetic Acid: Use for the Growth of Acetic Acid Bacteria

Material	Quantity
Malt extract	30.0 g
Peptone	5.0 g
Acetic acid	5.0 g
Agar	15.0 g

Mix above ingredients except the acetic acid in 950 mL deionized water.
Autoclave for 10 min at 115°C.
Cool to 55°C and add 50 mL of a 10% solution of acetic acid.
Cycloheximide (syn. actidione) can be added at 100 mg/L to suppress yeast growth if required.
Warning: Cycloheximide is toxic. Avoid skin contact, aerosol formation and inhalation.

TABLE 6.9 *Acetobacter–Gluconobacter* **Agar (AGA) Medium: Use for the Growth of Acetic Acid Bacteria**

Material	Quantity
Glucose	50.0 g
Calcium carbonate	30.0 g
Yeast extract	10.0 g
Cycloheximide	100 mg
Agar	15.0 g

Mix above ingredients in 1 L deionized water.
Autoclave for 15 min at 121°C.
Warning: Cycloheximide is toxic. Avoid skin contact, aerosol formation and inhalation.

9.3 Acetic Acid Bacteria

Malt extract agar (MEA) plus acetic acid medium or *Acetobacter–Gluconobacter* agar (AGA) medium with added actidione (syn. cycloheximide, see Tables 6.8 and 6.9) is suitable for the isolation of acetic acid bacteria.

Consistent use of the agar media chosen is important to adequately compare microbial content between batches of wine and provide consistent interpretation of results.

10. MICROBIOLOGICAL CULTURE MEDIA

Most media are available as a commercially prepared powder. Prepare according to the manufacturer's instructions, which involve weighing a designated amount into 1 L distilled/deionized water then sterilizing in a domestic pressure cooker or by autoclaving at 121° C for 15 min. Media often need pH adjustment which can be done after autoclaving, before pouring the plates, but the batch must be cooled to 47° C in this instance and the adjustment made aseptically. Alternatively, the medium can be prepared and boiled to dissolve the agar, tempered to 47° C, the pH adjusted, and then autoclaved.

After autoclaving the liquid agar media must be tempered to around 47° C and carefully handled to prevent contamination prior to and during pouring into Petri dishes. Once set, agar plates must be refrigerated inverted. All microbiological media have a shelf-life beyond which they deteriorate through the effects of dehydration and light. The manufacturer can supply this information.

The size and shape of containers used influence the temperature profile within the autoclave. In any case, all microbiological media will only achieve sterility when the internal temperature reaches 121° C and remains there for 15 min.

If you are manufacturing liquid media for 9-mL volumes, required for decimal dilutions in Section 11.3, then the amount dispensed in the tubes before autoclaving must allow for volume loss during autoclaving so that the final volume is 9 mL ± 2%. The amount of volume loss varies in different autoclaves and also depends on the shape and type of the container. Trials must be run to establish the correct initial dispensing volume to achieve the correct final volume.

11. TECHNIQUES

11.1 Methods

11.1.1 Sample Preparation

Gently mix the sample collected, using 25 up and down movements of about 30 cm over a period of about 12 s. Allow gas bubbles to dissipate. Sanitize the lid and upper container by wiping with tissue soaked with 70% alcohol before opening the sample for testing.

11.2 Melting Culture Media

11.2.1 Hazards

- Burns
- Glass breakage
- Explosion of agar.

WEAR SAFETY GLASSES OR PREFERABLY A FACE SHIELD AND HEAT-PROOF GLOVES WHEN PERFORMING THIS PROCEDURE.

Use either a boiling water bath or a microwave oven. Always loosen the lid on the bottle of agar before placing in the heating unit, to prevent explosion. You should not be able to remove the lid by lifting off but it should be loose enough to allow steam to escape. Never place a bottle of agar in a water bath or microwave straight from the refrigerator; bring to room temperature before heating. Avoid overheating and remove it as soon as the agar has melted.

To check that the entire agar is melted, lift the bottle and gently shake to observe any solid agar particles that may be present.

11.2.2 Notes

- Keep the agar culture medium in a molten state in a thermostatically controlled bath at 47 ± 2° C until such time as it is to be used. Agar starts to solidify below 45° C.

- Never use agar culture medium at a temperature higher than 50° C as this may kill the organisms present in the inoculum.
- The maximum working life of a molten agar medium is about 8 h.
- Discard. Do not resolidify unused medium for subsequent use.

11.3 Preparation of Dilutions

11.3.1 Hazards

- Burns from flame.

The purpose of preparing serial dilutions of the sample is to obtain as uniform a distribution as possible of the microorganisms contained in the test portion and to allow the growth of isolated colonies of the microorganisms within a countable range on the agar plates. Too many microorganisms in the sample tested will lead to too many colonies on the agar plate so that counting the colonies becomes impossible and the count inaccurate or even meaningless. Diluting the sample through serial dilutions is done so that some of the dilutions made will contain a countable concentration of microorganisms.

- **Serial dilutions** are prepared in a 10-fold series: −0, undiluted; −1, diluted 10-fold; −2, diluted 100-fold, etc. For convenience they are commonly referred to by the index to the dilution factor; that is, −1 represents 10^{-1} (Figure 6.1).
- **Decimal dilutions** (serial dilutions) are suspensions or solutions obtained by mixing a measured volume of the wine sample with a nine-fold volume of sterile diluent and by repeating this operation with further dilutions until a decimal dilution series, suitable for the inoculation of culture media, is obtained.

If, for example, the sample is expected to have 1,000,000 (10^6) yeasts/mL, at least four serial dilutions should be prepared in order to obtain counts within an acceptable range on the agar plates (i.e. to 1/10,000th). The ideal number of colonies to count is between 15 and 300. Counts outside this range are less accurate and considered to be an estimate.

Numerous diluents are acceptable for use, including peptone salt solution (Table 6.11), buffered peptone water (Table 6.10) and Ringer's solution.

TABLE 6.10 Buffered Peptone Water (ASA, 2002)

Material	Quantity
Enzymatic digest of casein	10.0 g
Sodium chloride	5.0 g
Disodium hydrogen phosphate (anhydrous)	3.5 g
Potassium dihydrogen phosphate	1.5 g
Deionized water	1000 mL

Mix to dissolve and adjust to pH 7.0 ± 0.2.
Dispense into tubes to give a final volume of 9 mL.
Autoclave for 15 min at 121°C.

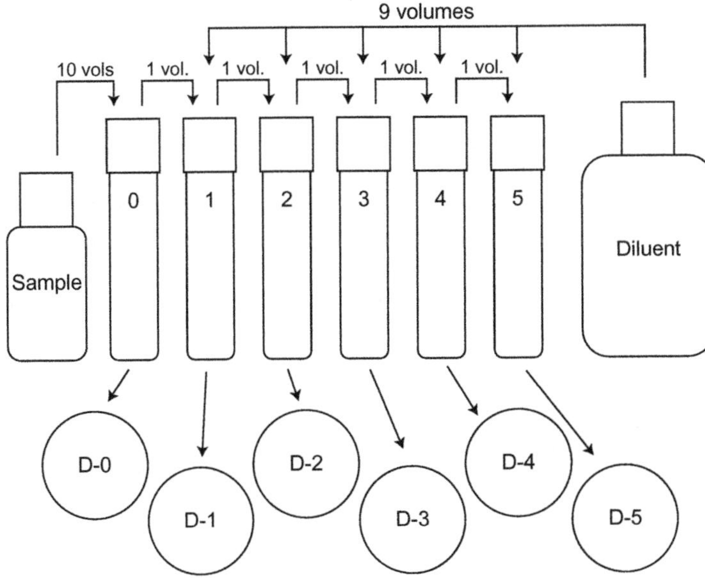

FIGURE 6.1 Example of a 10-fold dilution series including the undiluted original sample. First, 9 volumes of diluent are dispensed to each of the labeled dilutions from 1 to 5 or as chosen; the undiluted sample is then dispensed into the 0 tube and then 1 volume is dispensed to the next dilution; it is mixed thoroughly, and then 1 volume of that is dispensed using a fresh pipette to the dilution, et seq. An equal volume of each is then plated on to each plate for incubation. This example gives 10^0, 10^{-1} to 10^{-5}, i.e. 1 down to 1/100,000th of the original. All tubes are capped to minimize the risk of contamination.

11.4 Pour Plate Technique

11.4.1 Hazards

- Burns from flame.

The pour plate technique involves mixing of the sample inoculum with tempered molten agar and cooling and solidifying on a horizontal surface. Label the required number of empty Petri dishes with appropriate identification of the sample and the dilutions to be plated. Use 90-mm plates and always label on the base of the plate.

Agar plates should always be stored and handled in the inverted position (Figure 6.2). This prevents condensation droplets falling on the agar surface and the lid separating from the base of the plate.

As each decimal dilution is prepared, pipette 1 mL inoculum, using the sterile pipette that has prepared the dilution, into the center of the empty Petri dish(es). For example, when 1 mL of the −2 dilution is pipetted to prepare the −3 dilution, the same pipette is then used to transfer the −2 dilution aliquots into the empty Petri dish(es). Repeat the same procedure with the further dilutions, using a new sterile pipette for each decimal dilution. Upon removal of the molten agar from the water bath, wipe the drips of water from the outside of the bottle with some paper towel without agitating the agar.

Remove the lid from the bottle of agar aseptically, and flame the neck. Lift the lid from the base of the Petri dish for the minimum amount of time required to pour the agar into the base. Pour each of the Petri dishes containing sample inoculum with required molten agar medium so as to obtain a thickness of at least 2 mm (for 90-mm diameter dishes, 12−15 mL of agar is required).

TABLE 6.11 Peptone Salt Solution (ASA, 1994)

Material	Quantity
Enzymatic digest of casein (peptone)	1.0 g
Sodium chloride	8.5 g
Deionized water	1000 mL

Mix to dissolve and adjust to pH 7.0 ± 0.2.
Dispense into tubes to give a final volume of 9 mL.
Autoclave for 15 min at 121°C.

After replacing the lid, gently swirl the Petri dishes to mix using five clockwise, five anticlockwise, then five up-and-down movements. Allow the plates to stand until the agar has set.

For surface inoculation techniques (spread plate, membrane filtration) agar plates prepared in-house may need to be dried before use. Dry with the lids in place and the agar surfaces facing downward, in an oven set at a temperature between 25 and 50° C, until the droplets have disappeared from the surface of the medium. This will prevent aerial mold contamination on the surface of the plates.

11.5 Spread Plate Technique

11.5.1 Hazards

- Burns from flame.

This is a surface inoculation technique which involves spreading of sample inoculum over the surface of the prepoured agar medium and is the preferred method for the isolation of yeasts and molds.

Prepare the total number of serial dilutions required for the test sample. Starting with the highest dilution to be plated and using a sterile pipette, transfer 0.1 mL of inoculum on to the surface of the agar in the center of the prelabeled plate(s). Repeat the same procedure with the further (lower) dilutions, using the same sterile pipette for each decimal dilution.

Note that in order to minimize errors, the highest dilution is pipetted first, followed by each dilution in turn back down the dilution series to the lowest dilution, i.e. −4 to −1.

Spread the inocula uniformly and as quickly as possible on the surface of the medium using a sterile glass or plastic spreader. While spreading the inoculum, remove the Petri dish lid in such a way that it does not interfere with the spreading of the inoculum but provides a cover to the agar from air contaminants.

Begin rotating the spreader in a continuous circular direction while rotating the base of the dish with the hand holding the lid in the opposite direction. Make sure the spreader does not leave the agar surface and avoid touching the edges of the Petri dish. Keep the movements smooth.

If multiple dilutions are plated, one sterile spreader can be used. Start spreading on the plate(s) with the most dilute solution first and then on each dilution in turn back

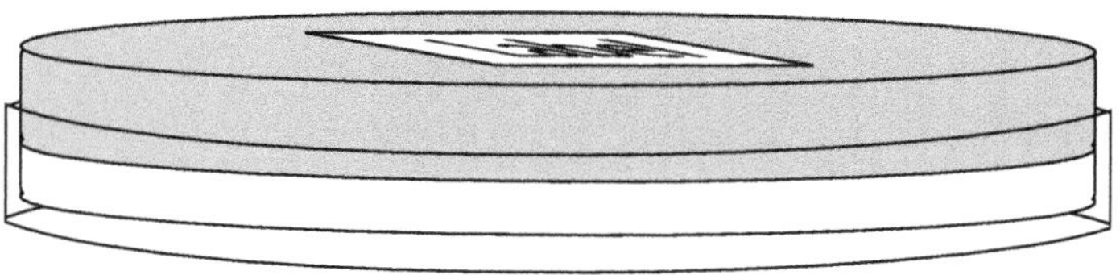

FIGURE 6.2 Diagram of culture plate illustrating the correct orientation, inverted, and the correct placement of the lid, on the base. Ensure that inadvertent swapping of covers does not lead to incorrect labeling.

down the dilution series to the lowest dilution, e.g. from −4 to −1.

Allow the plates to stand until the inoculum on each has been completely absorbed, which should occur within 15 min of spreading.

11.6 Membrane Filtration Technique

11.6.1 Hazards

- Burns from flame.

This technique is suitable for the analysis of liquid samples that can pass through a 0.45-μm pore size filter membrane. Filtration allows larger volumes of samples to be tested and hence provides results with higher sensitivities. Samples that are expected to have very low levels of microbial flora are tested effectively with this technique. The microorganisms are retained on the filter surface, allowing the liquid to pass through.

When the filter is placed on the agar surface the essential nutrients for growth are absorbed through the filter and colonies grow on the filter surface. The filtration for such samples is ideally performed in a laminar flow clean workstation. A filtration technique also enables the removal of antimicrobial agents such as SO_2 in the sample by postfiltration washing of the filter membrane and allows microorganisms, if present, to grow.

When the wine is tested by this method before sterile filtration and bottling, sediment may remain on the surface of the filter and interfere with the growth of microorganisms.

Label the required number of Petri dishes containing growth medium with appropriate identification of the sample and the sample volume or dilution to be filtered. The filtration assembly and stopper are to be sterilized as a unit prior to use.

Check that all tubing coming out of the filter manifold and into the collection reservoir and from the collection reservoir to the pump is securely connected and in good working condition. Before placing funnel assemblies, rinse the manifold with 70% alcohol by turning on the vacuum and dispensing alcohol into each manifold port from a squeeze bottle. Unpack the filter assembly and securely place on the manifold.

Sterilize flat-tipped forceps by immersing in 70% alcohol, then draining off the excess alcohol and flaming. Aseptically open a sterile membrane and grasp the outer part of the membrane with the forceps, taking care not to tear or damage the membrane. Open the filter funnel and center the membrane with its grid-marked side upward on to the funnel's membrane support (Figure 6.3). Replace the filter funnel and lock into place. (Note: filtration units complete with membranes are commercially available.)

FIGURE 6.3 Placing a sterile filter membrane on a filter funnel support. *(From Silliker Aust P/L with permission).*

Prewet the filter by pouring a few milliliters of sterile diluent into the filter funnel. The diluent generally used is 0.1% peptone water. Always flame the necks of bottles of diluent.

Add the sample or initial suspension into the funnel while the manifold valves are turned off. If the volume to be filtered is less than 10 mL, pour a minimum of 10 mL of diluent into the cup before adding the sample. A sterile pipette is used for volumes of less than 50 mL. Volumes of 50 mL or greater can be poured directly into the funnel using the graduated markings.

Turn on the manifold valve so that the vacuum pulls the sample through the filter. Turn the manifold valve off. If rinsing of the filter is required this is usually carried out with sterile diluent of choice. Use an amount of diluent that takes into consideration the amount of the sample or initial suspension filtered. For smaller amounts, rinsing with 30 mL is enough. Filtration of a larger volume requires rinsing with the same amount of diluent.

Turn off the manifold valve, open the funnel and, using sterile forceps, remove the membrane from the support base, taking care not to tear or damage the membrane. Place the membrane, grid side up, on to the surface of the required agar using a rolling motion to avoid trapping air bubbles between the membrane and the agar. There should be no excess moisture forming a ring around the filter on the agar plate.

Use a fresh sterile filter funnel assembly for each sample analyzed. However, if various amounts and/or multiple dilutions of one sample are filtered, the same filter funnel assembly can be used for one sample. The filtration assembly in such a case is used on the highest dilution first and then on each dilution in turn back down the dilution series to the lowest dilution, then to the sample volume in increasing order, e.g. −2 to −1 to 10 mL to 100 mL.

Always empty the effluent collection reservoir when it is nearly full. Turn off the vacuum control tap and vacuum pump. Remove the collection vessel and pour the effluent into the laboratory sink. Avoid splashing the effluent and creating aerosols.

11.7 Most Probable Number Technique

The most probable number (MPN) method is a statistical technique for estimating the most probable number of a bacterial species per specified unit of the material under test. Multiple test portions of appropriate dilutions of the sample are inoculated into tubes of liquid culture medium. The combination of growth-positive and growth-negative tubes in each set is used to cross-refer to statistical tables, from which can be calculated, with a known degree of certainty, the most probable number of bacteria per unit of sample. This technique is not recommended in routine wine microbiology as it requires more time, is more labor intensive and requires more laboratory incubation space than other methods.

12. INCUBATION OF PLATES

12.1 Yeasts

Incubate plates upright at 25°C, for 5 days, under aerobic conditions. If slow-growing fungi, such as *Brettanomyces*, are suspected, after examination of plates at 5 days reincubate for a further 5 days and examine again.

12.2 Lactic Acid Bacteria

Incubate plates inverted at 25°C for 4–10 days under microaerophilic conditions (i.e. under reduced oxygen, usually 3–5%). This can be achieved by flushing a plastic bag or desiccator with dry nitrogen from a cylinder and sealing before incubating, or by purchasing a commercially available atmosphere generation system.

12.3 Acetic Acid Bacteria

Incubate plates inverted at 25° C for 2–4 days under aerobic conditions.

12.4 Counting and Calculation

After incubation, examine the dishes, if possible, immediately. Otherwise they must be stored refrigerated until they can be read.

Examine the plates under subdued light. Use a tally colony counter to register the count and use a marking pen to mark the counted colonies on the Petri dish to avoid recounting them.

It is important that pinpoint colonies should be included in the count, but it is essential to avoid mistaking particles of undissolved or precipitated matter in dishes for pinpoint colonies.

Examine doubtful objects carefully, using higher magnification, in order to distinguish colonies from foreign matter.

Spreading colonies or spreaders occur on the plate as chains or clumps of colonies (Figure 6.4). They occur on wet plates, when the agar and sample are not adequately mixed, and when the sample sticks or adheres to the Petri dish, or sometimes the organism itself is a vigorous grower that outgrows the other colonies on the plate.

If all the colonies appear in a clump and the clump is where the sample was added, then it is likely that the cells have stuck or adhered to the Petri dish or the agar was not properly mixed with the sample. The problem can usually be reduced by adding the agar immediately after pipetting the sample and mixing. If the problem is sample specific, a spread plate procedure may be necessary.

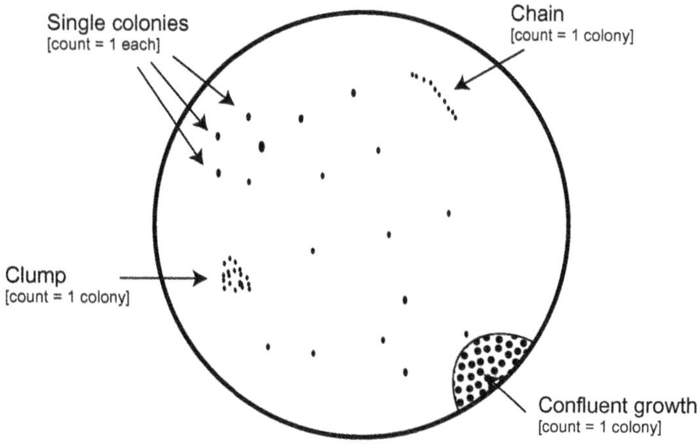

FIGURE 6.4 Guide for counting colonies. Do not attempt to distinguish between colonies that are touching one another, whether as groups, chains or a coalescent mass. *(From Silliker Aust P/L with permission).*

Spreading colonies shall be considered as single colonies (Figure 6.4). If less than 25% of the medium is overgrown by spreaders, count the colonies on the unaffected part of the medium and calculate the corresponding number of the entire dish.

For spread and pour plates, counting is to be carried out on those plates containing fewer than 300 colonies. For membrane filtration plates, counting is to be carried out on plates containing fewer than 80 colonies. After the incubation period, count the colonies on all of the plates from each dilution plated and record the count along with the dilution on the laboratory worksheet. Plates appearing to have a count over 300 can be recorded as >300.

If none of the dilutions has a count less than 300, efforts should be made to estimate the count from at least one of the dilutions with the lowest count, by counting subsections of the plate, to a maximum of six subsections, and multiplying by the number of sections to give an estimate of the count.

If the plate is divided in half, count one half of the plate and multiply the colony count by two. If the plate is divided into quarters, count two, diagonally opposed quarters and multiply the number of colonies by two to give the count. If the plate is divided into six, count two diagonally opposed sections and multiply the number of colonies by three to give the count.

Calculate the count per milliliter or gram from the dilution having counts between 15 and 300 for spread and pour plates and between 20 and 80 for membrane filtration plates. The count per milliliter or gram is calculated by multiplying the mean of the colony count by the dilution factor and reported to two significant figures; for example, 36 colonies at the -4 dilution calculates as: $36 \times 10{,}000 = 360{,}000$ cfu/mL.

Colony-forming units (cfu) are used as each colony may have been derived not from a single cell but rather from a group of cells.

Chapter 7

Harvest Protocols

Chapter Outline

1. SAMPLING

1.1 Terminology

- **Vineyard:** A set of vines and land under common management or ownership.
- **Block:** A contiguous set of rows and panels under a single management system and usually planted to a single clone or cultivar; sometimes referred to as a 'patch'.
- **Harvest unit:** An area of vines, the fruit of which will be combined in a single ferment.
- **Vigor unit:** An area defined as having common yield and vegetative growth (biomass) attributes based on aerial imaging or soil mapping, and winemaker preference.

1.2 Preamble

The goal of sampling fruit in the field is to estimate, within an acceptable error, the final maturity values observed at crushing. This is difficult to achieve because ripening is asynchronous within a cluster, within a vine or between vines. Furthermore, yield varies from vine to vine and thus a value from a high-yielding vine or section of a vineyard should be given more weight than a value from a low-yielding vine or section unless, as is usual, the samples are combined (e.g. Figures 7.1 and 7.2 and Tables 7.1 and 7.2). Importantly, the sampling protocol needs to be practical and economic. One of the best studies of sampling was conducted in the early 1960s and this study first put values to statistical measures of variance that could be used to balance sampling and to determine economic limits to sampling intensity (Rankine et al., 1962).

An excellent recent review of variation in berry composition and wine quality is that of Coombe and Iland (2004), but see also Blouin and Guimberteau (2000) for a thorough summary of the composition of ripening fruit for a number of cultivars over many seasons in France and factors that determined harvest date.

Variation in maturity and yield in a vineyard block contains both systematic and random elements. Sampling efficiency is greatest if the systematic sources of variation are mapped. Maps may be provided by a commercial 'precision viticulture' provider and based on either a satellite image or multichannel digital aerial video data: normalized difference vegetation index (NDVI) or plant cell density (PCD). Variation in biomass may also be judged

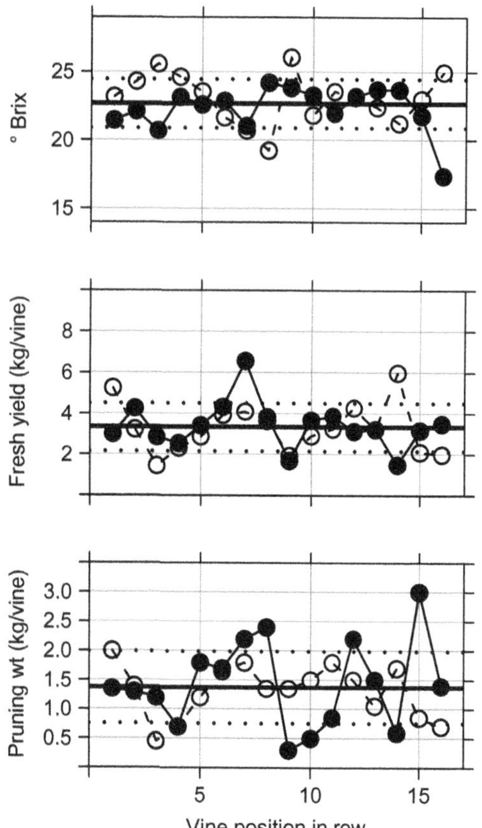

FIGURE 7.1 Plot of ° Brix, yield and vine biomass (pruning weight) values for 30 clusters harvested from individual vines in a single row from two adjoining blocks of Cabernet Sauvignon. ° Brix values are a weighted mean calculated from the individual bunches (Table 7.1). Closed circles: row 1; open circles: row 2 (adjacent). *(J. Wisdom, unpublished).*

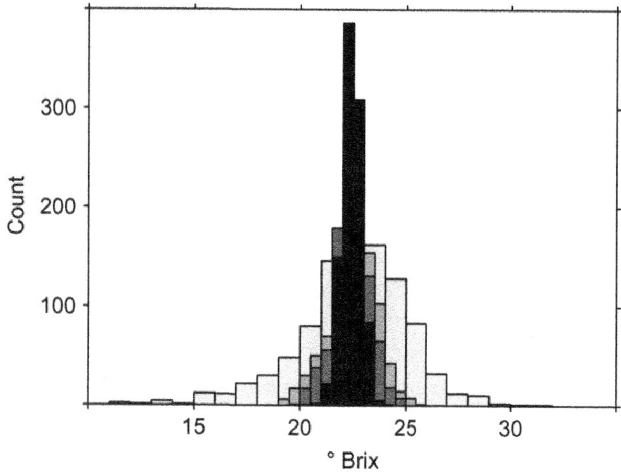

FIGURE 7.2 Histograms of ° Brix values for 960 bunches harvested from 32 vines (Figure 7.1). Reading from the outside inward, 960 single bunch values, then 960 random samples of 5-, 10-, 30- or 100-bunch subsamples from the 960-bunch set of data (cv. Cabernet Sauvignon). Bin ranges are 1.0 for the single bunch samples, 0.5 for the 5–30-bunch samples and 0.25° Brix for the 100-bunch sample. *(Data courtesy of J. Wisdom, unpublished).*

from aerial photographs, soil maps, landscape topographical maps or in smaller blocks by physically sampling pruning weights, yields or measuring butt circumference (good exercises for students).

In a typical vineyard, one-third to one-half of the total variation may be systematic and related to soil type or topography, while the remainder will be equally distributed as vine-to-vine variation within a particular classification of site within the block/patch and within a single vine. Variation also occurs within a cluster (Coombe and Iland, 2004). We recommend whole-cluster sampling to avoid this problem but this practice leads to large sample weights which may not be appropriate for experimental or hobby vineyards. Alternatively, a standardized cluster sampling procedure may be developed as variation within a cluster is mainly systematic, top to bottom and exposed to shaded. A recommended practice is to take two berries from the top, two from the middle and one from the bottom (Hamilton and Coombe, 1992). An advantage of berry sampling techniques is that more vines are sampled and generally this more than offsets the within-cluster error (Roessler and Amerine, 1958, 1963). Berry sampling is, however, more costly and time-consuming than whole-cluster sampling.

TABLE 7.1 Variation in Maturity of Whole Cabernet Sauvignon Bunches Harvested from 32 Vines, 30 Bunches per Vine

Measure	*n*	Mean	Range (min.–max.)	*sd*
° Brix	960	22.6	11.7–31.9	2.59
pH	960	3.78	3.00–4.40	0.218

Note: See also Figure 7.1.
(Data from J. Wisdom, unpublished).

TABLE 7.2 Sample Statistics for the ° Brix Data in Figure 7.2

Sample Size	Mean	*sd*	CV %	$-CL^{0.95}$	$+CL^{0.95}$
1-bunch sample	22.6	2.59	11.5	17.5	27.7
5-bunch sample	22.5	1.17	5.2	20.2	24.8
10-bunch sample	22.4	0.835	3.7	20.8	24.1
30-bunch sample	22.4	0.490	2.2	21.5	23.4
100-bunch sample	22.4	0.269	1.2	21.9	22.9

Note: CV = coefficient of variation; CL = confidence limit = $t^{0.05} \times sd$ (i.e. 95% of values will be within this range and 5% outside it). If this variance were applied to the whole of a block of vines, then about a 100-bunch sample would be required to provide a value of ±0.5° Brix of true mean with a probability of 95%.

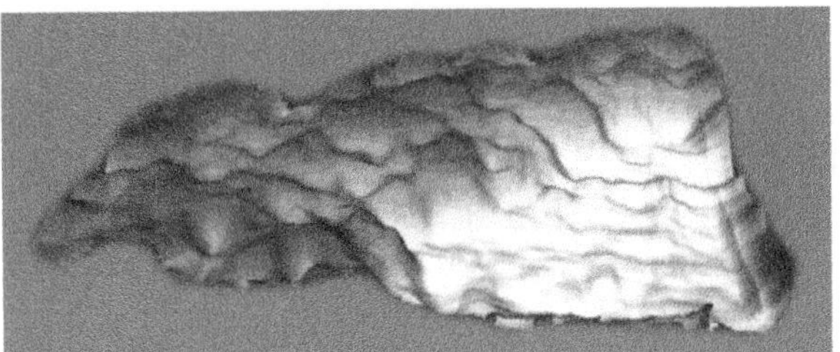

FIGURE 7.3 Three-dimensional reconstruction of depth to ferricrete base from data obtained by ground-penetrating radar. The site was a seemingly level 3-ha Chardonnay plot in the Margaret River region of Western Australia. *(Burgos et al., 2008).*

The importance of mapping was demonstrated in a study which showed that harvest maturity varied by nearly 2 weeks across a single block on a seemingly flat landscape (3 ha). The variation was location related and not random. Soil depth and consequently vine vigor was the most important factor, and fruit from vines growing on the shallowest soils (lowest canopy density, Figure 7.3) matured first even though the entire vineyard was irrigated regularly and the vines were of a uniform age (Burgos et al., 2008).

Usually, it is not economic to sample to high precision. The viticulturist and winemaker rely on regular, repeated sampling to develop a trend line which usually is more reliable than a single sampling point (Figure 7.4). Note that the ° Brix rise is approximately linear over the range but the tartaric acid values would be better fitted by a curve. The ° Brix values are anticipated to plateau before rising again as the berry becomes overripe and begins to shrivel. Should an exceptional climatic event occur, such as a rainstorm, then more emphasis should be placed on single sampling points as the trend will be disrupted.

An adequate sampling protocol will provide a value that is $\pm 0.5°$ Brix of the final value after pressing. Because sampling precision varies with $\sqrt{1/n}$, a four-fold increase in sample number (size) is required to double the precision (i.e. halve the variance). Examples of the expected variance and sample sizes are presented in Table 7.1 and Figure 7.5. The values presented here are probably conservative as they were taken from a small, contiguous set of vines. In Figure 7.2, the data are presented as though samples of one to 30 bunches were combined (adjusted for bunch weight): the greater the bunch number per sample, the less the error. For this plot 100-bunch samples would have been required to meet the initial objective of value at the press $\pm 0.5°$ Brix. This underlines the importance of defining harvest units and sampling within blocks of similar vines (vigor and yield).

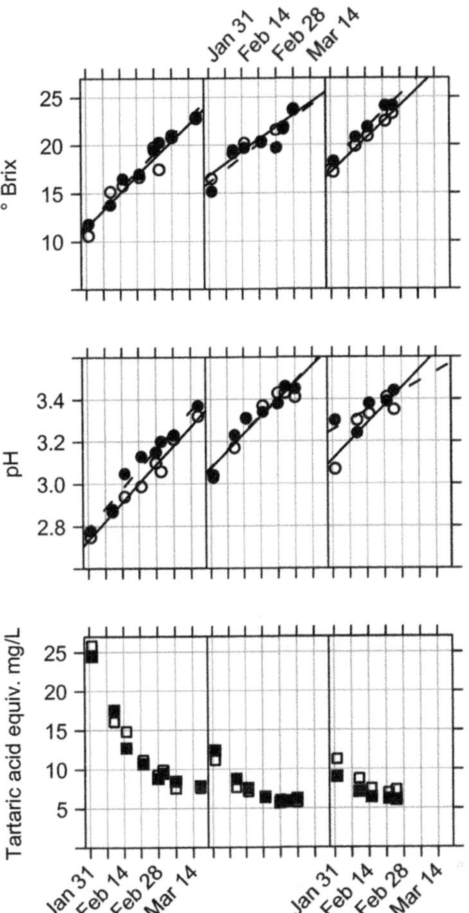

FIGURE 7.4 Plot of maturity values for samples of one bunch per vine from two plots of four panels (four vines per panel) on four adjacent rows of Chardonnay vines on each of three locations within the Margaret River regions, showing the trend in maturity: ° Brix, pH and titratable acidity (TA). Closed circles: plot 1; closed circles: plot 2. Note the diversity of maturity date and rate (slope) of ripening, and that the rise in pH closely follows that of °Brix, more so than of TA which plateaus (presumably all malic acid has been consumed and only tartaric acid remains at the final sampling date). *(Considine, unpublished).*

Figure 7.5 represents repeated sampling from six, four-panel, four-vine per panel plots in a 2-ha plot. It shows that there is an important degree of spatial variation in the rate of ripening in even such a relatively small plot; plots 1–3 were ripening more slowly than the average, plot 4 was close to the average, and plots 5 and 6 were ripening faster than the block average. However, note that the individual values (closed circles) give a misleading impression of the differences between the plots.

Note that in calculating a final value, it will be necessary to determine a weighted average that accounts for both the relative area in each subsampling unit and

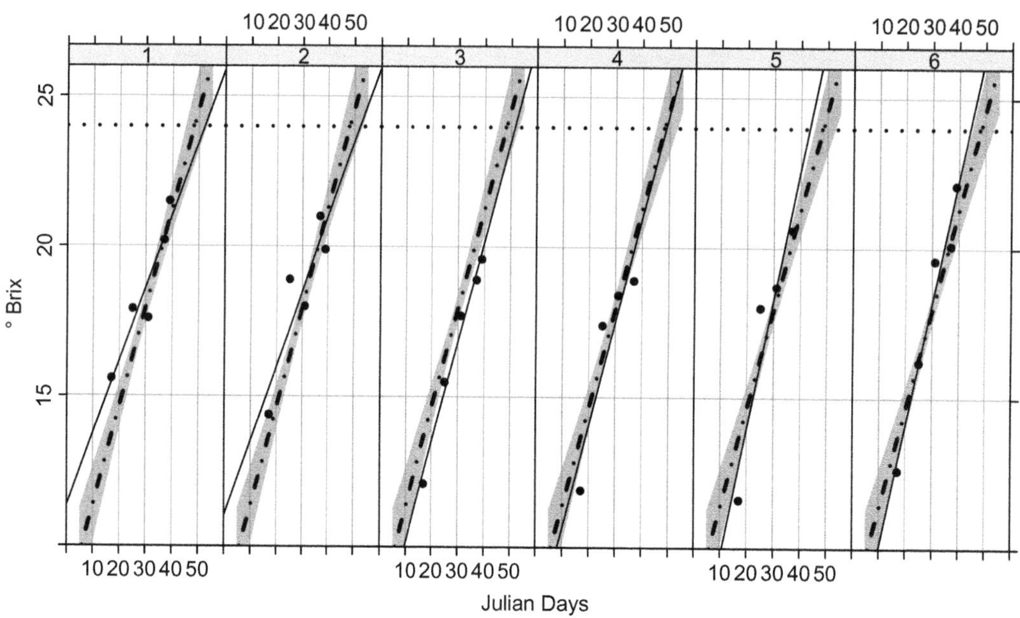

FIGURE 7.5 **Plot of ° Brix values from six 24-vine plots located at random within a 3-ha site (cv. Chardonnay).** One bunch was harvested at each sample time from each vine and combined for each plot. The horizontal dotted line represents the target harvest maturity, the shaded zone is the 95% confidence limit for the combined regression line (dashed line), and the solid regression line is that for the individual plot. Predicted harvest dates vary from day 43 to day 52, with the earlier value being associated with the shallow soil zone. *(Considine, unpublished).*

the relative yield per vine (or meter of canopy) (Rankine et al., 1962). A fault common to all published studies is that they are conducted on relatively small plots, and those of Roessler and Amerine (1958, 1963) confound sampling with site. Nevertheless, these studies serve as useful guides to the development of sampling protocols; especially those of Rankine et al. (1962) and the review of Coombe and Iland (2004) but also those of Clingeleffer (2001) and Krstic et al. (2002).

Finally, any sampling protocol must relate directly to harvest protocols. A mapped vineyard may have arbitrary boundaries drawn in a block/patch that is practical for mechanical harvesting operations but which may not truly represent the underlying variance boundaries. Each harvesting unit should be sampled separately as a means of 'ground-truthing' the vineyard.

1.3 Occupational Health and Safety

- High-visibility vest
- Sign-in–sign-out protocol for all field activities
- Nature of bird/pest deterrent strategy adopted, e.g. may involve gunshot activity or use of carbide cannons (in which case hearing protection is appropriate)
- Climatic events, e.g. risk of lightning strike in a storm: vineyard wires act as an excellent lighting conductor! Leave immediately if there is a risk of lightning strike.

1.4 Quality Assurance Records

- Person, date.

1.5 Procedure

1.5.1 Vineyard Block Mapped for Systematic Variation

- Define harvest units by chosen method (PCD, soil conductivity, soil depth, etc.).
- Estimate the relative area (vine number, row length) in each defined vigor unit within the harvest unit [e.g. high, medium and low biomass (vigor) units].
- Optional: estimate the anticipated yield at harvest in each unit (Clingeleffer, 2001).
- Calculate the sample size required to obtain an aggregate sample that represents the crush (see below).
- Define a minimal path through the vineyard that traverses the defined unit: commonly this involves a zigzag walk through the harvest unit, sampling each side of the row.
- Sample whole bunches (or berries) using the 'grab' technique but ensure that a balanced sample is obtained to represent the variation present within a vine (side to side and exposed to enclosed being the primary variables) (Rankine et al., 1962). Sample at about the same time each day and avoid sampling when fruit is wet or hot (best to sample early to mid-morning).

1.5.2 Unmapped Vineyard Block

- Divide the block into sub-blocks of no more than 2 ha area each.
- Proceed as for mapped vineyard.

1.6 Equations and Calculations

In devising sampling regimens it is assumed that the distribution of the samples conforms to a predictable pattern. In most instances it will be a normal distribution such as that shown in Figure 12.2 (see Chapter 12). While the actual data do not quite conform to a normal bell-shaped distribution, they are pretty close (this data set may have included some second crop bunches). In this instance, the value measured (e.g. of ° Brix) on two-thirds of all bunches will fall within the mean value ± the standard deviation (*sd*). This is a measure of the width of the distribution curve. Usually, however, we consider a 95% probability spread as being the acceptable risk range, i.e. there is a 1 in 20 risk that the value will fall outside the confidence limits each time a sample is taken (±CL, Table 7.1).

For example, for the data in Table 7.2, the mean ° Brix is 22.6 ± 0.08 *se*. This is far more precise than is required and would leave nothing to harvest! To achieve the stated precision of ±0.25° Brix a sample of 90–100 bunches would have been needed (three per vine). This is not practical as repeated sampling of the same vine would remove half the crop. One bunch per vine would yield a precision of ±0.5° Brix and one per three vines that of ±0.8° Brix. Neither is satisfactory and it is thus unsurprising that the winemaker must often deal with fruit that differs significantly from that which he or she was expecting. In this instance the sampled blocks were not particularly uniform (Figure 7.1).

To calculate the sample size (*n*), the following values are required (Cochran, 1962):

ae: acceptable error (°B, e.g. 0.5)
sd: standard deviation of the sample (obtain from historical records, repeated measurements or guess (in our data it was about 10% of the mean of all bunches for single bunch samples; Table 7.1).

$$n = 4 \times sd^2/ae^2$$

The way to minimize this problem is to plan harvest units that are relatively small and uniform, thus reducing the range of maturity found within the fruit, and to use regression analysis of sequential samples (e.g. Figures 7.4 and 7.5). Some variation may be desirable from a wine-making viewpoint as it may provide a greater range of flavor and aroma characteristics. However, the winemaker still needs to know that the fruit being pressed is within a known range of values. Yield or volume is another matter of probably even greater consequence and accurate estimation presents problems similar to those of sampling (e.g. Clingeleffer, 2001).

Finally, some sampling protocols adopted by industry are based on repeated measures of identified vines or panels of vines. This may be a sound approach for yields and developing relationships between particular vines/panels and harvest units, from one season to the next, but is far from ideal when sampling for maturity. Maturity samples are best taken at random from within the harvest unit and rerandomized each sample time. This will give a better impression of the unit as a whole than repeated sampling of a few vines: the samples will show more variation, but the trend line should, over the sampling period, yield a more accurate estimate of the (weighted) mean of the harvest unit than is possible with any single sample.

Good sampling protocols lead to good wines.

1.7 Equipment and Materials

See Analysis section for the laboratory requirements (Chapter 10).

- Labeled bags
- Picking secateurs or knives
- Laboratory-scale press (e.g. a 'cone' press, small basket press or pneumatic press); it is best not to use a blender as this may give a misleading view of the composition of the must produced by a wine press, unless you wish to extract the seeds as well as the flesh and skin (see below).

1.8 Benchmark Values

Benchmark values are provided in the analytical section (Chapter 10) but these should be negotiated with the winemaker. In many cases, the analytical chemical methods described here are used as a guide only: flavor will usually be the deciding factor unless there are some limits. For example, the middle graph in Figure 7.4 shows a vineyard in which the acidity has dropped well below the ideal; a winemaker may choose to harvest it a little earlier than flavor dictates in order to preserve a reasonable level of acidity. At the level displayed there may well be no malic acid remaining (note that pH values closely follow the rise in ° Brix) (Figure 7.4).

2. MEASURING BERRY MATURITY

2.1 Introduction

Assessing the optimum time to harvest is a critical control point in the wine-making process. Failure to get this aspect right will forever limit the quality and value of the final product as some aspects, especially aroma potential and color, cannot be corrected in the wine-making process. For other aspects, the wine will always reveal to the discerning customer the impact of the adjustments.

Aspects that are routinely measured are sugars (° Brix/ ° Baumé) and titratable acidity (TA). Other aspects that may be assessed are berry weight and volume of extractable juice (Figures 7.6 and 7.7), as well as sensory analysis, whether formally (see next section) or informally. In addition, especially for red wine cultivars, phenolics and color may be measured. Such data are provided at an industry level by the Faculty of Oenology, Bordeaux (http://www.bordeauxraisins.fr) along with other measures that are only available to well-equipped research laboratories [e.g. level of herbaceous characters, methoxypyrazines, 2-methoxy-3-isobutylpyrazine (IBMP)].

While rough extractions may be made pressing with a well-placed foot on a sample, sealed in a tough plastic bag, sugars for red and white cultivars are best assessed when extracted using a standardized laboratory pneumatic press. This should give a result that is closely related to that achieved in the wine-making process. This extract is also appropriate for the assessments of acids and phenolics in white and rosé-style wines. Fruit from red wines should be homogenized to extract the skin and the seeds, and the resulting juice filtered or centrifuged and used to measure pH, acids and phenolics.

Although, commonly, the whole sample will be crushed and analyzed, it is better practice to conduct a more detailed analysis on subsamples.

2.2 Preparation for Analysis

- Weigh the entire sample and record bunch number and weight.

2.2.1 *Simple Analysis*

- Record bunch number, weigh and press in a standardized fashion. Record weight of marc.

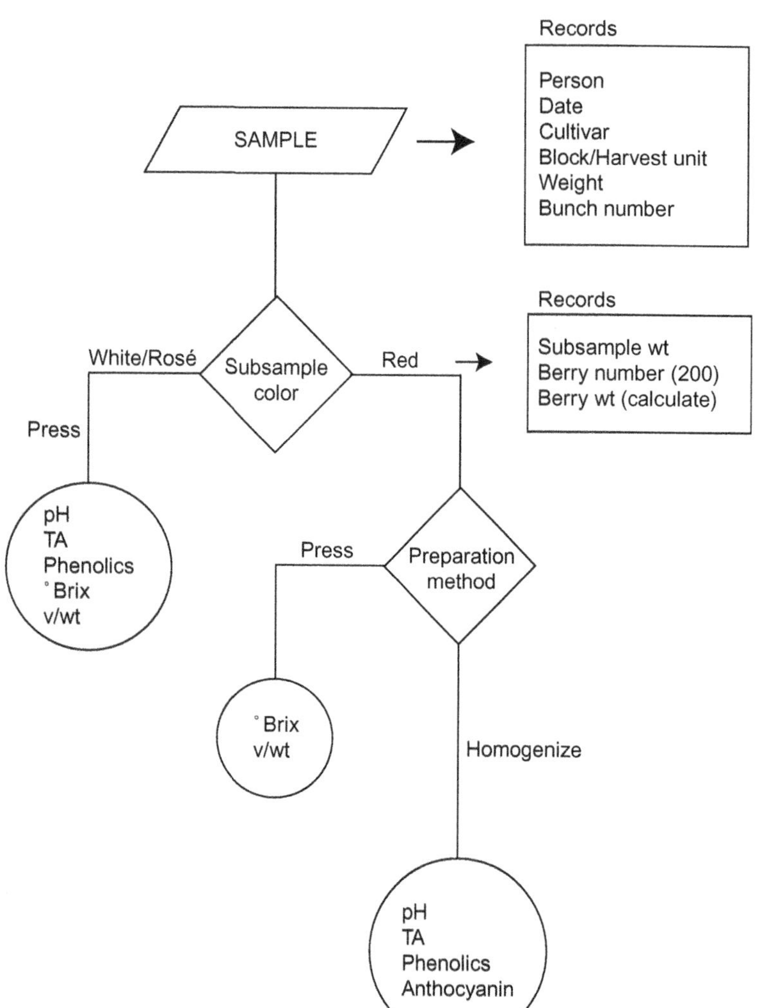

FIGURE 7.6 Flow diagram for processing and analyzing sample data for assessing maturity. TA = titratable acidity. See also sections for specific methods in Chapters 10 and, especially, 11 for further details.

- Conduct sugar and acid analyses (see Chapter 10) and calculate and record fraction of juice obtained (= sample weight − marc weight).
- Plot data.

2.2.2 Detailed Analysis

- Collect sample by an appropriate protocol and proceed immediately to analysis. If sample is damp due to dew or rain, dry with absorbent paper before analysis. Record sample weight and bunch number (Figure 7.6).
- Using a pair of pointed secateurs, chop the sampled bunches into small units (10–20 berries per unit). Place all in a large plastic bag, close and roll the samples, gently, back and forth to obtain a uniform mix.
- Weigh a subsample of about 0.5 kg and then cut off or pull off (carefully, e.g. using a slotted spoon) individual berries. Count and weigh 100 berries and record the weight.

Red cultivars:

- Prepare two subsamples of 200 berries each.
- Use one for sugar and acid analysis after pressing in a standard fashion; calculate juice extraction and measure sugar content (° Brix).
- Then homogenize the other and use for phenolics and anthocyanin assays (see Chapter 11).

White and rosé cultivars:

- Prepare one sample of 200 berries and, after pressing in a standard fashion, calculate juice extraction and then measure pH, titratable acids and ° Brix (see Chapter 10 for all) and, optionally, phenolics (see Chapter 11).
- Record and plot all data immediately.

2.3 Comprehensive Analysis Example

The examples shown here refer to aggregated regional samples for Merlot and Cabernet Sauvignon in Bordeaux

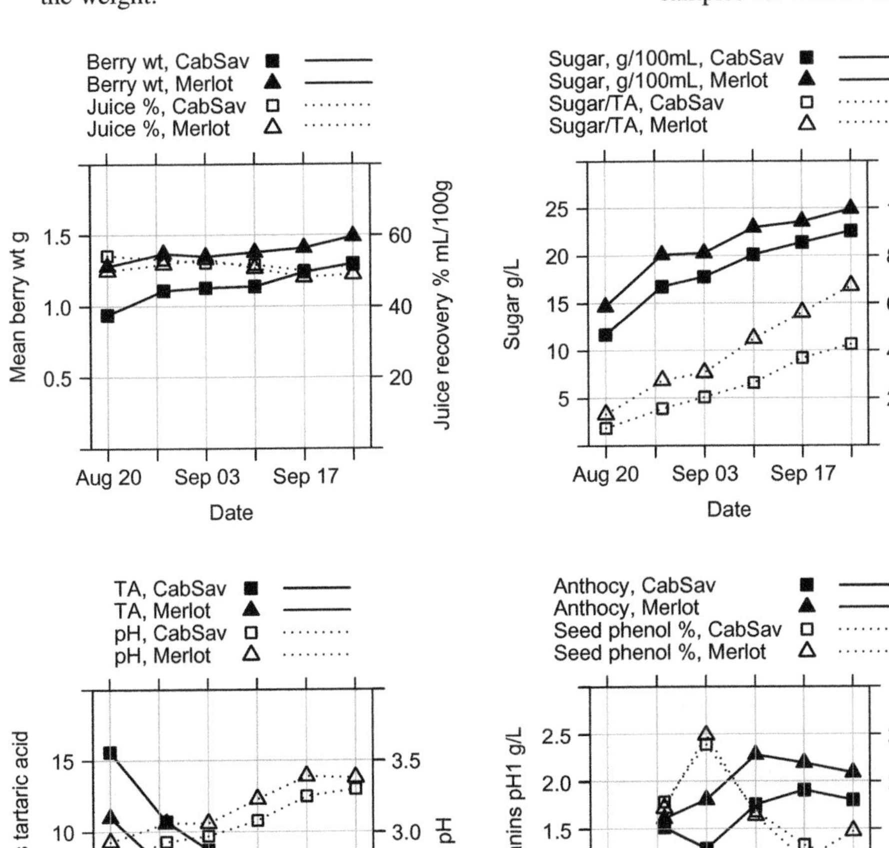

FIGURE 7.7 Some examples of maturity data obtained using the Yves Glories protocol for phenols and anthocyanins in red grapes and data for Cabernet Sauvignon (CabSav) and Merlot for the 2012 season. TA = titratable acidity. See also Chapters 10 and 11. *(Source: www.bordeauxraisins.fr, with permission).*

for 2012, an unusually warm and early harvest season (Figure 7.7). The data show a marked difference in berry weight, sugar, acid and seed-derived phenolics. While none of these shows a clear point of maturity, seed-derived polyphenols were at a minimum on about 17 September, while anthocyanins may be declining and sugar–acid ratios becoming unacceptably high beyond this point. These data are compiled from regional analyses and a clearer indication would be available at the subregional level, Graves versus Bordeaux, etc. Sensory assessment is required to determine optimum stage of maturity for harvest. Technical assessments such as these are rarely done.

3. BERRY MATURITY BY SENSORY ANALYSIS

Protocol author: Jacques Rousseau, ICV, La Jasse de Maurin, 34970 Lattes, France, www.icv.fr.

3.1 Introduction

Determining the state of the berry and its readiness for harvest is a key factor in wine making or indeed in achieving quality outcomes for table grapes. Key sensory factors are sugar and acid levels, pH, flavanones and tannins—especially tannin ripeness. While winemakers have routinely tasted berries before harvest, few would bother with such a formal procedure as this. The method is intended to train individuals to recognize the key ripeness characters; or actually, the unripe characters (Rousseau, 2001). The descriptors were developed for French circumstances and deal principally with unripe to ripe fruit. In Australia and even in the south of France, overripe fruit is also an issue, so consider descriptors and scales that may be used to define those attributes (e.g. Winter et al., 2004). A cool climate approach has also been developed (Pedneault et al., 2012).

Climatic conditions affect the coordination of ripening of berries: in hot climates, acidity levels may decline to overripe before polyphenolic and flavor attributes mature; in cold climates, flavor maturity may peak before sugar and tannin maturity. The winemaker will need to compromise and adjust the outcomes as allowed by legal constraints governed by national food and beverage laws. In Australia, it is common to add acid to a wine, while in the colder parts of Europe, sugar may be added. This process should be adapted to your own region, allowing for season-to-season differences, as has been attempted for Australia (Winter et al., 2004).

Few varieties exhibit in the berry the full spectrum of flavors present in the mature wine because such flavors are present as precursors; usually bonded to sugars, and not volatile (see Chapter 3). A recent study has examined these relationships in detail for Cabernet Sauvignon, with promising results (Forde et al., 2011). Exceptions are Riesling, the 'muscat' clone of Chardonnay (1:1 free to bound) and Muscat-related cultivars (1:5 free to bound). Fruit of these cultivars usually exhibit a proportion of free aroma compounds of the terpene class. However, even in such cultivars, the vast majority of the aroma is present as bound forms. Acid hydrolysis is the primary cause of their release. Pectinase enzymes can help once the free sugar levels have been reduced by fermentation. Some yeast β-glucosidases are also effective even in the presence of free sugars (Otero et al., 2003). Saliva is pretty ineffective at releasing these aromas because the principal enzyme is an α-amylase, suited to starch.

The purpose of this protocol is to train vineyard and winery staff and to develop a consensus approach to the determination of maturity for harvest and for preliminary grading of fruit as suitable for wines of a particular quality segment: 'vin de table' to 'Grand Cru' (super premium). The methods consider the whole bunch, the intact berry, the pulp, the skin and the seed. The tables are based on those devised by Rousseau (2001). An example of the comparisons that may be made is presented in Figure 7.8, which shows the change in values as fruit matures from marginal for harvest to fully mature and ripe for harvest.

This methodology should always be complemented by quantitative analyses: ° Brix or ° Baumé, TA and some form of colorimetric analysis for red wine cultivars. However, its value lies in the fact that some aspects of berry quality cannot be readily assessed in the laboratory and that decisions often need to be made in the field and need to be well judged. This takes practice.

3.2 Training and Research Procedure

3.2.1 *Materials*

- Fresh berries with pedicel attached, 10 per person per assessment lot (variety, harvest block, etc.)
- Plastic-backed laboratory sheeting to protect benches; tissues or napkins
- Small containers; Petri dishes or beakers, at least one per person
- Spittoon or other sealed container for wet wastes; one per pair
- Bowls of clean water for rinsing hands; one per pair
- Record sheets and tables; pens/pencils.

Match the collection of samples to your sampling procedure, including harvest and 'vigor' units. If your field sampling is of berries, ensure that the sampler collects berries with the pedicel attached. Berries should be freshly harvested, preferably in the cool of the early morning; bring to room temperature if chilled (do not hold, even overnight, in a cold room).

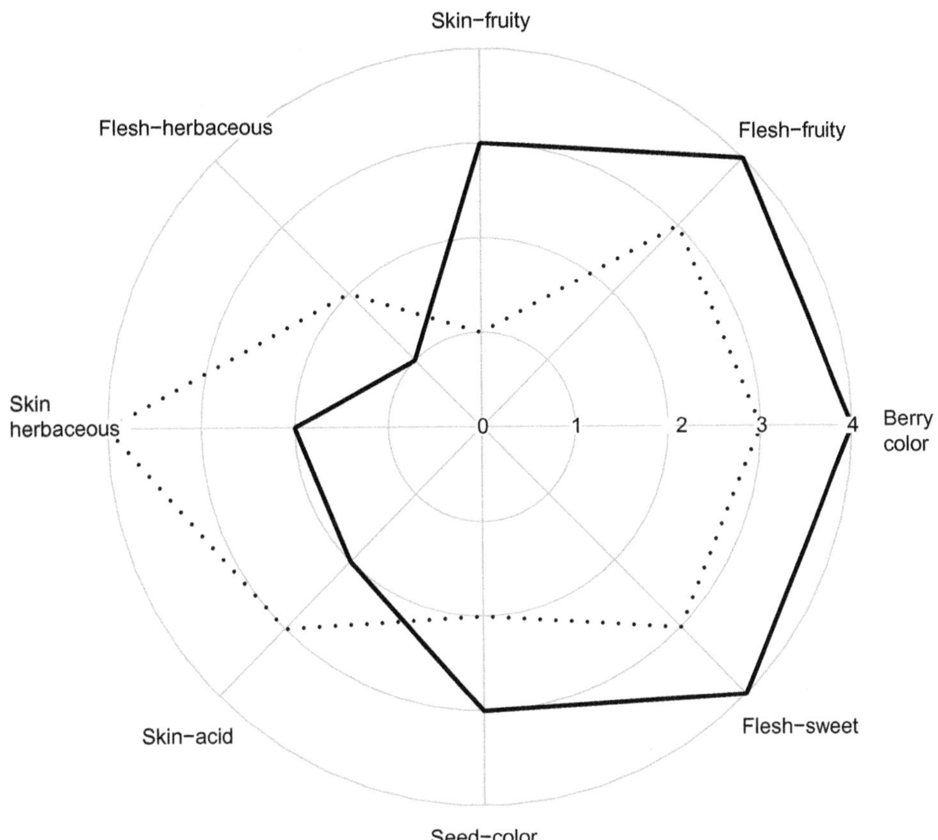

FIGURE 7.8 Radial plot of selected quantitative descriptive sensory analysis (QDSA) values for cv. Grenache. Dotted line: immature, minimum maturity for harvest; solid line: fully mature for harvest. *(Data from Rousseau, 2003).*

While this procedure can be conducted in the field, it is best done in the tasting room or winery to ensure constant tasting conditions when comparing fruit from diverse cultivars and in differing climatic conditions (not in the laboratory, owing to restrictions on eating foods in such areas).

Choose three lots of three berries at random from a container holding the grapes. Fill in the corresponding Table 7.4. Record the full identity of the variety tasted (vineyard, variety, lot and date picked). Begin tasting the berries. Each taster should taste at least three lots of three berries from the same lot: tick the box that corresponds to the rating given for all of the descriptors, defined according to the procedure below and in Table 7.3. The taster may record the three samples of three separately or give an integrated impression. The former is preferable but perhaps not practical in all circumstances. Spit out seeds, skin and pulp when finished (as for wine-tasting procedures).

3.2.2 *Physical and Visual Assessment*

For each assessment, record the rating from low to high (1 to 4) in the Quantitative Sensory Analysis (QSA) form (Table 7.4). The table reads left to right regardless of whether the measure rises or falls as the berry matures. Sugars, color and aroma rise while acidity, astringency, vegetativeness and firmness fall. Thus, for example, a level of 1 is high for some parameters but low for others. Note that the order varies from the original in some cases to minimize the risk of mis-entry (Rousseau, 2001).

Note also that the ranking is reversed for parameters that fall with ripening. This is intended so that the table gives an immediate visual impression of fruit maturity for wine making.

- Before removing the berry from the pedicel, squeeze each lightly by exerting the same pressure each time in order to evaluate the firmness (Coombe and Bishop, 1980).
- Then, pull the pedicel from the berry to evaluate the ease with which it can be removed.
- Finally, examine the skin, particularly the base, and record its color.

3.2.3 *Tasting the Flesh*

- Put three berries in your mouth; extract the flesh by successively crushing each berry between the tongue

TABLE 7.3 Physical and Visual Descriptors for Quantitative Descriptive Sensory Analysis (QDSA), Institut Coopératif du Vin (ICV)

Component Action	Characteristic and Score			
	1	2	3	4
Berries				
Physical attributes				
Squeeze berry between finger and thumb	Hard berry, bursts under strong pressure	Elastic berry, changes shape slightly under pressure but returns to its original form	Berry changes shape easily, plastic, returns slowly to its original shape	Soft berry, crushes under slight pressure
Pedicel attachment				
Pull pedicel from berry	Berry strongly attached, pedicel tears the skin	Pedicel comes off with some pulp attached	Pedicel readily removed with a little uncolored flesh (the 'brush')	Pedicel easily removed with very little flesh
Berry color				
Red or black cultivar Examine the pedicel end	Pink to pale red	Red	Dark red	Dark red or black
White cultivar Examine the whole berry	Green to pale yellow	Yellow	Straw yellow	Amber yellow
Flesh				
Skin–flesh adhesion				
Crush 3 berries against the roof of your mouth and gently separate flesh from skin. Put the skin and seeds aside, but do not discard				
Chew the skins	Flesh adheres strongly to skin and seeds	A thin layer of flesh adheres to the skin	Flesh layer only slightly present but juice is released when skin is chewed	No flesh attached to skin and no release of juice when skin is chewed
Sweetness				
Release the juice from the flesh	Not sweet	Some sweetness	Sweet	Very sweet
Acidity				
	Very acidic	Acidic	Moderate acidity	Mild acidity
Vegetative flavor/aroma				
	Very strong	Strong	Mild	Absent
Fruity/floral aromas				
	Nil	Weak	Intense	Jam-like, cooked
Skin				
Toughness				
Chew the skins 10–15 times	Very difficult, large pieces	Difficult, small pieces	Fairly easy, mixture almost homogeneous	Easy, homogeneous mixture
Tannic intensity				
Run the tongue over the palate	Tongue slides effortlessly	Tongue sticks slightly	Tongue slides with difficulty	Tongue slides with great difficulty

(*Continued*)

TABLE 7.3 (Continued)

Component Action	Characteristic and Score			
	1	2	3	4
Acidity				
Analyze the crushed, chewed skins	Highly acidic	Acidic	Moderately acidic	Not very acidic
Astringency				
	Lip slides easily over gums	Lip sticks slightly to gums	Lip slides with difficulty over the gums	Lip slides with great difficulty over the gums
Dryness of tannins				
Run the tongue over the palate and evaluate the time necessary to resalivate	Tongue slides easily over the palate—no difficulty in resalivating—fine and silky grains	Tongue sticks slightly—a little difficulty in resalivating—medium-size grains	Tongue slides over palate with difficulty—difficult to salivate for several seconds—coarse grains	Tongue slides over palate with great difficulty—difficult to salivate for >5 s
Grassy aromas				
Analysis of crushed chewed skins	Very intense	Intense	Weak	Absent
Fruity aromas				
Analysis of crushed, chewed skins	Absent	Weak	Intense fruit	Intense jam
Seeds				
Color				
	White, yellow–green	Brown–green	Grayish brown	Dark brown
Strength				
Chew the seeds between the incisors	Outside layer is soft, seed crushes under strong pressure from the incisors	Outside layer is fine, seed is still moist, crushes under strong pressure from the incisors	Almost no outside layer, seed is hard, slightly crisp	No outside layer, seed is very crisp
Flavor				
Analysis of crushed chewed seeds	Do not taste	Grassy, green	Grilled	Roasted
Astringency				
Analyze the crushed chewed seeds	Lip slides easily over the gums	Lip sticks slightly	Lip slides with difficulty	Lip slides with great difficulty
Tannic intensity				
Run the tongue over the palate	Tongue effortlessly slides across the palate	Tongue sticks slightly	Tongue slides with difficulty	Tongue slides with great difficulty

(After Rousseau, 2003).

and the palate; keep the three pulps in your mouth; spit out the skins and the seeds into your hand and set aside; extract the juice, completely crushing the pulp (flesh) between the tongue and the palate.

- While separating the flesh from the skin, evaluate the adherence of the flesh by:

a. observing the ease with which the flesh detaches itself from the skin

b. examining the seeds and the presence of attached flesh

c. at the time of the final chewing of the skin, monitoring the release of juice, if any

TABLE 7.4 Berry Sensory Analysis Form (QSA)

Name:		Date:											
Vineyard/Sample:		**Sample 1**				**Sample 2**				**Sample 3**			
Part	**Measure**	**Score**				**Score**				**Score**			
Berry	Softness	□ 1	□ 2	□ 3	□ 4	□ 1	□ 2	□ 3	□ 4	□ 1	□ 2	□ 3	□ 4
	Pedicel attachment	□ 1	□ 2	□ 3	□ 4	□ 1	□ 2	□ 3	□ 4	□ 1	□ 2	□ 3	□ 4
	Skin color	□ 1	□ 2	□ 3	□ 4	□ 1	□ 2	□ 3	□ 4	□ 1	□ 2	□ 3	□ 4
Flesh	Attachment	□ 1	□ 2	□ 3	□ 4	□ 1	□ 2	□ 3	□ 4	□ 1	□ 2	□ 3	□ 4
	Sweetness	□ 1	□ 2	□ 3	□ 4	□ 1	□ 2	□ 3	□ 4	□ 1	□ 2	□ 3	□ 4
	Acidity	□ 4 ←	□ 3	□ 2	□ 1	□ 4 ←	□ 3	□ 2	□ 1	□ 4 ←	□ 3	□ 2	□ 1
	Grassy flavors	□ 4 ←	□ 3	□ 2	□ 1	□ 4 ←	□ 3	□ 2	□ 1	□ 4 ←	□ 3	□ 2	□ 1
	Fruity flavors	□ 1	□ 2	□ 3	□ 4	□ 1	□ 2	□ 3	□ 4	□ 1	□ 2	□ 3	□ 4
Skin	Toughness	□ 1	□ 2	□ 3	□ 4	□ 1	□ 2	□ 3	□ 4	□ 1	□ 2	□ 3	□ 4
	Tannins	□ 4 ←	□ 3	□ 2	□ 1	□ 4 ←	□ 3	□ 2	□ 1	□ 4 ←	□ 3	□ 2	□ 1
	Acidity	□ 4 ←	□ 3	□ 2	□ 1	□ 4 ←	□ 3	□ 2	□ 1	□ 4 ←	□ 3	□ 2	□ 1
	Astringency	□ 1	□ 2	□ 3	□ 4	□ 1	□ 2	□ 3	□ 4	□ 1	□ 2	□ 3	□ 4
	Dryness	□ 4 ←	□ 3	□ 2	□ 1	□ 4 ←	□ 3	□ 2	□ 1	□ 4 ←	□ 3	□ 2	□ 1
	Grassy	□ 4 ←	□ 3	□ 2	□ 1	□ 4 ←	□ 3	□ 2	□ 1	□ 4 ←	□ 3	□ 2	□ 1
	Fruity	□ 1	□ 2	□ 3	□ 4	□ 1	□ 2	□ 3	□ 4	□ 1	□ 2	□ 3	□ 4
Seeds	Color	□ 1	□ 2	□ 3	□ 4	□ 1	□ 2	□ 3	□ 4	□ 1	□ 2	□ 3	□ 4
	Hardness	□ 1	□ 2	□ 3	□ 4	□ 1	□ 2	□ 3	□ 4	□ 1	□ 2	□ 3	□ 4
	Flavor	□ 1	□ 2	□ 3	□ 4	□ 1	□ 2	□ 3	□ 4	□ 1	□ 2	□ 3	□ 4
	Tannins	□ 1	□ 2	□ 3	□ 4	□ 1	□ 2	□ 3	□ 4	□ 1	□ 2	□ 3	□ 4
	Astringency	□ 1	□ 2	□ 3	□ 4	□ 1	□ 2	□ 3	□ 4	□ 1	□ 2	□ 3	□ 4

Note: The data in the highlighted cells are reversed in order to give an immediate visual impression of the maturity state.
(After Rousseau, 2003).

TABLE 7.5 Field Assessment Interpretation Characters and Scores

	Maturity Rank				Abnormal Characters
	1	2	3	4	
Technical maturity	Pulp not sweet, acidic, adheres strongly to skin and seeds	Pulp slightly sweet and slightly acidic, some pulp adheres	Sweet pulp, little acidic, almost no pulp adheres	Pulp very sweet, very low acidity, no pulp adheres	Pulp both little sweet and little acidic, pulp like gelatin, adheres strongly to seeds (water deficit stress); pulp both very sweet and acidic (cold climate)
Aromatic maturity of pulp	Strongly grassy	Neutral or slightly grassy	Slightly fruity	Very fruity, jam	Moldy or muddy taste (*Botrytis*); acetic acid, tea, infusion (water deficit stress), fruit in alcohol, overmature fruit
Skin maturity	Very hard, grassy, acidic, dry skin with little tannin intensity, green tinges on white varieties, pink, on red varieties	Hard skin, grassy to neutral, not fruity, rather acidic and dry; some green (white varieties) or clear red (red varieties) near the pedicel	Skin rather crushable, slightly acidic and dry, neutral to fruity with no grassy aromas, uniform color at the base	Very crumbly skin, without acidity or dryness, very fruity, strong color extraction when pressing the skin between fingers on red varieties	Moldy or muddy taste (*Botrytis*); very thin skin (important, risk of damage), skin both crumbly and grassy
Seed maturity	Seeds green or bright yellow, without brown	Seeds brown and green; strong astringency when licking	Brown without any green; grassy and astringent, crushes without cracking	Dark brown or gray, toasted aromas, little astringency, cracks entirely between the teeth	Brown seeds and hard grassy acidic skin; green seeds and soft skin (cold climate)

(After Rousseau, 2003).

TABLE 7.6 Brief Sensory Analysis Form

Vineyard/Sample					Person	Date
	Score					
Component	1	2	3	4	Decision	Abnormalities
Technical maturity	☐	☐	☐	☐		
Flesh flavor	☐	☐	☐	☐		
Skin maturity	☐	☐	☐	☐		
Seed maturity	☐	☐	☐	☐		

(After Rousseau, 2003).

d. then, assess the juice from the three berries; observe the sweet and acid tastes and the type and intensity of flavors.

3.2.4 *Tasting the Skin*

After having spat the juice and the pulp out, and recorded your observations, put the three skins back in your mouth and keep the seeds in your hand. Record the following observations as you proceed:

- Chew the skins 10–15 times. While chewing, evaluate the toughness of the skin.
- Then, rub the crushed mixture obtained on to the palate, the cheeks and between the gums and the lip membrane. Run the tongue twice from the back to the

front of the palate and evaluate the friction force the second time to rate the tannic intensity.

- Evaluate the acidity of the skin.
- After spitting out the crushed mixture, run your lips over your gums and observe the astringency and the dryness of the skin.
- Finally, observe the type of flavors of the skin and their intensity (this evaluation can also be done with the crushed mixture in the mouth).

3.2.5 *Appearance and Taste of Seeds*

Examine the color of the seeds. If green traces remain, do not taste them; pronounced astringency due to low molecular weight tannins could saturate the sensory cells of the mucous membranes and deprive you of tasting powers for several minutes.

- In the event of the absence of green traces, chew the seeds between the incisors and observe the crushability of the seeds.
- Next, chew the seeds 10–15 times and observe the flavors, tannic intensity and astringency.

These tables can be subject to statistical analysis and the data presented graphically as an aid to presentation of the data (Figure 7.8).

3.3 Practical or Field Protocol

Use the summary berry analysis forms (Tables 7.5 and 7.6).

In a vineyard, or when a quick decision is required, it is not practical to make a complete assessment for each of the 20 criteria. In these circumstances, and indeed for routine use, four characters only need be assessed:

- Technical maturity; balance between sugar and acidity of the pulp
- Aromatic status of the pulp
- Skin maturity
- Seed maturity.

It is easy, with some training, to give a summary rating of these four levels. Table 7.5 gives a brief interpretation scale. Values of 1 or 2 are below the average, which means that the grapes have not reached a sufficient level for wine making; rankings of 3 or 4 are above the average, meaning that the fruit has reached a level sufficient for wine making (even if not optimal).

The sensory profiles can be stored as data to enable comparisons between vineyards or years.

Chapter 8

Winery Protocols

Chapter Outline

A Complete Guide to Quality in Small-Scale Wine Making.

1. CLEANING, HYGIENE AND MAINTENANCE

1.1 Purpose

Cleaning is a vitally important part of wine making. Proper cleaning processes minimize the risk of contamination by spoilage organisms and taints from wastes and general contaminants.

1.2 Occupational Health and Safety

- Corrosive alkali and acids
- Heat—hot water
- Hazardous substances—e.g. metabisulfates, oxidants (chlorine, bromine, iodine, peroxides)
- Environmental hazards—alkali, phosphates, surfactants.

1.3 Quality Assurance Records

- Person, date
- Items
- Method applied.

TABLE 8.1 **Cleaning Agents and Their Relative Effectiveness in Removing and Killing Biofilm (Gel)-Forming Types of *Brettanomyces bruxellensis***

Agent	Rank	Type	Active Agent
Sanibac®	4	Oxidant	Chlorine
Acryl Aquaclean®	3	Detergent, ketones	Surfactants
Soda ash	4	Alkali	Sodium carbonate
Caustic soda	1	Alkali	Sodium hydroxide
Oakite 62®	2	Alkaline detergent	Sodium hydroxide and surfactants
Biocidal ZF®	2	Detergent, quaternary ammonia	Surfactants

Note: Rank 1 = kills organism in film or free (attached to surface); 2 = mildly effective at killing gel-embedded and largely effective for free organisms; 3 = mildly effective for killing gel-embedded and free; 4 = ineffective.
(Joseph et al., 2007).

1.4 General

Hygiene is a critically important aspect of winery operations. The goal is to minimize the risk of spoilage. While it is virtually impossible to maintain an aseptic operation, it is important to minimize the risk of a high load of endemic spoilage microorganisms and to ensure that, as far as is possible, the containers and lines comply with health authority guidelines and rules. Many older, wooden structures harbor not only desirable strains of yeast but also undesirable forms that are difficult if not impossible to eradicate (e.g. *Brettanomyces* spp.). Minimizing access to pest animals, rodents and birds is important in maintaining hygiene during and between vintages.

Devising and conducting a cleaning protocol should be based on minimizing the use of chemicals that pose an environmental risk—caustic and sodic chemicals in particular. If these must be used then consider how they will be recycled or disposed of according to local regulations. In general, the cycle is: wash, sanitize, rinse, use; wash, sanitize, allow to dry, then store.

The objectives are to remove wastes and sediments including tartrates, tannin–protein complexes and lees, and to eliminate contaminating microorganisms. Cleaning agents may be classified as antimicrobial (heat, alkali, sanitizing agents), surfactants (detergents) or physical (pressure) (Table 8.1).

Sanitizing agents are usually oxidants, either alone or combined with a surfactant or an alkali. Common oxidants are potassium metabisulfate, hydrogen peroxide, chlorine and iodine. The alkali may be ammoniacal or caustic. It is usually better to avoid the caustic forms as they pose an environmental risk and therefore a waste disposal issue. All agents are potentially hazardous, either through corrosive effects on the skin and internal membranes or because of their potential to cause allergies—check the material safety data sheet (MSDS) for each and adopt safe practices. It is also worth remembering that chlorine is a precursor to TCA (cork taint, 2,4,6-trichloroanisole) and should be used with caution, if at all.

Cleaning is the first priority as biocides need to reach the target organism in order to be effective. Sediments, scales (e.g. bitartrates and tannin–protein complexes) and biofilms (microbial colonies embedded in a 'slime' matrix) formed during fermentation are highly insoluble and must be removed before proceeding with any secondary hygiene process. Living microbes, yeasts and bacteria are killed at 80° C and this is normally sufficient for cleaning small batch fermenters and equipment, especially if visually clean. This temperature may not kill spores of yeast, fungi or bacteria. These may even resist boiling water (thus laboratory microbiological processes involve high-pressure steam at 121° C).

Cleaning of highly contaminated vessels may be achieved through the use of hot caustic soda solutions to dissolve and hydrolyze the precipitates before neutralizing the surface with citric acid. Such a severe treatment also lyses and effectively sterilizes the treated surface. However, these treatments are hazardous to personnel and to the environment and are best minimized. High-pressure hot water/steam treatment can achieve similar goals on most surfaces and is also suitable for timber (e.g. barrels).

The use of surfactants is normally restricted to cleaning inaccessible components, lines, pumps, filters, complex fittings, etc. Often these are based on alkaline sodic chemicals that pose an environmental risk. The potential of a range of cleaning agents and surfactants to remove and kill *B. bruxellensis* that had formed a biofilm has been assessed, and only caustic soda (NaOH) proved efficient (Table 8.2 and Figure 8.1). However, its use should be minimized in the interest of limiting risks.

1.5 Care of Stainless Steel

While particular grades of stainless steel are resistant to corrosion, they are not immune. In addition, untreated stainless steel may be a source of hydrogen sulfide (H_2S) and should be washed with citric or tartaric acid to remove any manganese sulfide that may have formed on the surface (Rankine, 1989). Commercial tanks and vats should be treated, 'passified' or 'pickled' before use, either chemically or electrochemically. Citric acid (e.g. 10% at 65° C for 30 min) is commonly used after thorough cleaning and rinsing, but damaged and heavily scaled surfaces may require more aggressive treatment (e.g. phosphoric or nitric acid), which will be carried out by a registered commercial agent (e.g. ASTM, 2006).

1.6 Care of Barrels

Barrels are one of the single largest expenses in the production of fine wines, accounting for perhaps as much as half of the total cost. They are fragile and susceptible to contamination because they are constructed of abutted staves, and timber is porous and subject to shrinking and swelling. Cost, value and susceptibility to contamination all mean that a lot of care must be invested in maintaining this resource. New oak should be stored in its shrinkwrap in a humid environment until required (e.g. in a barrel store cellar). Old oak should be washed, dried, sulfited and held in a humid environment.

TABLE 8.2 Table of Cleaning and Sanitizing Agents According to Use in the Winery

Agent	Rinse (Required)	Items	Indicator	Comment
Steam/high-pressure hot water	No	All	General cleaning and sanitizing	May not be fully effective in lines and complex fittings
Caustic	Citric	Stainless steel tanks, plastics	Hard tartrate scaling	Waste disposal issue, take care to neutralize; full protective clothing necessary
Carbonate	Citric			
Hypochlorite	Water	Usually smaller items but also effective for general area, not suitable for barrels	Good general purpose disinfectant	Sodium and chlorine residues may be a waste disposal issue; chlorine is highly toxic. Chlorine may corrode stainless steel and provide a substrate for TCA if used in wooden barrels
Iodine	No			As for hypochlorite but fewer waste issues; toxic; stains clothing and skin
Metabisulfite	No			Allergies, strong irritant
Peroxide	No	Lines and fittings		
Sodium carbonate peroxyhydrate ($2Na_2CO_3.3H_2O_2$)	No			Contaminated barrels and general use, e.g. pumps and lines
Peroxide/alkali	Citric/ water			Pumps and lines
Chlorinated trisodium phosphate	Water			
Quaternary ammonium	No			

Note: TCA = 2,4,6-trichloroanisole (cork taint).

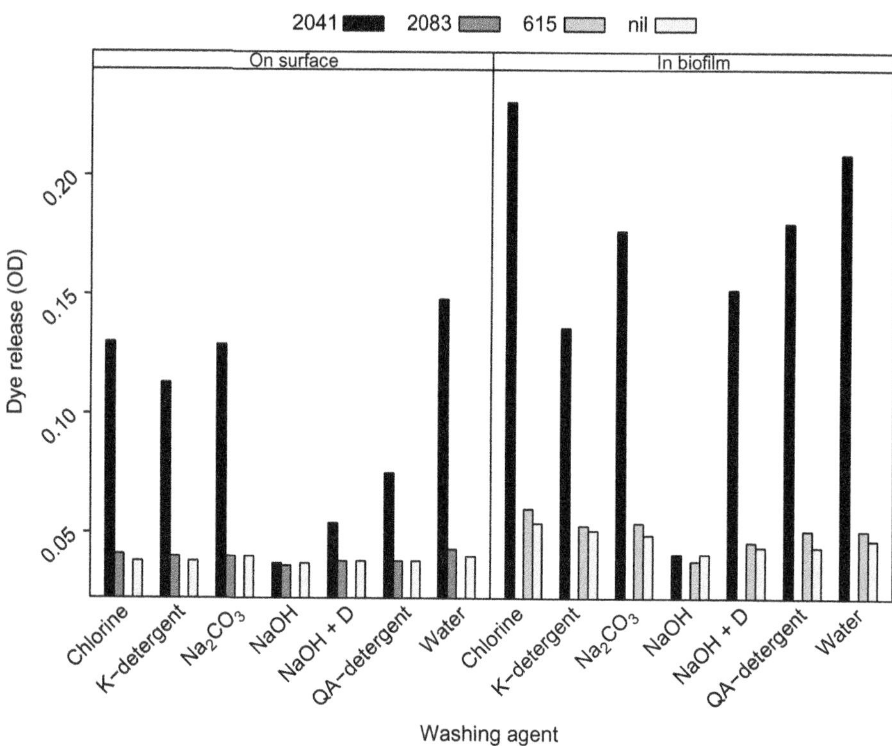

FIGURE 8.1 Comparison of the effectiveness of a variety of cleaning agents (Table 8.2) in removing particular strains (615, 2041, 2083) of *Brettanomyces bruxellensis* either adhering to the wall of a polystyrene culture plate (surface) or within a biofilm formed on the plate surface. The adhering or film-forming yeasts were stained with a dye (acridine orange in the adhesion study and methylene blue in the biofilm study). The higher the absorbance, the higher the level of yeast cells (not necessarily living) remaining after washing. *(Redrawn from Joseph et al., 2007, with permission).*

1.6.1 New Oak

New barrels should be inspected for uniformity, flaws and sensory and physical cleanliness before use. They should then be hydrated.

- **Method 1:** Fill one-third full with clean, cold water (not chlorinated or treated to remove chlorine so as to minimize risk of taint) and stand for 3–4 h, fill to two-thirds full and hold for a further 3–4 h, then top up and hold until seeping stops. Use immediately or sulfur. If the barrel continues to seep it will need to be replaced or repaired by a cooper.
- **Method 2:** Add about one-tenth volume of hot water, replace the bung and roll to ensure all inner surfaces are wetted. Stand on its end, fill the head with hot water and stand for about 15 min. Invert and fill the other end with hot water and stand. Then turn on its side, remove the bung and drain. You may need to test for a seal by refilling with cool water or by assessing whether a vacuum has developed as the hot water cools. Use immediately or sulfur. This method conserves water and is more rapid but perhaps less reliable, especially if the barrel has been stored dry for a long period.

Do not hold barrels with the same water for more than 2 days. If necessary add potassium metabisulfite (PMS) to protect against microbial contamination, but be careful not to oversulfur and risk other problems.

1.6.2 Used Barrels

Used barrels should be washed with hot, high-pressure water to remove deposits, then with cold water, and either used immediately, or sulfured and used or stored. Note that sulfuring at this stage will add sulfur dioxide (SO_2) to the wine (5 g S will on burning give 10–20 mg/L SO_2 subsequently) (Ribéreau-Gayon et al., 2006b).

Spoiled, contaminated or tainted barrels are difficult to treat and the chemical treatments required to eliminate the spoilage usually destroy the character of the barrel. Thus, prevention is far better than the cure (see specialized literature).

1.7 Recipes and Methods

1.7.1 Vessels, Processing Equipment and Containers

- **Citric metabisulfite:** Dissolve 7.5 g of PMS and 30 g of citric acid in 10 L of water or multiples thereof.
- **Caustic wash:** Dissolve 5 g of sodium hydroxide per liter of water.
- **Citric (neutralization) wash:** Dissolve 5 g of citric acid per liter of water.

1.7.2 Pumps and Lines and Fittings

These are best treated by rinsing with clean water. Rinse with an alkaline–oxidative cleaner (sodium percarbonate, syn. sodium carbonate peroxyhydrate) until the solution runs clean, and then circulate for a period before rinsing with a citric, then a water rinse. Usually a 10 g/L solution is appropriate, dissolved in warm water before dispersing in cold (but concentrations may vary from manufacturer to manufacturer, so check the label).

1.7.3 Barrels

- **Barrel sulfuring:** Burn a sulfur stick or candle (5–10 g) per 22-L barrique. The barrel should be fully drained and preferably dry (if appropriate, use dry, hot air to dry the inner surface, but don't overdo). The candle or wick or ring should be held in the center of the barrel so as not to burn the timber. If necessary, the bung may be replaced for a period to ensure that all surfaces are affected. This must be carried out in a well-ventilated area.

If the barrel is to be stored, the sulfuring should be repeated after 5–6 days, once the barrel has dried fully, and the process repeated at 8–12 week intervals. A barrel should, at all times, have a sulfurous aroma internally. Note that barrels which have been treated in this manner should be soaked in cold water for 2–3 days before use to avoid taints.

2. PH ADJUSTMENT

2.1 Purpose

pH adjustment serves a number of purposes in wine making. Primarily, pH adjustment optimizes the environment for the growth of desirable microorganisms and discriminates against growth of undesirable organisms–it is a first line in the defense of wine quality (sulfur is probably primary). Secondly, pH has an important impact on the palate and the perception of the balance of sweet and sour flavors. Thirdly, pH affects the release of aromas and the color of the wine, especially red wines. pH is important.

2.2 Occupational Health and Safety

No particular issues.

2.3 Quality Assurance Records

- Person, date, acid (and source), amount per volume.

2.4 Procedure

- Calibrate pH meter (ideally against a saturated solution of potassium bitartrate).
- Pipette 50 mL of wine or clarified must into a beaker—do not dilute with water.
- Titrate to the desired pH with a solution of 10.0 g/L of the chosen acid (normally L-tartaric) or base (normally $KHCO_3$).
- Calculate the weight of acid or base to be added to the wine (in the case above, 5 mL equates to 1 g/L of wine or must).

  (See Figures 8.2 and 8.3.)

2.5 Notes

1. Options, apart from direct adjustment, include the use of a malolactic fermentation and blending (use a Pearson square to estimate volumes; see calculators or any one of a number of websites, e.g. NTWG, 2007).
2. It is difficult to calculate the correct amount of acid/alkali to add for pH adjustment (but see Devatine et al., 2002). Regulations permit the addition of tartaric and a range of other acids (ascorbic, sorbic, citric, malic,

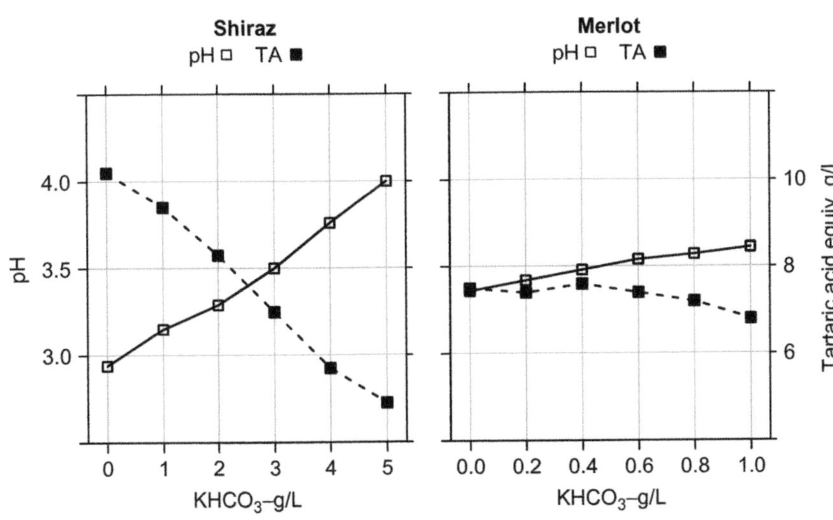

FIGURE 8.2 Plot of changes to pH and titratable acidity (TA) in two wines in response to the addition of $KHCO_3$. *(Data from AWRI, 2011a, with permission).*

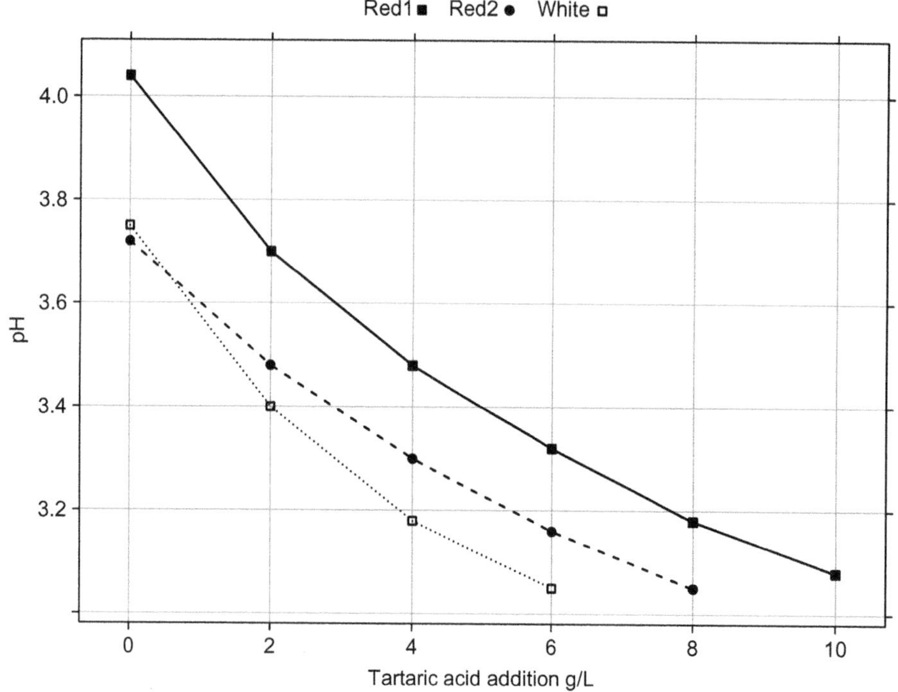

FIGURE 8.3 Plot of change in pH for three wines in response to successive additions of tartaric acid. Note the differences in slope between the red and the white wine due to the smaller buffering capacity of the white wine (less potassium and other salts) and the reduction in slope generally as the pH approaches the pK_{a1} of tartaric acid (3.01). *(Data from Rankine, 1989, with permission).*

TABLE 8.3 Organic Acids Added to Wine and Their Characteristics

Acid	*M*	Protons	pK_a	Metabolizable
L-Tartaric[a]	150.09	2	2.93, 4.23	No
L-Malic	134.09	2	3.40, 5.11	Yes
Citric	192.12	3	3.14, 4.77, 6.39	Yes
Ascorbic[b]	176.12	1	4.17, 11.57	Yes
Sorbic[c]	121.14	1	4.8	No
Erythorbic[b]	176.12	1	cf. Ascorbic	No

M = molar mass.
[a]*D and L refer to the effect that a solution of the substance has on polarized light, turning it to the right hand (dextro) or the left hand (levo). This property led to glucose or glucose polymers being termed 'dextrose' or 'dextrans', and fructose and fructose polymers being terms 'levulose' or 'levans', respectively. A mixture of the two forms is termed 'racemic'.*
[b]*Antioxidants, erythorbic is an isomer of ascorbic acid but with low vitamin C activity.*
[c]*Antimicrobial agent used to stabilize sweet wines.*

lactic, erythorbic and metatartaric acid) (Table 8.3). Also see, for examples, the TTB (Alcohol and Tobacco Tax and Trade Bureau), Organisation Internationale de la Vigne et du Vin (OIV), Wine Australia (AWBC), Australian Wine Research Institute (AWRI) or Food Standards Australia New Zealand (FSA) web sites for current regulations, which are defined by region.

3. The simplest way to estimate the amount needed is to titrate a known volume of wine with a known concentration of the acid of choice—e.g. a 10 g/L solution. Figures 8.4 and 8.5 provide a guide to the order of the amount that will need to be added. However, treat all wines individually as each will differ in its buffering capacity and thus in its response to the addition of acid or alkali.
4. In white wines it may at times be necessary to increase the pH and either calcium or potassium carbonate is used for this purpose. These compounds may also be used to reduce the concentration of organic acids in the wine, and thus its buffering capacity, making it easier to acidify the wine (Ribéreau-Gayon et al., 2006b). See also section on tartrate stability (Chapter 10, Section 8).

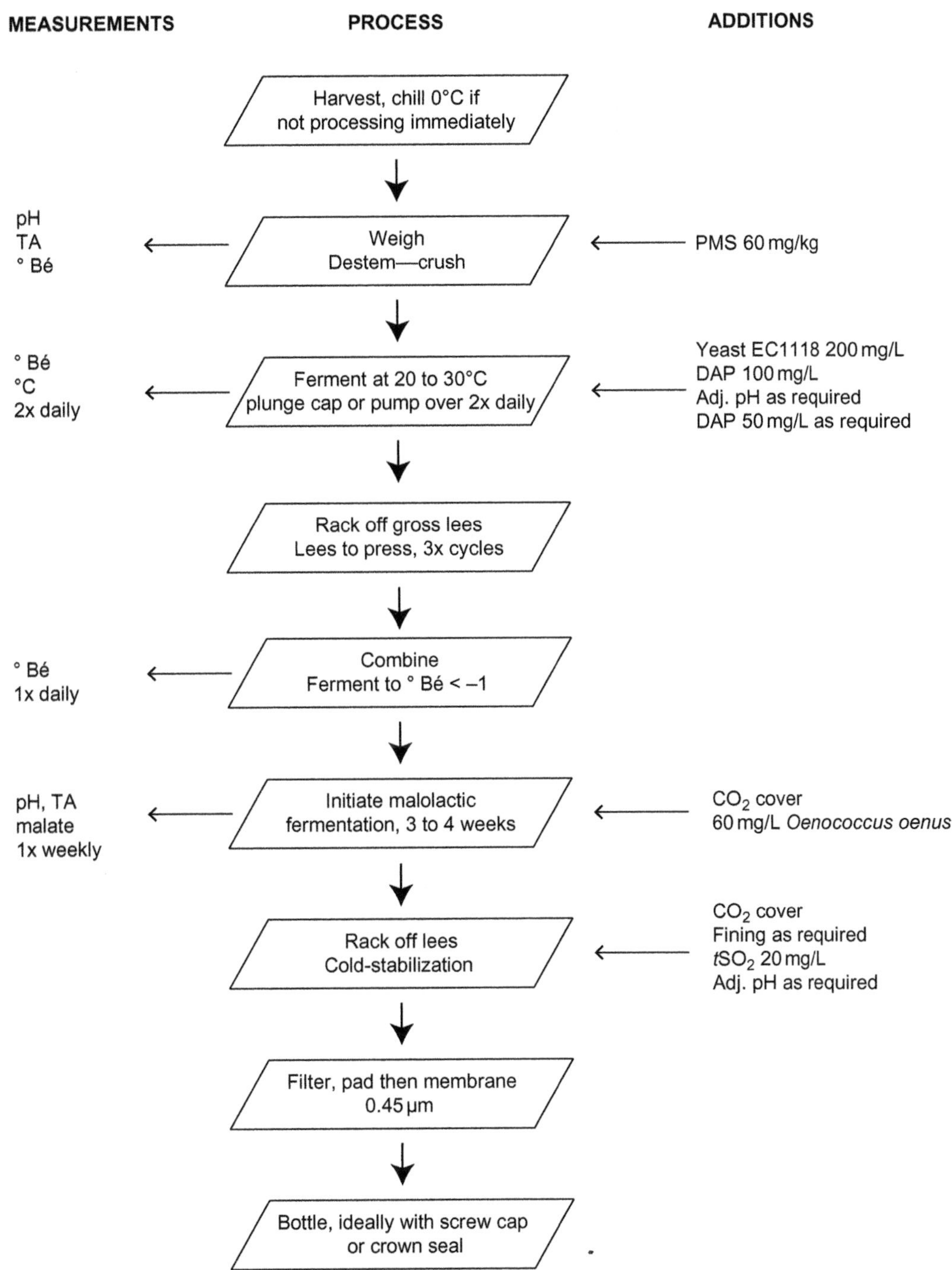

FIGURE 8.4 General protocol for the production of small-scale red wines. PMS = potassium metabisulfite; DAP = diammonium phosphate.

5. Major adjustments should be carried out before fermentation commences to provide optimal conditions for discriminating between competing beneficial and fault-causing microorganisms. Minor adjustments to enhance palate balance only should be made prebottling.
6. While a number of acids may be used, L-tartaric is probably the acid of choice as it is not metabolizable. Malic and citric acids may be used, and usually the latter is the acid of choice for final adjustments, but both may be metabolized by microorganisms and therefore sterility in the bottle is essential. Citric acid, in particular, may give rise to diacetyl, which may be good but in modest quantities only.
7. A mathematical model has been developed to predict the outcome of pH adjustment based on knowledge of

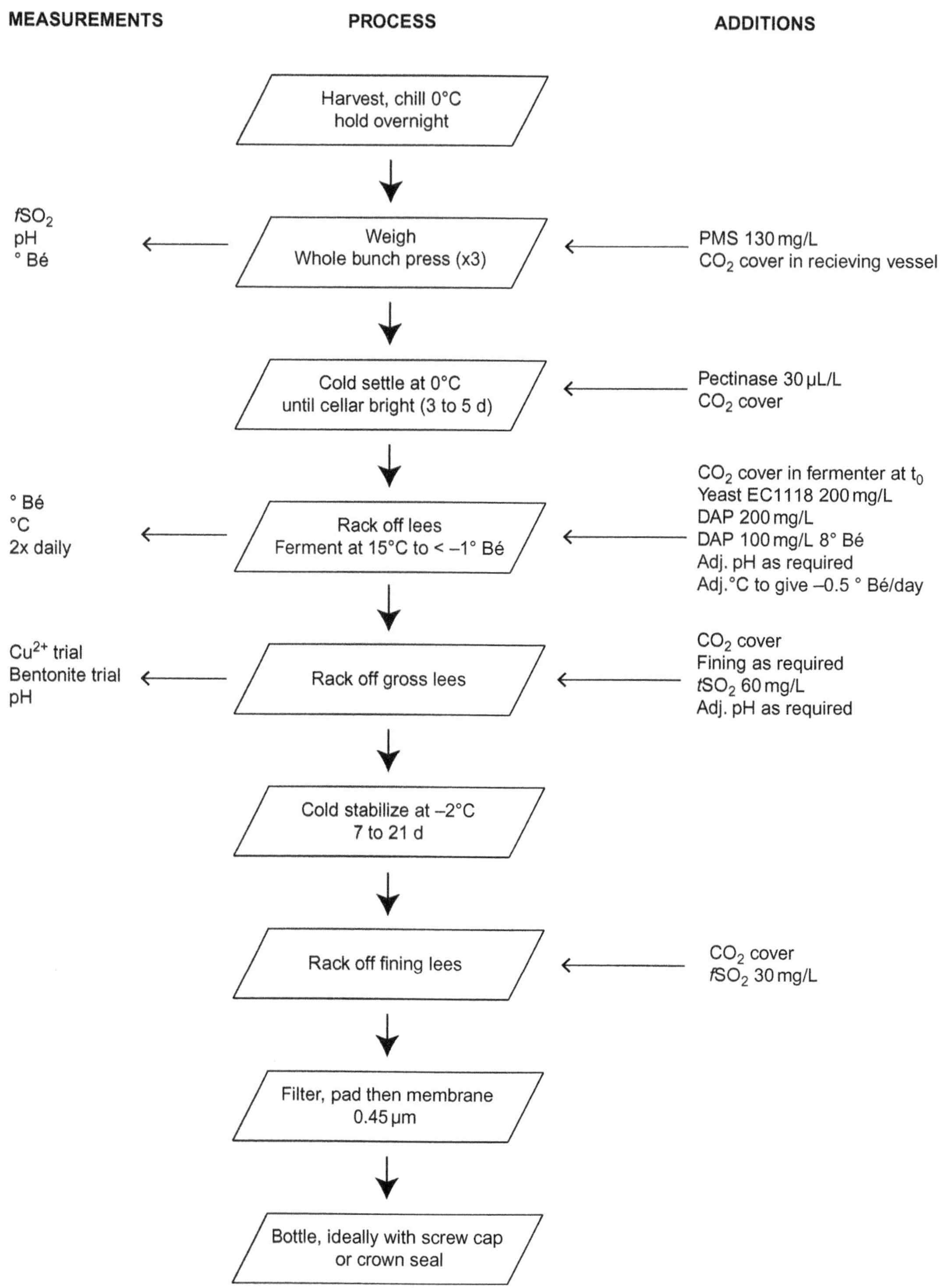

FIGURE 8.5 General protocol for the production of small-scale white wines. PMS = potassium metabisulfite; DAP = diammonium phosphate.

pH, titratable acidity, tartaric acid and potassium. At the time of publication the authors suggested that it required further verification in practice (Devatine et al., 2002).

2.6 Cautions

1. Use only L$^+$-tartaric acid (natural), not the cheaper DL (racemic) form as the mixture of the two forms may complicate cold-stabilization of the wine.
2. The pH of red wine may rise during fermentation owing to the release of K^+ from skins during maceration.
3. If adding DL-malic acid before a malolactic fermentation be aware that only the L form will be metabolized, i.e. the pH may be lower than otherwise.
4. Do not adjust with tartaric acid following cold-stabilization as this may destabilize the wine.
5. Do not adjust with citric acid before a malolactic fermentation as this may give rise to excessive volatile acidity and excessive levels of diacetyl (buttery aroma).
6. pH may rise or fall following cold-stabilization depending on pH at the start: adjust below pH 3.5 to ensure that pH falls (see tartaric acid stabilization Chapter 10.8).
7. Do not adjust with malic acid following a malolactic fermentation as this may lead to an in-bottle fermentation unless rigorously filter-sterilized.
8. If treating 'must', chill for an extended period first to encourage precipitation of KH tartrate.
9. If deacidifying, beware frothing may occur due to carbon dioxide (CO_2) production.
10. Always taste as flavor is the final arbiter.

2.7 Equipment and Materials

- pH meter and standards
- Magnetic stirrer and Teflon-coated bar
- 50 mL burette
- 150 mL beaker
- 10 g/L solution of tartaric acid (or other selected acid in the form being added to the wine, i.e. either anhydrous or hydrate as available)
- $KHCO_2$ (alternatively, $CaCO_3$, K_2CO_3 or Acidex®: $CaCO_3$ + malate–Ca–tartrate double salt).

2.8 Benchmark Values

- White table wines: approximately pH 3.3, but adjust to palate test
- Red table wines: approximately pH 3.4, but adjust to palate test.

3. CULTURE PREPARATION AND INOCULATION

3.1 Purpose

While wine has been produced since the establishment of culture in the human race, only since the role of microorganisms was discovered has the production of quality wine become routine. The use of industrial or purpose-selected strains of yeast and bacteria enables a winemaker to 'mold' the flavor and aroma profile of a given wine style. Perhaps more than any other invention, the use of selected strains of wine yeasts and lactobacilli enables the production of quality wine, routinely.

3.2 Yeast

Yeast must be rehydrated and activated before use. Usually the supplier will provide a protocol (e.g. YeastRehydration_Lallemand.pdf from Scott Laboratories). The protocol provided here is based on a general recommendation (Soubeyrand et al., 2006). Some recommend the use of a yeast nutrient solution rather than water. The objective is to achieve approximately 5×10^6 colony-forming units (cfu)/L. This will rise as high as 10^9 during the ferment.

The standard protocol is to rehydrate 1 g of active dry yeast per 10 ml of glucose solution (50 g/L) at 37° C for 30 min. Allow the yeast to cool to room temperature during the holding period to reduce temperature shock on inoculation.

Inoculate at a rate of 0.25 g/L (2.5 mL/L) under a cover of CO_2. In the case of white wine, ensure that there is some head space to accommodate the frothing that will occur, so that the trap does not become contaminated. In the case of red wine, ensure sufficient head space to allow the cap to rise (it may double the depth of the must).

Temperature control is critically important. A ferment that is unregulated will quickly heat up, even to levels that affect viability and impair the sensory characteristics of the wine. A good starting temperature is about 18° C. In the case of a white wine, this will be progressively reduced to about 15° C, while a red wine may be allowed to run to 25–27° C or even a little higher. This heating can occur within 24 h, so be vigilant.

3.3 Oenococcus

Freeze-dried bacteria may be inoculated directly into a wine at the end of fermentation, although successful inoculation may be carried out at the start, the middle or the end of the ferment (Rosi et al., 2003). Dried cultures contain on the order of 5×10^9 cfu/g. One should aim to achieve 10^6 to 5×10^6 cfu/L. That is, 1 g of dried inoculum should suffice to inoculate 1000 L, three hogsheads or about five barriques.

At the time of inoculation the total sulfur dioxide (tSO_2) should be below 5 mg/L, pH between 3.2 and 3.5, temperature between 18 and 21° C and alcohol less than 15%. If these conditions are not met then specialist strains may need to be used. Inoculations should be made under CO_2 cover.

The progress of a malolactic fermentation can be followed by paper or thin-layer chromatography, by enzymatic analysis of malic acid or by following colony-forming unit counts. At the end of the ferment, malic acid should be measured enzymatically or by another sensitive analytical technique (high-performance liquid chromatography or gas–liquid chromatography) to confirm completion.

4. A RED WINE PRODUCTION PROTOCOL

In the vast majority of cases, red wines for teaching, experimental and semi-commercial purposes will be prepared in open fermenters, and that is assumed in the text that follows. Be sure to allow sufficient space above the level of the liquid for formation of a cap (at least plus one-third of the depth of juice).

4.1 Harvesting and Transportation

Small lot wines are usually made from hand-harvested fruit. All fruit, whether hand or machine harvested, should be picked when near its lowest diurnal temperature: it will follow air temperature, approximately, with a lag of about 1 h. Thus, in many regions it will be necessary to pick from first light until about mid-morning only. Fruit to be transported should be treated with a light dusting of $K_2S_2O_5$, chilled before transport and held in a cold room on receipt. Mechanically harvested fruit should have its full amount of $K_2S_2O_5$ added to the container at picking to reduce the action of the enzyme polyphenol oxidase, which may spoil the fruit (Boulton et al., 1996; Jackson, 2008; Ribéreau-Gayon et al., 2006a).

4.2 Initiating Fermentation

If possible, bring the juice to 20° C or lower and the yeast culture to a similar temperature. A large difference in temperature between juice and culture will cause a thermal shock to the yeast cells and may seriously impair the initiation of the fermentation. The speed with which the yeast cells adapt and initiate the ferment is determined by temperature, density of yeast cells, free sulfur dioxide (fSO_2) concentration and nutrient status.

4.3 Monitoring Fermentation

Plunge the cap, and record and graph at least twice daily the ferment temperature, ° Baumé and aroma (paying particular attention to sulfurous, H_2S aromas). Regular monitoring is especially important during the early to middle phase of fermentation to ensure a steady reduction of ° Baumé of about 1.5–2 ° Bé/day. Be alert for any decrease and change in sulfurous aromas that may signal nutrient deficiency and the onset of a sluggish ferment. The onset of a sluggish ferment is frequently apparent as a relatively slow rate of fermentation from the outset, so act early: assess nitrogen status and if necessary add more, and check the microbiology of the fermentation, yeast count and culture for contaminating microorganisms.

4.4 Racking off Gross Lees

A good rule of thumb is to do this at about 3–4° Bé, but ideally a palate assessment for astringency and bitterness is used to determine the optimum time. The wine is dropped/transferred into a press and run through three cycles with remixing the skins between each and run or pumped into a closed fermenter under CO_2 cover (dry ice chops). It is a good idea to sample each pressing and to conduct a palate test on each.

Complete the fermentation with an airlock. As before, continue to follow the course of the ferment by ° Bé and when 'dry' (< − 1.0° Bé, sugar <2 mg/L) take a sample for sugar and for SO_2 analyses.

4.5 Initiating Malolactic Fermentation

Commercially, *Oenococcus oeni* may be introduced at racking, but it is better for teaching and research and the small-scale winemaker to separate the two ferments to minimize complications and to aid problem solving should the fermentation be slow or fail to finish.

It is critically important that the total sulfur concentration be below 5 mg/L, although some sulfur is beneficial in helping to minimize the growth of spoilage lactobacilli (here pH is also important: it should be less than 3.5 and probably a little less than 3.4, allowing for a rise in pH as the malic acid is converted to lactic acid).

Unlike yeast, there is no need to culture the freeze-dried microbes before inoculation; simply spread the requisite weight/unit volume over the surface. It is likely that some lactobacilli will already occur in the wine and it is important to ensure that the selected strain prevails by adding the correct amount of fresh inoculum. Commercial additives, mainly of vitamins, are available as an adjunct. However, a successful yeast primary fermentation should, on autolysis of the yeast cells and with the remaining residual sugars, provide adequate nutrition. The remaining factor is temperature, which should be close to room temperature: 18–20° C is generally regarded as suitable.

This fermentation should be monitored. At first this can be done by observing the release of CO_2 via bubbling of the solution in the airlock. Completion should be checked by chromatography (malic and lactic acids) or enzyme analysis for malic acid (see Chapter 11). Thin-layer chromatography on a microscope slide or aluminum backing is useful as a quick test.

4.6 Racking and Aging

Small-scale fermentations are usually aged in the bottle unless the scale is at barrel level. It is possible, however, to keep the wine on oak chips until palate tests indicate that sufficient extraction has occurred—add PMS before this step to protect the wine. Otherwise, rack the wine off the gross lees by siphon or racking plate into a fresh, closed container, with the addition of a small pellet of dry ice at the outset, for final adjustments, fining and stabilizing. Add PMS to bring the total SO_2 to about 40 mg/L and hold in a lightly closed vessel and with minimal ullage (space between wine and seal), e.g. 5–10 mm. Do not fill completely as this will encourage contamination. Reassess the free SO_2 and if necessary adjust to 25–30 mg/L.

4.7 Fining and Stabilizing

The choice of fining and stabilizing procedures depends on sensory analysis and on chemistry (Table 8.4). Tests that should be done at this stage include, pH, TA, free and total SO_2, a copper fining trial and a bentonite fining trial (optional). If sensory analyses indicate volatility then volatile acids should also be assayed (regrettably, the only way to remove these may be by reverse osmosis; not presented in this text).

Once the appropriate additions have been made, finely ground potassium bitartrate should be added as per the protocol and the wine brought down and held at the desired temperature until settling is complete (Chapter 10.8).

4.8 Filtering and Bottling

Prebottling tasks include final adjustments to pH and SO_2, and usually blending. As with white wines, it is important to avoid risks associated with taints from natural cork in experimental wines and thus to close with either a screw cap or a crown seal. Every item that the wine is exposed to must be sterile.

The process of bottling at the small scale is labor intensive and involves first transfer to a keg under nitrogen or CO_2 cover, and then through the coarse filter, then the fine and into the bottle. Ullage in the bottle should be 25–30 mL. If using a screw-cap machine, be sure to conduct trials to obtain a good seal. If using cork, be sure to sterilize the cork in 10 g/L PMS and rinse before use. Use only good-quality corks, or use a synthetic cork or a crown seal.

5. A WHITE WINE PRODUCTION PROTOCOL

5.1 Harvesting and Pressing

For white wines, as for red, fruit should be harvested as cool as the environmental conditions allow. This not only saves energy, but enables a rapid transition from picking to pressing and to adjusting and settling. The fruit should be weighed to enable the volume of juice to be estimated and an appropriate quantity of PMS added at each pressing to achieve the final target value of about 50 mg/L tSO_2.

It is advisable to conduct a trial run to determine the yield of juice per kilogram of fruit, and the yield of juice and its quality (sensory and chemical) for each pressing cycle. While, normally, pectinase would not be added until after pressing, it may be added at crushing if it is intended to have a skin extraction treatment, as may be

TABLE 8.4 Red Wine: Common Fining and Stabilizing Agents and Their Characteristics

Agent	Range	Heat Stability	Comments
Egg albumen	1–2 egg whites/hL (30–60 g/hL)	Not heat stable	Reacts with high molecular weight polyphenols, best for softening and polishing red wines prior to bottling (Margalit, 2004). Recommended agent
Blood albumen	10–20 g/hL	Not heat stable	A traditional fining agent. Fresh blood (cow) is now proscribed for health safety issues, and indeed many countries now ban the use of blood products. Regarded as excellent for softening stalky, young tannic red wines and fine wines (Ribéreau-Gayon et al., 2006b)
Gelatin	5–10 g/hL	Heat stable	Softening and tannin stabilization, especially in young red wines, e.g. pressings; choose type according to wine phenol content (Ribéreau-Gayon et al., 2006b)

practiced for Semillon and Sauvignon blanc (see Figure 5.4 in Chapter 5). Usually, if included, this is of short duration to avoid excessive extraction of bitter skin and seed tannins and phenolics. This, however, would make for a good student exercise with times varied and sensory evaluations carried out on the pressings and the finished wines.

The receiving vessel, if glass, should be precooled to a temperature close to that of the fruit and CO_2 added in the form of 'dry ice' chips.

5.2 Initiating Fermentation

In the vast majority of cases, white wines for teaching, experimental and semi-commercial purposes will be prepared in a closed fermenter with an airlock. Be sure to choose a fermenter with a volume that will provide sufficient space above the level of the liquid for the frothing that may occur at the outset (ca 20% of depth of liquid) but not so large that there will be a risk of oxidation. It is important to maintain a low temperature and CO_2 cover at all times to preserve aromatic volatiles and to prevent oxidation.

Yeasts are prepared as for red wine but the temperature may be lower and the fermentation will be much slower to initiate. Yeast nutrition is, however, even more important and use of a proprietary yeast nutrient medium is recommended to minimize the risk of a sluggish or stuck fermentation (see Section 3.2). If the temperature is raised to promote the initiation of fermentation, it is critically important that it be monitored three or even four times daily and, once active fermentation is evident, that the temperature be lowered progressively to the target temperature (13–15° C) to achieve the target rate of sugar utilization (ca 0.5° Bé/day). An 'explosive' onset of fermentation is often the prelude to a stuck or sluggish ferment.

5.3 Monitoring Fermentation

Record and graph, at least twice daily, ferment temperature, ° Baumé and aroma (paying particular attention to sulfurous, H_2S aromas). When sampling, ensure that the sample is representative of the ferment as a whole. Regular monitoring is especially important during the early phase of fermentation to ensure that temperature is adjusted to maintain a managed ferment. Act early if there is evidence of sluggishness developing.

If signs of slowing of the rate of fermentation occur, then the temperature should be raised a little and other steps taken as necessary to promote completion (see Chapter 6.2). If necessary, check the microbiology by conducting yeast cell counts and culturing to check for the presence of contaminating microorganisms.

5.4 Racking, Fining and Cold-Stabilizing

In small-scale fermentations racking-off is usually accomplished with a racking plate or siphon. The fermentation lees are usually discarded. Commercially, the wine from the lees would be recovered by filtration or centrifugation but kept separate from the racked wine as it may be bitter or carry other flaws. However, in some wines, sparkling and Chardonnay, the wine is left on the lees for some months to encourage the development of yeast autolysis characters (Boulton et al., 1996; Jackson, 2008; Ribéreau-Gayon et al., 2006a).

Cold-stabilization is more important generally in white than in red wines because white wine is held and served chilled. Testing for stability against the formation of hazes and crystalline deposits due to excess tartrates is important. Also, as this process will usually cause a change in pH and palate balance, it should be done before final adjustments are made to acidity and pH prior to bottling.

Other palate adjustments are made at this stage, to reduce sulfurous aromas, bitterness and protein instability (Table 8.5). Tests should be conducted for each of these and appropriate additions made before chilling commences. Protection against oxidation should be maintained by a CO_2 cover and by ensuring that the fSO_2 is about 30 mg/L (total of about 60 mg/L). In commercial situations, acids other than tartaric may be used at this stage, including citric and ascorbic acids. Caution is needed with these additions as citric acid is fermentable and ascorbic acid has been shown to ultimately increase rather than reduce browning, in-bottle, in white wines (Peng et al., 1998).

A sequence for clarifying and stabilizing may be as follows: add bentonite at the selected concentration, mix and hold at room temperature for a few days before adding, e.g. casein at the selected concentration (50–100 mg/L) and hold for 2–3 weeks, before racking and cold-stabilizing by the addition of finely ground tartaric acid and chilling to the desired temperature. Alternatively, silica gel (30 g/100 mL) may be combined with gelatin and possibly isinglass in the range of 25–100 ml: 1.25–10 g: 1.5–3 g/hL (Bearzatto, 1986; Ribéreau-Gayon et al., 2006b). Add the proteins first, then the silica gel.

5.5 Filtering and Bottling

White wines are always sterile filtered before bottling, to achieve not only sterility and therefore microbial stability in the bottle, but also clarity and brightness. In commercial and perhaps teaching situations, the membrane used for this final filtration may be checked directly for viable bacteria and yeasts and placed on a culture plate as a quality assurance process (fluorescent vital dyes such as fluorescein isothiocyanate are particularly good but a

TABLE 8.5 White Wine: Common Fining and Stabilizing Agents and Their Characteristics

Agent	Range	Heat Stability	Comments
Isinglass	1–5 g/hL	Heat stable	Reacts with monomeric phenols, must use immediately; can use alone to enhance brilliance. Difficult to prepare and 'true' isinglass is rarely available due to source limitations (sturgeon fish). Can be used alone with white and even with red wines. Difficult to rack and probably best used with silica gel to reduce this problem
Casein and milk	5–20 g/hL	Heat stable (?)	Reported to refresh color and flavor and to protect again oxidation; milk has ca 30 g/L casein and 10–15 g/L other proteins which are not heat stable; use low-fat milk and note that use of milk is not legal in Europe (e.g. Ribéreau-Gayon et al., 2006b)
Blood albumen	10–20 g/hL	Not heat stable	A traditional fining agent. Fresh blood (bovine) is now proscribed for health and safety issues, and indeed many countries now ban the use of blood products (Ribéreau-Gayon et al., 2006b)
PVPP	10–20 g/hL		Reduction of bitterness, e.g. in pressings, use in preference to gelatin as removes less flavor (e.g. Margalit, 2004). Helps to reduce browning in white wine and remove discoloration and bitterness (Ribéreau-Gayon et al., 2006b)

Note: PVPP = polyvinylpolypropylene.

wide range of DNA/RNA and vital dyes is available that will suit the purpose—see specialized literature).

6. BENTONITE PROTEIN STABILITY TEST

6.1 Purpose

The primary purpose of this treatment is to remove heat-unstable proteins from wine and thus to prevent the formation of a haze of coagulated proteins in the bottle on storage. Secondarily, it may be used to assist in the removal of proteins added to remove bitter and excessively astringent phenolics, and proteins that may cause allergies, such as egg white or fish products.

6.2 Occupational Health and Safety

No particular issues.

6.3 Quality Assurance Records

- Person, date, supplier, batch, quantity.7

6.4 Sources of Error

1. Use of a tainted batch of bentonite.
2. Using different batches for the stability test and for the subsequent adjustment.
3. Adjusting pH after the settling the bentonite.

6.5 Procedure

- Prepare a stock solution of 5% w/v bentonite by heating ca 80 mL distilled water to ca 60° C and while stirring, slowly sprinkle 5 g of bentonite on the surface and mix it in thoroughly. Allow to cool, stand overnight, then bring to volume in a 100 mL measuring cylinder or volumetric flask (50 mg/mL). Repeat heating and mixing if necessary as it must be thoroughly dispersed.
- Pipette 50 mL of wine into each of seven 25 mm diameter test-tubes.
- Pipette 0, 0.5, 1.0, 1.5, 2.0, 2.5 and 3.0 mL of the stock into the tubes, mix thoroughly and allow to settle for at least 30 min.
- Centrifuge about 10 mL of each and then further clarify by filtration through a 0.45-μm filter into a narrow test-tube (10 mm).
- Place a marble on each and heat to 80° C for 6 h in a water bath or heating block. Allow to cool (ideally hold at 4° C overnight, 6 h) (see Sarmento et al., 2000, for other variants).
- Assess turbidity using a collimated light beam in the dark or (optionally) use a turbidity meter or a spectrophotometer at 700 nm (see Chapter 10.9).
- Repeat if necessary, using a smaller range.

6.6 Chemistry

Bentonite is a clay mineral based on montmorillonite, a swelling clay mineral, with a negative surface charge and a high surface area. It is available as either Ca^{2+} or Na^{+} forms, usually the latter. It is a colloid with a particle size of less than 1 μm and as such tends to form

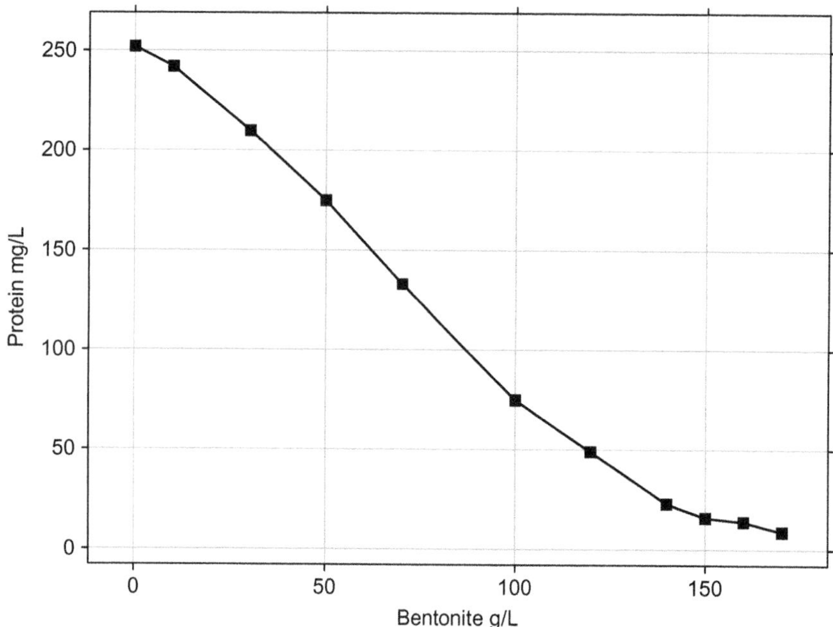

FIGURE 8.6 Plot of change in protein content, measured by Coomassie brilliant blue assay, of a wine following bentonite addition. *(Data from Boyes et al., 1997, with permission).*

stable suspensions. In the presence of a polycation, e.g. a protein, the particles aggregate (flocculate) to form larger particles that will settle under gravity, not only clarifying the solution but removing the protein (Figure 8.6) (Boyes et al., 1997). This is a complex phenomenon in a matrix as diverse as a wine. The efficiency of the process is subject to the presence of competing cations (e.g. K^+, Na^+, Ca^{2+} and H^+), so trial and error is the only way to determine the optimum quantity.

6.7 Notes

1. Wines contain many proteins and glycoproteins. Some of these are metabolic enzymes but many are so-called thaumatins and chitinases which have antifungal properties. On heating, proteins coagulate and form hazes. The heating test is to determine whether they are substantially removed (i.e. is a haze formed?).
2. Bentonite will also remove other components that bind to its surface such as complex tannins and anthocyanins, so its use is generally restricted to white wines. However, low concentrations are frequently used with red and rosé wines, especially red wines with unstable anthocyanins (low in tannins, e.g. Grenache).
3. Care needs to be taken to ensure that there are no off-flavors associated with particular batches of bentonite (it is, after all, a soil-forming mineral) and that the wine is not diminished in its flavor profile through the excessive use of this (re)fining agent.
4. Bentonite and most protein fining agents are more effective at low pH. Therefore, make any pH adjustments before conducting fining trials.

6.8 Equipment and Materials

- Wine (food)-grade bentonite (be sure to assess sensory qualities before use)
- Distilled water
- 25-mm diameter test-tubes (seven per wine) and racks
- 10-mm diameter test-tubes (seven per wine) and racks
- 100-mL measuring cylinder or volumetric flask
- Variable autopipette (0–10 mL) and tips
- 50-mL bulb pipette
- Bench centrifuge and tubes
- Disposable syringes and 0.45-μm filters
- Collimated light source and dark room
- Spectrophotometer and disposable cuvettes (optional)
- Water bath or heating block (80° C)
- Glass marbles (seven per wine).

7. COPPER FINING TRIAL

7.1 Purpose

Copper may be added to remove excess (stinky) H_2S or monomercaptans arising from metabolism of sulfur amino acids (methionine, cysteine, cystine), and attain a balanced aroma profile in a wine, especially for white wines.

7.2 Occupational Health and Safety

- Heavy metal toxicity, especially if using cadmium for diagnostic purposes.

7.3 Quality Assurance Records

- Person, date.

7.4 Procedure

- Prepare a stock solution of 2 g/L hydrated copper sulfate ($CuSO_4.5H_2O$) or 0.5 mg Cu^{2+}/mL.
- Measure an equal amount of wine into five screw-cap vials or jars (50–100 mL).
- Bring the wines to 0.0, 0.5, 1.0, 1.5 and 2.0 mg/L of Cu^{2+} by adding an appropriate volume of the stock solution (if 100 mL, then 0, 100, 200, 300, 400 μL of the stock solution).
- Mix well and then pour into a wine glass and assess the sensory impact of the additions.
- Repeat using a narrower range to locate the minimum amount of copper to add.

7.4.1 Option

- Reseal samples and set aside for 1 week, then retest.

7.5 Chemistry

Copper salts react with sulfides to form insoluble copper sulfites. A small excess of such a salt will rapidly remove all sulfites such as H_2S.

Cupric ions (Cu^{2+}) are highly soluble in aqueous solutions (e.g. 30 g $CuSO_4.5H_2O$ can be dissolved in 1 L of cold water, whereas only 3.0 μg of the cupric ion will dissolve from the copper sulfite (CuS). Copper will, however, also react with mercaptans (thiols, R-SH), some of which are important aroma volatiles whereas others are fault producing, causing rubbery, onion, cabbage, sewage-type aromas.

7.6 Notes

1. Gentle aeration may be sufficient to remove the more volatile components and therefore err on the side of caution.
2. Mercaptans are also important flavor compounds. This adds to the need to act cautiously when using copper: it is better to avoid the problem through careful monitoring of the ferment (for a discussion of the complexity see Ugliano et al., 2011). Copper addition has also been reported to remove the aroma associated with the critically important mercaptans in Chenin blanc and Colombar (Roland et al., 2011).
3. It is common, particularly in white wines which are fermented under strictly anaerobic conditions, for small amounts of sulfides to be produced (e.g. H_2S). This can become severe if yeasts are forced to use sulfur-containing amino acids as a nitrogen source, either from free amino acids in the juice or from hydrolyzed proteins. It is thus common to remove excess amounts of such compounds by treating the wine with mineral salts that form insoluble sulfides.
4. Copper in the form of copper sulfate is the most commonly used salt, although others such as cadmium and even mercury are used in laboratory tests (these are highly toxic heavy metals and must be treated with great caution). An alternative rarely used in practice is to immerse a brass solid that will react with the H_2S and form reduced CuS (black) on the surface.
5. Cadmium reacts preferentially with H_2S and can be used to determine whether this is the cause of the problem, whereas copper reacts with both H_2S and monomercaptans. Neither reacts with polymercaptans at the concentration used and these must first be converted to monomercaptans with ascorbic acid (Iland et al., 2004; Zoecklein, 1995). If this is the case then an additional fining with bentonite or activated charcoal may be required. Commercial test kits are available (e.g. http://www.pros.co.nz).
6. Copper citrate (usually on a bentonite base, 2 g/100 g) has been approved in a number of jurisdictions as an alternative to $CuSO_4$ with a number of benefits, notably a greater reactivity with sulfides (e.g. FSA, 2007). This article also discusses the formation of sulfide hazes (casse) and their treatment.
7. The recommended maximum level is 0.5 mg/L to avoid casse formation (this is also the legal limit in many countries, but greater than in some others).

7.7 Equipment and Materials

- 2 g/L hydrated copper (cupric) sulfate
- Five screw-cap jars or bottles per wine
- Sensory glasses, five per wine.

7.8 Storage

Indefinite at room temperature.

8. OTHER FINING TRIALS

8.1 Purpose

The targets of other fining trials are commonly phenolic compounds that produce a bitter or excessively astringent palate. These are either proteins or synthetic agents such as polyvinylpolypropylene (PVPP). Other fining processes include the use of charcoal to remove aroma taints and/or color, PVPP to remove color, and specialized processes to remove acetic acid or alcohol (high-pressure membrane filtration). These are beyond the scope of this text, and if the guidelines are followed then these more exotic processes should not be required.

Also, in practice, a winemaker will combine various fining, clarifying and stabilizing treatments and the trials given here may need to be modified to reflect that. Thus, the wine used as the base for a secondary addition may need to be that already treated by the selected primary agent (e.g. bentonite). The finer aspects of the chemistry and technology of fining are reviewed by Ribéreau-Gayon et al. (2006b).

8.2 Occupational Health and Safety

No particular issues, although individuals may be allergic to particular agents such as isinglass (fish product) or egg albumen.

8.3 Quality Assurance Records

- Person, date, acid (and source), amount per volume.

8.4 Procedure

- Prepare stock solutions as below.
- Dispense 100-mL volumes of wine into capped containers (note that there should be minimal ullage to avoid oxidation).
- Dispense and mix the appropriate stock solution volume to produce the desired concentration including a blank (Tables 8.4 and 8.5 provide guidelines).
- Stand until the sediment from the treatment has settled. This may take several days (alternatively, the wines may be filtered, but take care to avoid oxidation that may impair the sensory assessment).
- Carefully dispense aliquots into tasting glasses and conduct duo–trio tests against the control (the untreated wine).
- Conduct the sensory assessment and if necessary repeat the activity with a narrower range of agent.

8.5 Stock Solutions

8.5.1 *Egg Albumen (Fresh)*

Break eggs and separate the white (albumen) from the yolk. Filtering through a single layer of cheese cloth may assist in the separation process. Pour the albumen into a tared container (e.g. a large beaker). Prepare 10× the weight of 0.5 g/100 mL of NaCl (or phosphate buffer) and add to the beaker. Mix gently to disperse and denature the proteins. The final solution will contain 10 g albumen/100 mL (100 mg/mL). Thus, a series might be 0, 0.3, 0.4, 0.5 and 0.6 mL stock additions, each to one of five 100-mL wine samples.

8.5.2 *Casein*

Disperse 20 g of casein in 1 L of 0.1 M Na_2 or K_2 hydrogen phosphate buffer, pH 8.9. Stir continuously until a non-settling suspension has developed.

8.5.3 *Blood Albumen*

Do not prepare this agent freshly. Purchase a commercial preparation and follow the manufacturer's instructions if intending to trial this agent.

8.5.4 *Gelatin (Oenological Grade)*

Dissolve 10 g in 1 L of distilled, warm water (<40° C, 0.1 g/mL) or purchase a commercial preparation in liquid form. A series might be 0, 5, 6, 7, 8, 9, 10 g/hL or 0, 5–10 mg/100 mL or 0, 50 to 100 μL/100 mL. Note that it is best if this reagent is combined with a silicate (silica gel) to minimize the risk of overfining (e.g. 1 part gelatin to 5 parts silicate).

8.5.5 *Isinglass*

Disperse 5 g of finely chopped isinglass in 1 L of 10 g/L of citric acid in distilled water. This is difficult to prepare and it may need to be ground in a mortar and pestle or blended once moistened. Allow to soak overnight (see also Watts et al., 1981). A series might be 0, 10, 20, 30, 40, 50 mg/L, which is produced by adding 0, 0.2, 0.4, 0.6, 0.8 or 1 mL/100 mL.

Chapter 9

Principles of Analysis

Chapter Outline

1. PRIORITIES

All analyses, whether sensory, biological or physical–chemical, have common goals: to ensure quality for the consumer and profits for the producer. They are intended to assist the winemaker to make consistent and reliable judgments during wine production, judgments that minimize the risk of failure or suboptimal outcomes. Analyses aid learning. Systematic recording and analysis is a prerequisite for all continuous improvement and quality assurance schemes. The right analysis helps to resolve problems that inevitably occur. Analyses also ensure that legal and legislative requirements are met.

The definitive source of information regarding analytical methods is the *Compendium of International Methods of Wine and Must Analysis* [Organisation Internationale de la Vigne et du Vin (OIV), 2006a, 2006b]. This also contains a guide to quality assurance accreditation for wine laboratories, although its guide will need to be supplemented by organizations established for this purpose in particular member countries. Wines sold to the public or exported will usually need to be assessed for issues regarding food safety and assurance of origin and composition by local authorities (e.g. Standard 4.5.1 Wine Production Requirements, http://www.comlaw.gov.au). Also, local authorities will oversee such matters as label design and content and trademarks.

This being said, few wineries maintain more than a basic laboratory, and most defer the complex, sophisticated and legally binding assays to contract laboratories that can achieve considerable cost benefits through investment in technologies that are not considered here: infrared spectrometers that can be calibrated to measure many aspects of must and wine composition simultaneously, e.g. pH, color, alcohol and even sensory quality (Cozzolino et al., 2008, 2011), and microtiter plate reader spectrophotometers that can both analyze large numbers of samples quickly and minimize sample and reagent volumes (important if using enzymes and specific substrates). Nevertheless, the analyses selected here remain at the heart of wine chemistry, and are vital in calibrating and checking the newer methods which rely on calibration and consistency of the matrix (the presence of other substances in a solution under test).

2. QUALITY ASSURANCE

The very term 'quality assurance' implies a known relationship between the attribute being assessed or graded in terms of quality and the value of its rank, from unacceptable, to acceptable, desirable and even highly desirable, is known and can be assessed or measured, reliably and repeatedly. For some aspects of berry and wine composition these values are well known and readily available (e.g. Table 9.1), but for others, long experience and rigorous recording may be required to develop an in-house database of values that are a sound guide to quality in the experience of the potential consumer at some future date.

Most quality assurance activity is designed to minimize the risk of problems that may reduce wine quality below the potential in the fruit at harvest. Higher order quality assurance intended to guide the selection of fruit and the subsequent management of the fermentation and aging is often an in-house aspect guarded closely by the owner. In such circumstances there is no shortcut to experience to gain a sufficient history to guide the production of consistent super-premium wines. The analyses provided here will support the production of commercial wines, consistently. Only art and experience can take the wine to the next level; that is up to the winemaker—to you and your viticulturist or fruit supplier. It is for this reason that, generally, the winemaker determines viticultural policy.

Timeliness of an analysis often determines whether a problem can be resolved. This is especially the case during a fermentation where options to correct faults are highest at the early stage and may be irreparable if left too long.

TABLE 9.1 Maximum Acceptable Limits for Particular Substances

Substance	Limit	Comment
Sulfur dioxide	150 mg/L	Red wines
	200 mg/L	White and rosé wines
	300 mg/L	Still wines >4 mg/L reducing agents
	400 mg/L	Some sweet white wines
Sulfates	1 g/L	
(as K_2SO_4)	1.5–2.5	If stored in oak or sweetened
Methanol	400 mg/L	Red wines
	250 mg/L	White and rosé wines
Propylene glycol	150 mg/L	Still wines
Ethylene glycol	<10 mg/L	
Diethylene glycol	10 mg/L	
Propylene glycol	300 mg/L	Sparkling wines
Arsenic	0.2 mg/L	
Boron	80 mg/L	As H_3BO_3
Bromine	1 mg/L	Exception: sodic soils
Cadmium	0.01 mg/L	
Copper	0.5 mg/L	
Fluoride	1 mg/L	
Sodium	<60 mg/L	
Zinc	5 mg/L	
Ochrotoxin A	2 mg/L	
Citric acid	1 g/L	
Gluconic acid	<300 mg/L	
Volatile acidity	20 mg/L	
Malvidol diglucoside	15 mg/L	From *Vitis* hybrids

Note: This is not a definitive list and values may vary according to circumstance and particular regulations.
(OIV, 2006a).

2.1 Legal Requirements

Most nations have enacted laws to protect consumers from inept, greedy or antisocial providers of foods and beverages. There are many cases dating back to the phylloxera epidemic in Europe of attempts to extort wine consumers with artificial or adulterated wine, not merely bad wine—there is no law against that!

Use of juices from other fruits can be detected by analysis of the anthocyanins, sugars and or acids, as these are frequently diagnostic. Methanol has been a common adulterant in artificial wine and wines made from immature grapes have been sweetened with diethylene glycol to give them body and to balance the excessive acidity. These are highly toxic and life-threatening additives and it is important that consumers feel secure when consuming wine to a responsible level, that consumption is not life-threatening!

There is a temptation to add artificial sweeteners in regions where the laws forbid the addition of sugar (e.g. Australia) or acids (e.g. France; although climate warming is leading to dispensations). There is a requirement also for fortified sweet wines to contain a minimum of natural alcohol produced by fermentation of the must (4%), otherwise the 'wine' should be labeled a fortified must, not a wine. This too can be assessed by analysis of the products of fermentation.

Labeling laws also have been enacted to ensure that the wine described on the label is the wine in the bottle and that, in Australia and several other countries, it has a stated alcohol content. Sophisticated analyses of rare earth elements in wine or even near infra-red spectroscopy (NIR) can be used to assure wine origin and the method of production of sweet wines can also be assessed. Analysis can go well beyond quality assurance in the winery.

2.2 Hazard Analysis and Critical Control Point

The approach most commonly adopted to monitor and control quality (food and beverage safety) internationally is hazard analysis and critical control point (HACCP).

In many food industries there are legislative requirements to ensure that it is applied.

There are seven principles:

- Conduct a hazard analysis.
- Identify critical control points.
- Determine critical limits for each critical control point.
- Identify critical control point monitoring requirements.
- Establish corrective actions.
- Develop record-keeping procedures.
- Establish verification procedures.

These are seemingly self-evident in nature. However, along with identifying risks and responses, the purpose is to ensure that all materials can be traced to origin, that individuals responsible for each activity can be identified and that an independent audit can be conducted subsequently (if done under legislative or a national or internationally recognized standards, e.g. http://www.iso.org, ISO 9000, Quality Management Systems).

Therefore, this is not a trivial matter, but once completed and undertaken, risks to quality are surely minimized. Larger organizations frequently conduct such analysis to include all areas of risk. It is a good idea, even if not opting to adopt a formal process, to include an informal analysis to be aware and prepared to act to protect your product, your profitability and your reputation and, not least, the personal safety of yourself and your staff.

3. CONCEPTS

The OIV lists methods under four categories of chemical analysis:

I. Criterion benchmark method: The method used to determine the 'accepted' value, i.e. that by which all other methods are calibrated.
II. Benchmark method: Applies when Class I methods are not available. Is suitable for calibration and dispute resolution.
III. Approved method: Approved by the OIV and suitable for monitoring, inspection and regulatory purposes.
IV. Auxiliary method: Conventional or new, yet to be approved technique.

In addition, individual countries have agencies that will approve particular methodologies with respect to their individual regulatory requirements. In Australia this is the National Association of Testing Authorities (NATA), which oversees accreditation to the International Organization for Standardization/International Electrotechnical Commission (ISO/IEC) 17025 and ISO 15189 international standards. Such organizations usually provide guidelines for best practice maintenance of laboratory equipment and standardization (e.g. http://www.nata.asn.au). This organization works with Standards Australia and New Zealand (http://www.standards.org.au), which assesses and publishes standards governed by legislation and which covers such matters as laboratory and winery design. Wine composition is governed by Food Standards Australia New Zealand (http://www.foodstandards.gov.au). Equivalent organizations will be found in all countries and should be consulted before determining your own operational practices.

In all laboratories, it is good practice to write down a standard operating procedure for all equipment and for all analyses. Such a procedure should be complete, including not only the technical but also the safety and the regulatory issues. It should be part of the development of an HACCP approach to quality assurance.

The procedures provided here have been, wherever possible, checked against original references and against industry and national standards where appropriate.

3.1 Terminology for Solutes in Solution

Terms and abbreviations in this text conform to the National Institute of Science and Technology (NIST) reference on *Système international* (SI) unit rules and conventions (Thompson and Taylor, 2008), except that the unit for volume will be the liter (L) rather than the cubic meter (m^3) as per strict usage of the SI system and generally mass will be in grams (g) rather than kilograms (kg) (numerically, 1 kg/m^3 is equivalent to 1 g/L). In order to avoid ambiguity, we will always state the units rather than referring to percent or other units.

3.1.1 Chemical Names

Chemists abbreviate the names of inorganic substances in a manner that is consistent worldwide and defined by the International Union of Pure and Applied Chemistry (IUPAC), and which largely reflects the original names applied by their discoverer (usually German). See the Periodic Table of the elements (RSC, 2013).

3.1.2 Weights and Volumes

Weights and volumes practices in the general community are not rigorous and are open to misinterpretation. Thus, adding 20 g of sucrose to 1 L of water is not the same as making up a 20 g/L solution, although, for practical purposes, the difference will be minor (i.e. the sucrose addition will cause a small increase in volume). However, when considering analytical chemistry, it is important that rules regarding the preparation of solutions are strictly adhered to.

Chemists also use the convention of standard (laboratory) temperature and pressure (STP) because density and volume change with temperature and pressure (combined gas law). For all practical purposes we are concerned only with the temperature that can have a bearing on an assay. We regard 20° C (68° F) as 'normal' and adjust readings according to the measured temperature when necessary; however, this is not universal and some regard 15° C (60° F) or 25° C (77° F) as the standard—it is best just to state the conditions. Pressure is usually the mean atmospheric pressure (101.3 kPa) but, as a rule, this is not material in wine chemistry.

Solutions may be prepared on a weight for weight (mass for mass, m/m, kg/kg), weight for volume (kg/L) or volume for volume (L/L) basis.

Thus, a 20% (g/100 g) solution of sucrose would be prepared as 20 g of sucrose dissolved in 80 g of water (1 mL of water at 20° C weighs 0.99823 g or about 1.0 g).

A 20 g/L (g/L) solution of sucrose would be prepared by dissolving 20 g of sucrose in a volume somewhat less than 1 L, and then making the solution up to 1.00 L in a volumetric flask, which, strictly speaking, will have been calibrated for a temperature, and the careful analyst will measure the temperature of the solution adjusting for any temperature difference (usual only when preparing primary standards).

A 20% (ml/100 mL) aqueous ethanol solution would be prepared by adding 20 mL of pure ethanol to 80 mL of water.

When preparing solutions it may be important to correct for the actual composition, which for many reasons is frequently less than 100%. This may be due to the presence of water, fillers, stabilizers or impurities. Check the label.

Note: Never leave regent jars open for longer than is necessary to remove the quantity you require for the solution, because the substance may absorb water or carbon dioxide, or may oxidize. Never return excess quantities to a reagent jar.

3.1.3 *Stoichiometry and Equations*

Definitions

To further enhance the ability of an analyst to make valid comparisons, to compare 'apples with oranges', chemists use standard units of mass, concentration (mass per unit volume) and temperature (SI units). The terms 'molar' or 'molarity', 'equivalence' and 'normality' are redundant and are not used in this text. The terminology used is based on the structure of matter, of atoms and molecules and their molecular mass relative to the stable isotope of carbon (^{12}C), which is defined as having a relative atomic mass of exactly 12. The relative atomic mass of an atom varies from 1.008 (hydrogen) to 238 (uranium), and higher.

The weight of a molecule is the sum of the weights of the atoms that comprise the molecule. Thus, water has a molecular composition of HOH or H_2O, with the subscript indicating the number of copies of an element in the molecule, in this case two hydrogen atoms and one oxygen atom.

The **mole** (mol, *n*) is one of the seven base units in the SI system. It is defined as the amount of substance of a system that contains as many elementary entities (atoms, molecules) as there are atoms in 0.012 kg of carbon-12. Also, by definition, the number of molecules present in one mole of any substance is the same. This number is Avogadro's constant (6.022×10^{23}).

Molar (atomic) mass (*M*, formerly molecular weight, *A*, atomic weight) is numerically equivalent to the relative molar mass, which is defined as the ratio of the average mass (*m*, kg) of a substance, molecule or atom to 1/12th of the mass of an atom of the nuclide ^{12}C. Molar mass is formally defined as kg/mol but conventionally as g/mol. For example,

$$M(\mathrm{NaCl}) = A(\mathrm{Na}) + A(\mathrm{Cl}) = 22.990 + 35.453 = 58.443\ \mathrm{g/mol}$$

Also, weight (mass) can be converted to moles for the purposes of calculation. Thus,

$$100\ \mathrm{g\ NaCl} = 100/58.443 = 1.711\ \mathrm{mol}$$

Concentration (*c*) is the number of moles per unit volume of solution (conventionally, mol/L). A solution that contains 1 mole of a substance dissolved in 1 L is termed a 1 molar mass solution.

$$\mathrm{c} = M/V = \mathrm{mol/L}$$

Stoichiometry is a term for the proportions in which chemicals react, e.g. the neutralization of sodium hydroxide (NaOH) with hydrochloric acid (HCl) is 1:1, whereas that of NaOH with tartaric acid is 2:1.

$$\mathrm{NaOH} \leftrightarrow \mathrm{Na^+} + \mathrm{OH^-}$$

$$\mathrm{HCl} \leftrightarrow \mathrm{H^+} + \mathrm{Cl^-}$$

These ions react in solution to form sodium chloride (salt) and water (by addition):

$$\mathrm{Na^+} + \mathrm{OH^-} + \mathrm{H^+} + \mathrm{Cl^-} \leftrightarrow \mathrm{NaCl} + \mathrm{H_2O}$$

Note that when the equation is balanced, i.e. the number of NaOH molecules exactly equals the number of HCl molecules, the solution is neutral, neither acid nor basic. This is the equivalence or neutralization point. For strong acids and bases such as NaOH and HCl, the equivalence point is at neutral pH (pH 7); for weak acids and bases the equivalence point may be greater or less than pH 7, respectively.

In preparing chemical solutions we take account of this knowledge and prepare solutions of known molecular mass. Thus, in the example above, a solution containing 40 g of NaOH in 1 L of solution is said to be a molar mass solution. Such a solution will react exactly with a 1 mol/L solution of hydrochloric acid (36.5 g in 1 L of solution) and will neutralize it completely, producing a solution of salt, i.e. the acid will neutralize the 'caustic' base.

Standard international conventions apply and 1 mmol/L represents 1/1000th of a mole or 0.001 mol/L and 1 μmol/L is 1/1,000,000th of a mole or 0.000,001 mol/L.

One molecule of tartaric acid will react with two molecules of NaOH and thus a balancing solution of tartaric acid would contain half the molecular weight of tartaric acid dissolved in 1 L (75/2) (Figure 9.1).

In summary:

- Prepare g/g and mL/mL solutions by adding the components to produce either a percent (g/100 g, %) or per 1000 (g/1000 g %) solution or mixture.
- Prepare g/L solutions by bringing the predissolved solution to the final volume (usually 1 L or a proportion thereof).

3.2 Volumetric Apparatus

Volumetric apparatus in common use in laboratories include:

- **Volumetric flasks**, which have a bulbous base and a long neck with a single graduation indicating the volume contained when filled to this point, such that the lower point of the meniscus is level with the graduation. They are usually designed to be accurate at a particular temperature that is engraved on the flask (20 or 25° C). These flasks are used for preparing 'stock' and 'standard' solutions.
  - A stock solution is a concentrate of a solution that will be diluted before use.
  - A diluent is the solution used to dilute the stock solution, usually distilled water for most analyses used in wine chemistry. These flasks need to be kept clean in order to be accurate (e.g. films of oils will prevent the development of a normal meniscus). Overheating during oven-drying may cause deformation which will impair their precision (glass is, after all, a liquid!).
- **Bulb pipettes** come in a wide range of volumes and are commonly used in basic wine and juice analyses. They have a bulb in the middle and a graduation midway toward the 'mouth' end. It is generally unsafe to use the mouth to draw up liquids, but wine and juice may be safely mouth-pipetted. All other solutions must be pipetted using a pipetting aid or syringe.
- **Graduated pipettes** also come in a wide range of sizes and are commonly used to deliver intermediate volumes, whereas bulb pipettes deliver whole-number volumes only. They are not as precise as bulb pipettes. They also come as 'touch' or 'blow-out' types. Touch pipettes drain to the volume, whereas blow-out pipettes must be fully emptied to deliver the marked volume. The latter were in common use for 'rough' work whereas the former are used for more precise work.
- **Measuring cylinders** also come in a range of sizes and are used for rough work rather than for analysis. Again, the lower edge of the meniscus is the guide to the volume. A glass or plastic transfer pipette is commonly used to adjust the volume when using one of these.
- **Burettes** are similar to large graduated pipettes but with a tap at the base. They are commonly 50 mL in volume and available as open burettes in which the initial and final values must be read, or as autozero burettes which have a reservoir and are filled to zero volume before each use. These are by far the most

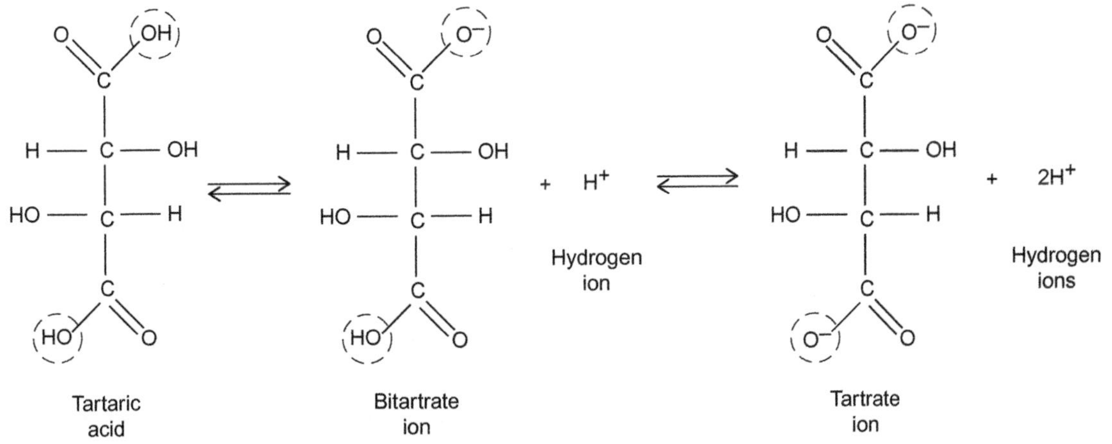

Tartaric acid ⇌ Bitartrate ion + H^+ (Hydrogen ion) ⇌ Tartrate ion + $2H^+$ (Hydrogen ions)

FIGURE 9.1 Drawings of a tartaric acid molecule showing the associated, partly dissociated and the fully dissociated forms (TA.H_2, $TA^{2-} + 2H^+$). The dashed circles show the acidic parts of the molecule.

convenient for routine work unless you can afford an autotitrator. A smaller volume burette (25 mL or less) is preferred for sulfur dioxide analyses.

- **Autopipettes** have largely replaced mouth pipettes for reasons of safety and precision. However, they are subject to abuse and must be used with good technique to work to specification. They are available as fixed or variable volume and from 10 to 10,000 μL. Each volume range is designed for specific tips which are color coded (yellow for P20 to 200, blue for P1000).

Rules for autopipettes:

- Never draw fluid into the barrel of the pipette itself.
- Never lay a pipette down while there is fluid in the tip. The fluid may accidentally find its way into the barrel.
- Never turn the adjustment scale below or above the full range settings.

To maximize precision, always use the smallest volume pipette and tube for a given total volume.

1. Set the desired volume. Turn the volume up just slightly past the desired setting, then down.
2. Attach a tip. Press it on firmly, with a slight twisting motion. The tip must make an airtight seal with the pipette barrel.
3. Depress the plunger to the first stop.
4. Insert the tip just below the surface of the liquid you want to transfer.
5. Release the plunger s l o w l y. (Releasing the plunger suddenly may cause the pipette to underfill.) Also, be aware that the liquid level in the tube will drop as you withdraw fluid. Ensure that it does not drop below the level of the tip or you will suck air.
6. As you withdraw the tip, touch it to the side wall of the tube to remove excess fluid from the exterior.
7. To dispense, depress the plunger slowly to the first stop; hesitate; then depress all the way. Never dispense a small volume into air. Always dispense into a liquid or on to the wall of a tube so that surface tension will draw the expelled liquid off the tip. Unfortunately, this may mean that the tip is contaminated and cannot be reused.
8. With the plunger still fully depressed, remove the tip from the liquid.
9. Consider what to do with the pipette tip. Should you discard it or use it again? If in doubt, throw it out.

3.3 Balances

Most analytical balances in use are electronic and robust. All require a vibration-free environment and leveling to ensure accurate and precise results. All should be kept clean and dust free. Better quality balances have built-in calibration facilities, which should be used regularly. At say 6-month or certainly annual intervals, a balance should be checked against a standard weight near the upper range of the instrument (e.g. 1.00 g or 0.100 g). Precision should be checked by repeated weighing of the standard, 10 times (these weights should only be handled with the tongs provided and never by hand). The standard deviation should be less than twice that which was supplied with the instrument. If a balance is faulty, it should be returned to a manufacturer's agent for service (see also NATA, 2010; this document provides tables as an aid to checking).

3.4 pH

pH is an inverse logarithmic scale designed to convert the concentration of hydrogen ions (acid, pH) or hydroxyl ions (alkali, pOH) to convenient numbers. The scale ranges from 0 to 14 and the actual concentration of H^+ ions varies from 10^0 to 10^{-14} moles per liter (1 to 0.000,000,000,000,01).

$$pH + pOH = 14$$

i.e. if pH is 3, then pOH is 11.

A wine at pH 3 has 10 times the concentration of H^+ ions as one at pH 4.

Note that this property applies to all ionizable compounds. Such compounds when dissolved in water have widely varying degrees of ionization and thus constants (i.e. the proportion that exists as a combined unit and that which exists as independent units).

3.4.1 pH Meters

- **Care of electrodes:** All manufacturers provide guidelines for safeguarding, cleaning and maintaining their electrodes (e.g. http://www.tps.com.au/handbooks/PH_ORP-INSTR_V3.2.pdf).

Electrodes consist of a fragile glass membrane separating the test solution from an electrolyte (silver chloride) and a silver electrode (simple type). Take care not to scratch it.

Condition a new or a dried electrode in pH 7 buffer or 4 mol/L KCl for 20 min. Rinse in distilled water. Dab dry with a lint-free paper. Check that there are no bubbles in the electrode bulb—if present, shake down as you would a medical thermometer. If it is a refillable type, check that there is sufficient reference solution in the electrode itself (actual solution depends on electrode type—check with the manufacturer). Loosen the refill cap when in use; close for storage. NEVER store in distilled or deionized water (this will ruin the electrode); use 4 mol/L KCl or a proprietary electrode storage solution.

If it is dirty, rinse in 0.1 mol/L HCl or detergent or dilute household bleach. Rinse thoroughly. Always calibrate before use, preferably with pH 4 and pH 7 buffer. Modern standards suggest the use of phosphate buffer pH 6.88 (20° C) rather than pH 7.0. The pH 6.88 buffer is available commercially and consists of equimolar (0.05 mol/L) monohydrogen/dihydrogen orthophosphate (e.g. 1.701 g KH_2PO_4 + 2.177 g K_2HPO_4 in 1 L).

A 10-fold dilution of this buffer can be used to test electrode performance. A pH of 0.2 ± 0.05 pH units higher indicates that the electrode is working normally. A lower result indicates that the electrode should be cleaned as per the manufacturer's instructions. If it still fails the test then replace it.

An appropriate choice of electrode is a double-junction electrode, which protects the electrolyte solutions from contamination by the wine (Howell and Vallesi, 2003).

3.5 Spectrophotometry

Definitions

- **Absorbance** is the proportion of light absorbed during passage through a path of particular length. It is measured in units of au (absorption units), usually defined by the path length of the cuvette (normally 10 mm). For precise work, absorbance may be calibrated using either neutral density filters or commercial reference solutions, but as long as standards are used this is generally immaterial.
- **Extinction** is, as for absorbance, used in the context of standardized values relating concentration to absorbance, hence extinction coefficient, ε (although absorbance is often used interchangeably).
- **Transmission** is the proportion (fraction) of light passing through a path of particular length (usually measured as a percentage).
- **Scatter** is light reflected from particulate material during passage through a path. Note that this is usually uniform, regardless of wavelength, but often estimated at 700 nm.

Many compounds absorb light of particular wavelengths and the absorbance varies directly with the concentration of the compound (at low concentrations). The method is non-destructive and quick, and can often be used to measure the concentration of particular compounds in complex mixtures, provided the wavelength at which the compound of interest absorbs light is unique. It is especially useful for phenolics, nucleic acids, proteins, anthocyanins and other compounds that absorb light in either the ultraviolet or visible range of wavelengths.

Modern spectrophotometers use ruled gratings and are usually robust and accurate with respect to wavelength. Calibration and checking of wavelength accuracy can be achieved using particular emission lines in the spectrum of mercury lamps (the usual light source in ultraviolet–visible light instruments). (Figures 9.2 and 9.3).

The Beer–Lambert law is used as the basis for the calculations, as follows:

$$I = I_0^{-\varepsilon c}$$

$$A = -\log[I/I_0] - \varepsilon x c$$

where I_0 is the initial light energy, I is the transmitted light energy (i.e. not absorbed or scattered), ε is the molar absorption coefficient of a 1 mol solution (g/L) of the chromophore under particular conditions, c is the concentration in moles, x is the path length (usually 10 mm) and A is the observed absorbance.

It is especially important that all surfaces are clean and that the solution is free of particulate matter and gas bubbles. Never touch the front or rear surface of a cuvette.

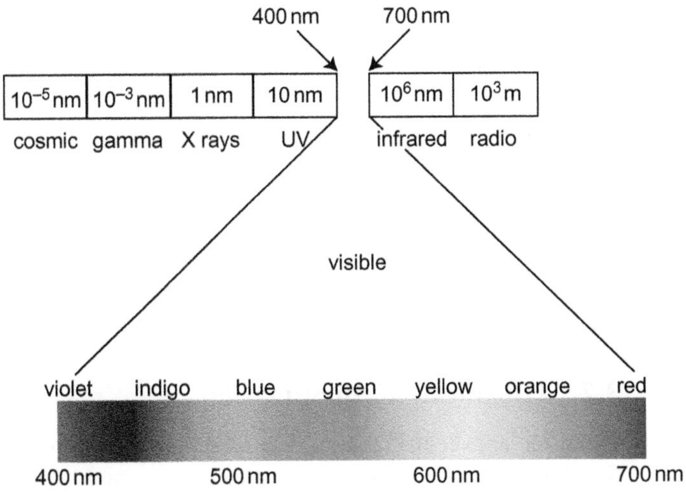

FIGURE 9.2 The spectrum of light energy arriving from the sun and the small part that is visible (400–700 nm).

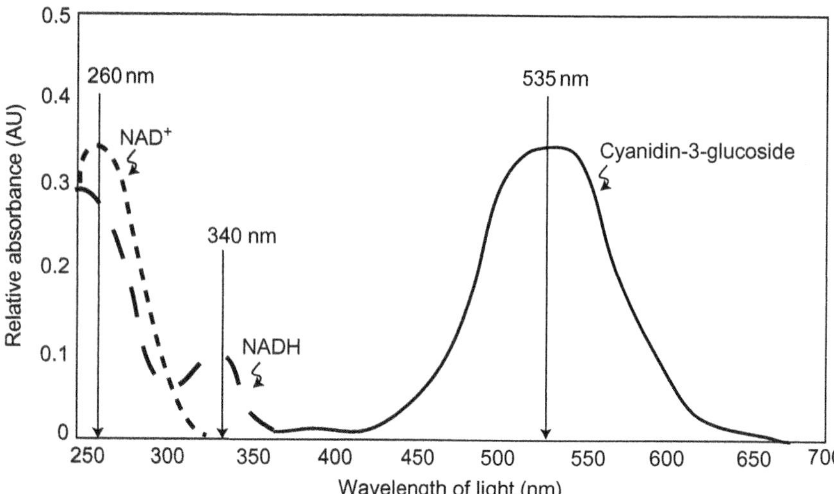

FIGURE 9.3 Absorption spectrum of three naturally occurring compounds. Cyanidin-3-glucoside is the principal anthocyanin found in red grapes, and NADH and NAD^+ are important in energy metabolism (and in some of the enzyme assays in this text). Note that at 340 nm only the NADH absorbs light (see Ammonium assay (Chapter 11.2) for an example).

Note that glass and most plastics absorb light in the ultraviolet range (<400 nm) and that either quartz (very expensive) or special plastics need to be used.

Modern spectrophotometers consist of a light source (mercury vapor for ultraviolet light and, typically, halogen for visible wavelengths). The light is split into its component spectra by means of a ruled grating (replacing the less precise glass or quartz prisms), collimated through a slit (to limit the range of wavelengths and to ensure that the light is traveling in a parallel band) and then passed through the cuvette and on to a photometer which measures the incoming light.

An important principle in spectrophotometry is that of always comparing the absorbance of the solution of interest with a blank that contains everything except the substance of interest. The method for 'Total phenols in white grape cultivars' provides a good example of this in that a specific absorbent, polyvinylpolypyrrolidone (PVPP), is used to remove most of the phenols so that the absorbance of other, otherwise interfering substances can be assessed and eliminated (Chapter 11.13).

In routine quality assurance tests, this principle may not always be strictly adhered to; provided the background absorbance is the same for all substances being compared, this may not be a serious issue. It is the ranking and experience rather than the absolute amounts that matter in that case. However, as you may find in the total phenol assay, that approach is not always satisfactory. This approach could be applied to the anthocyanin assay by using activated charcoal to absorb the anthocyanins and probably most of the phenols as well. If it did not, the PVPP could be applied as a secondary absorbent.

If you wish, you may assess some of these issues to check on the importance of the choice of correct reference or blank solutions. It is customary in the anthocyanin assay to use water as the reference solution, but 1M HCl and perhaps a volume of ethanol equivalent to that in the extract should be the minimal choice (Chapter 11.10).

3.6 Chromatography

Identification and analysis of the composition of natural products depend on the separation of the compound of interest from all other compounds within the very complex mixture that comprises a biological sample. Chemists use small differences in the properties of these compounds to separate one from another.

The principal properties used are differential solubility (between two non-miscible solvents), volatility (between a non-volatile solvent and a gas phase), reactivity (e.g. the reaction of a positive ion with a negative ion, such as in ion-exchange chromatography) and dimension (molecular sieving). In chromatography there is an immobile or stationary phase or component and a mobile phase. Passing a mobile phase with particular properties (e.g. a solvent or a gas) in which the compounds of interest are dissolved over a stationary phase with different properties will lead to those compounds that are, say, more soluble in the stationary phase than in the mobile, spending relatively more time in the stationary phase than in the mobile phase. These will travel more slowly than will compounds with a preference for the mobile phase.

This simple premise is the basis for thin-layer and paper chromatography, gas–liquid and high-pressure liquid chromatography.

3.7 Filtration–Centrifugation

Filtration and centrifugation are used to remove solids (insoluble particles) and to sterilize musts and wines. Particles of solid material above a particular size range will settle out under gravity, but the smaller the particle size the slower the rate of settling.

Of interest in wine making and analysis are coarse particles (fragments of skin, pulp and seeds), bacteria (>0.5 μm) and yeasts (>1 μm) and fine particles such as crystals of potassium bitartrate, denatured proteins and pectin gels.

Coarse particles (say >100 mm) may contain bitter and astringent substances as well as wild yeast, and extended contact with these will affect wine quality and risk the hygiene of the wine (i.e. wild ferments may occur because the solids react with sulfur dioxide and protect any yeasts embedded in the settled solids).

Fine particles (colloids) will affect the clarity of the wine and may contain wild bacteria and yeasts, thus affecting hygiene again but also interfering with any measurement based on light (they cause scattering of the light and give false readings). Very fine particles (<0.45 μm) will not settle unless they aggregate with a 'seed' particle or substance. These fine particles cause hazes in wine and affect spectrophotometric assays. They can be removed by the addition of fining agents such as bentonite (an aggregated clay; disaggregated clays will not settle) or denatured proteins such as egg white or fish protein that attract the particles, causing the haze to produce larger, aggregate particles that will settle or which can be filtered or centrifuged.

These relationships are apparent from the formal relationships indicated in the equations below, which apply in the general sense only to complex solutions and suspensions such as wine and must.

Relative centrifugal force (RCF) increases with the speed of rotation, squared, and directly with the diameter of the centrifuge:

$$\mathrm{RCF} = \frac{rv^2}{g}$$

$$v = 2.\pi.N/60 \text{ radian/s}$$

where r is the radius of the centrifuge (m), v is the velocity (radians/s), and g is the force of gravity (9.807 m/s^2). Thus, if you double the speed you reduce the settling time four-fold; halve the speed and you increase it four-fold.

The **settling rate (SR)** is often considered to be described by Stokes' law. This law relates the settling rate (v_s) to the radius of the particle, squared (r), the difference between the density of the particle and that of the fluid (ρ_p and ρ_f, respectively) and the viscosity of the fluid (μ, N s/m^2), and gravity (g, m/s^2; centrifugation multiplies this value):

$$v_s = \frac{2}{9} \times \frac{\rho_p - \rho_f}{\mu} \times g \times r^2 \text{ (m/s)}$$

Thus, settling increases with size of the particle (squared), gravity (or RCF) and the difference between the density of the particle and that of the suspending fluid.

4. PREPARATION OF GRAPE EXTRACTS

Great care must be taken when considering the preparation and storage of grape extracts because particular components are not distributed uniformly and some may not redissolve if frozen. This applies especially to potassium bitartrate, which is present in levels greater than are soluble in water (i.e. the juice is a super-saturated solution of potassium bitartrate). Presumably, tartrate is bound ionically to other components such as proteins, amino acids and tannins. This is a pH-dependent phenomenon: the lower the pH, the smaller the problem. It is a critical factor for winemakers wishing to produce cold-stable wines that do not form crusts of potassium bitartrate crystals once bottled (most important for white wines).

Thus, for fruit for red wines, which are extracted heavily during the fermentation process, a whole maceration of flesh and skin and seed may be appropriate. For white and rosé-style wines such a process would give a misleading result. For such fruit, it would be appropriate to extract only the flesh or pulp. For table grapes, the skin would be important, but any seeds that may be present must be excluded.

If intending to freeze grape samples prior to analysis, the extraction should be done before freezing to avoid the complications of mixing the materials from the various compartments in the berry. For organic acid, mineral ion and pH measurements, sample preparation and dilution should be done before freezing. Heating frozen samples (berries or juice) to 50° C for 1 h can bring the organic acids and salts back into solution. However, our experience has been that the values obtained may still differ significantly from the unfrozen values. Samples for organic acid or mineral analysis should be acidified or diluted to minimize the risk of error.

5. SAMPLING, DILUTIONS AND CALCULATIONS

The objective in all methods of analysis is to relate the value measured to the original material. Thus, it is essential to know the original values for berry number, weight and/or volume of wine, and then to keep track of the proportion taken for analysis. For some assays (e.g. ° Brix, pH) this is not a problem because the analysis is carried out on the whole sample, but for others (e.g. titratable acids, potassium, phenolics and organic acids) it is important.

5.1 Dilution Factor

The dilution factor is a dimensionless number that describes, unambiguously, the fraction of the sample in the final solution or mixture. It is equal to the volume of the original or stock solution used (V_0), divided by the total volume of working solution produced (V_1). It is the number that is used to determine the concentration or amount of a substance in the original sample, after an analysis that involves subsampling or dilutions.

The dilution factor also gives the relationship between solute concentration in the stock or original solution (c_0) and the working solution (c_1). A serial dilution series is illustrated along with the calculations in Figure 9.4.

5.2 Calculation Example: Titratable Acids

Desired result: grams of tartaric acid equivalent per liter of must (extract).

Sample volume of undiluted pressings or macerate or wine: 10 mL
Dilute with distilled or deionized water: 80 mL
Titration volume, 0.1 mol/L NaOH: 6.5 mL

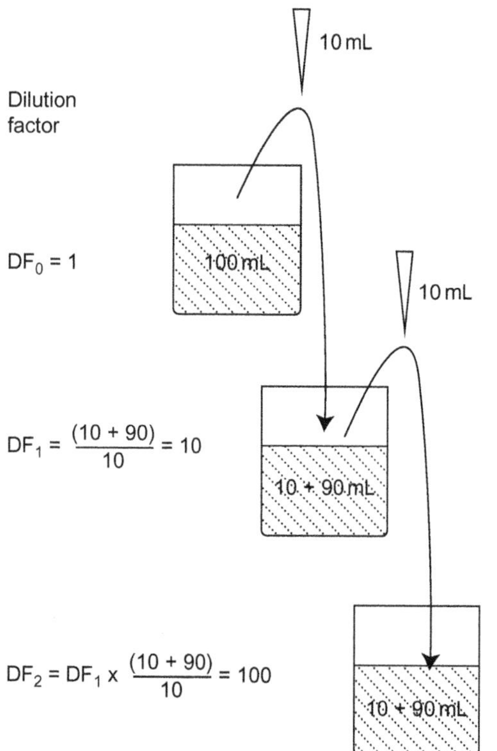

FIGURE 9.4 Diagram illustrating a 10-fold dilution series and calculations. The concentration of the substance in the final solution is 1/100th of that in the original.

Reactions:

$$\text{NaOH} \rightarrow \text{Na}^+ + \text{OH}^-$$

$$\text{Tartrate} - 2\text{H} \rightarrow \text{Tartrate}^{2-} + 2\text{H}^+$$

Therefore, one molecule of tartaric acid will react with two molecules of sodium hydroxide when the reaction is complete, i.e. when the Na^+ ions have swapped for the H^+ ions to form the disodium salt of tartaric acid (i.e. at about pH 8.2, not pH 7 because tartaric acid is a weak acid).

Calculation

$$2\text{Na}^+ + 2\text{OH}^- + \text{H}_2\text{Tartrate} \rightarrow \text{Na}_2\text{Tartrate} + 2\text{Na}^+ + 2\text{H}_2\text{O}$$

Note: This equation concerns equivalence in moles or molecules, whereas we want our answer in weight terms, grams. To convert one mole of tartaric acid to grams we need to multiply the answer by the molecular weight of tartaric acid, 150.

So the amount of tartaric acid in the 10 mL sample of our extract reacted with 6.5 mL of 0.1 mol/L NaOH.

Therefore, the moles of NaOH used (divide by 1000 because moles are defined as the molecular weight in grams dissolved in 1 L of solution and we have recorded our answer in mL—see Section 3.1.2, above):

$$= 6.5 \times 10/1000$$

and the moles of tartrate = half that amount because one molecule of tartaric acid reacts with two molecules of sodium hydroxide:

$$= 6.5 \times 10/1000 \times \{1/2\}$$

To convert this to weight we now need to multiply by the molecular weight of tartaric acid, 150:

$$= 6.5 \times 10/1000 \times \{1/2\} \times 150$$

Note that all values except for the volume of NaOH used in the titration are the same for all titrations, providing we use 10 mL and 0.1 mol/L NaOH. We can now simplify this equation:

$$= 6.5 \times 0.75$$
$$= 4.9 \text{ g/L}$$

Questions

Q1. We did not use the dilution volume (80 mL). Why not?
Q2. Can you write your own equation and factor now, using sulfuric acid (H_2SO_4) instead of tartaric acid?
Q3. Does it matter if the starting pH differs from one wine or grape to another?

Chapter 10

Essential Analyses

Chapter Outline

A Complete Guide to Quality in Small-Scale Wine Making.

1. ° BRIX

1.1 Purpose

° Brix is a calibration of the refractive index of a solution against dissolved sucrose (Table 10.1). The method used in the field is a hand-held analogue or digital instrument. These are commonly used to assess progress in fruit ripening in the field because the instruments are portable, reasonably robust and simple to use. The small sample volume, however, increases the risk of sampling errors unless samples are bulked before assessment or a thorough sampling procedure is adopted (see Sampling protocol, Section 1, Chapter 7). These instruments are not useful for fermentations. Precision laboratory instruments (Abbé) are available, are rarely used in viticulture but may be specified in 'standard' methods.

TABLE 10.1 Temperature Adjustment Values for °Brix

Temp. (°C)	Add to Value
10	−0.64
11	−0.58
12	−0.52
13	−0.46
14	−0.40
15	−0.34
16	−0.27
17	−0.21
18	−0.14
19	−0.07
20	0.00
21	+0.07
22	+0.15
23	+0.22
24	+0.30
25	+0.38
26	+0.45
27	+0.54
28	+0.62
29	+0.71
30	+0.79

Note: $adj = -1.1210 + 0.0407C + 0.0008C^2$.
More extensive correction charts are available (OIV, 2006a). See also Figure 10.2.

1.2 Occupational Health and Safety

No issues unless using an organic, primary refractive index standard for calibration, e.g. monobromonaphthalene or a preservative such as thymol is added to the standard sugar solution; if so, then check the material safety data sheet (MSDS) and handle accordingly.

1.3 Quality Assurance Records

- Person, date
- Instrument, standard(s), temperature.

1.4 Physics/Chemistry

The shadow boundary represents the critical angle at which light is reflected at the prism–liquid interface rather than being transmitted through the prism. This angle is dependent on the relative refractive index of the solution (or solid) and that of the prism, and on the wavelength of the incident light.

Low-precision refractometers use white light. Precision and electronic, digital, refractometers use monochromatic light (usually sodium D line 589 nm, yellow). In aqueous solutions, the refractive index is a function of the concentration of dissolved substances and their hydration clouds, i.e. it is related to water activity (a measure of the proportion of water that is bound more tightly to dissolved substances, ions and molecules, than to other water molecules). Thus, all dissolved substances contribute to the properties of the solution, not just those of sugars.

An infrared ° Brix meter will measure just the sugars and will give a lower value (correct) than a refractive index (Abbé)-type meter.

Analogue, telescopic, refractometers are rapidly being replaced with self-calibrating, temperature-compensating, digital instruments that are precise and robust. Care is required with these to avoid errors due to improper use,

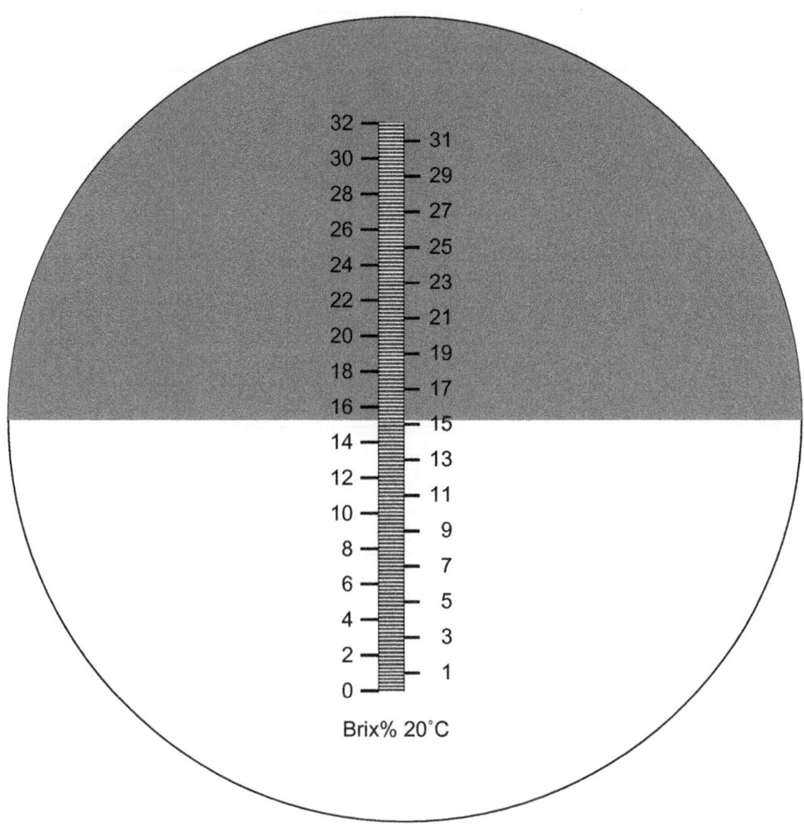

FIGURE 10.1 Image of graticule and shadow line for a conventional, hand-held, ° Brix analogue refractometer. The value is read at the interface between the shaded and the light regions; in this instance ca 15.3° Brix.

damage of surfaces, presence of gas bubbles and cloudy solutions.

1.5 Procedure

- Focus eyepiece diopter so that the graticule is sharply in focus (Figure 10.1).
- Check calibration with distilled water. The boundary between light and dark should intersect with the zero on the graticule; if not, set to zero (usually by turning a small knob on the side of the instrument or with a jeweler's screwdriver). Precision instruments will also enable the user to adjust the focus of the shadow–light boundary—check the manufacturer's instructions.
- Ensure that the solution fully covers the glass prism and lower the cover slowly to avoid air bubbles.
- Check calibration with an 18% (w/w) sucrose solution (18° Brix) or with a certified standard (preferably two standards to check linearity).
- Adjust the resulting value according to air (instrument) temperature, unless the instrument is temperature compensating (Table 10.2 and Figure 10.2). For precise work see Organisation Internationale de la Vigne et du Vin (OIV, 2006a).
- When finished, wash the prism surface with distilled/deionized water, dry carefully with an optical grade cloth and store in a dry environment.

1.6 Sources of Error

1. Failure to correct for temperature: note that it is the temperature of the instrument, not that of the original solution, that is important. The temperature of the thin film will quickly adjust to that of the considerably larger thermal mass of the instrument.
2. Failure to focus the shadow line and the graticule.
3. Careless closure of the chamber lid, trapping many air bubbles.
4. Excessive suspended particulate matter (turbidity causes light scattering, whereas dissolved substances do not, so color is not a cause of error).
5. Failure to check calibration against a standard.

1.7 Notes

1. The instruments generally used for fieldwork have a scale graduated at 0.2 g/100 g intervals and do not meet national standards criteria, which usually specify a readability of 0.1 g/100 g, similar repeatability (±0.1 g/100 g) and temperature compensation. Some

TABLE 10.2 Conversion Table for the Common Measures of Maturity in Fruit Musts

° Brix (Sucrose g/100 g)	Sucrose (g/L)	Refractive Index (20°C)	Density g/cm^3 (20°C)	Oechslé[a]	KMW[b]	Baumé[c]	Potential Alcohol (mL/100 mL) cf = 16.83 g %
14.0	125.1	1.35407	1.0557	55.7	10.9	7.7	7.43
15.0	136.0	1.35567	1.0599	59.9	12.0	8.2	8.08
16.0	147.0	1.35728	1.0642	64.2	13.1	8.7	8.73
17.0	158.1	1.35890	1.0685	68.5	14.1	9.3	9.39
18.0	169.3	1.36053	1.0729	72.9	15.2	9.9	10.06
19.0	180.5	1.36217	1.0773	77.3	16.3	10.4	10.72
20.0	191.9	1.36383	1.0817	81.7	17.4	11.0	11.40
21.0	203.3	1.36550	1.0862	86.2	18.6	11.6	12.08
22.0	214.8	1.36719	1.0906	90.6	19.7	12.2	12.76
23.0	226.4	1.36888	1.0952	95.2	20.8	12.6	13.45
24.0	238.2	1.37059	1.0998	99.8	22.0	13.2	14.15
25.0	249.7	1.37232	1.1049	104.9	23.2	13.8	14.84
26.0	261.1	1.37405	1.1095	109.5	24.4	14.3	15.51
27.0	273.2	1.37580	1.1144	114.4	25.6	14.9	16.23
28.0	284.6	1.37757	1.1190	119.0	26.8	15.4	16.91
29.0	296.7	1.37935	1.1239	123.9	28.0	16.0	17.63
30.0	308.8	1.38114	1.1288	128.8	29.2	16.5	18.35
31.0	320.8	1.38294	1.1336	133.6	30.4	17.1	19.06
32.0	332.9	1.38476	1.1385	138.5	31.6	17.6	19.78
33.0	345.7	1.38660	1.1437	143.7	32.9	18.2	20.54
34.0	357.7	1.38845	1.1486	148.6	34.2	18.8	21.25

Notes: A conversion factor of 16.83 g of sugar/L for each mL/100 mL of alcohol was used in this table (EU regulation) but it may in practice vary from ca 16 to 19.

[a]° *Oechslé:* = $(d_{20°C} - 1) \times 1000$.

[b]*KMW = Klosterneuberg must weight; similar to Brix but with an adjustment for non-sugar influences on final alcohol.*

[c]° *Baumé (20° C)* = $145 - 145/d^{20/}{}_{20°C}$ *or* $= 0.0181 \times 0.5532$° *Brix (Boulton et al. 1996).*

(Adapted from Blouin and Guimberteau, 2000, with permission).

instruments have more than one scale, usually Brix and/or alcohol ('potential', ° Baumé).

2. Rinse the prism with a little of the new solution before adding a sample for recording.
3. ° Brix is a calibration of refractive index of sugar solutions against the weight of sucrose as a percentage of the solution weight (e.g. 20 g of sucrose plus 80 g of water is a 20° Brix solution). Grapes contain mainly glucose and fructose with tiny quantities of sucrose. The refractive indices of glucose and fructose differ a little from that of sucrose and this difference is not included in the corrections (but is trivial in this context).
4. The relationship between ° Brix and alcohol yield varies with cultivar, year and location (Jones and Ough, 1985); but see also discussion in Boulton et al. (1996) and Figure 10.3.
5. The refractive index of water at 20° C is 1.3330. That of a 20° Brix solution is 1.3639, a difference of only 0.0309.
6. The precision of the measurement depends on a clean and scratch-free glass surface and a clear, non-turbid

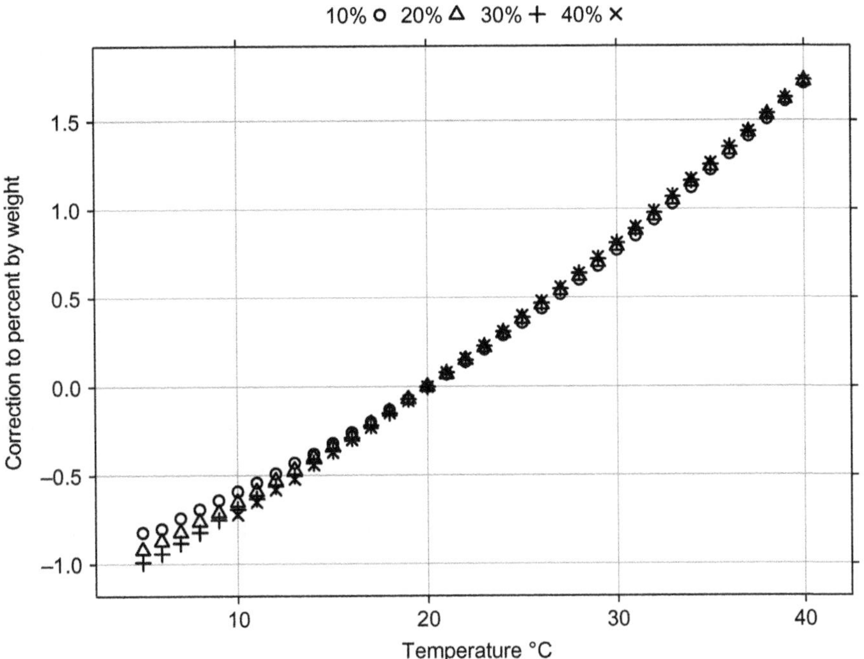

FIGURE 10.2 Influence of temperature and sugar concentration (%) on the correction factor required for ° Brix estimation. Note that the correction is greatest at temperatures below about 15° C: $cf = -1.382 + 0.0740T - 0.00349pc + 0.000104T.pc$, R^2: 0.995, where cf is the correction factor; T is the temperature (° C), and pc is the observed ° Brix value (g/100 g). *(Equation calculated from data presented in OIV, 2006a).*

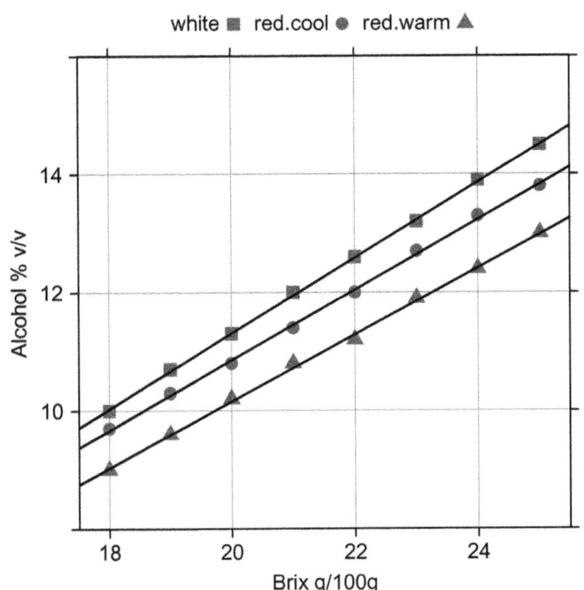

FIGURE 10.3 Effect of climate and cultivar on final alcohol content at particular ° Brix. $y = -1.50 + 0.64x$ (white), $y = -1.05 + 0.60x$ (red.cool), $y = -1.14 + 0.57x$ (red.warm); white: white cultivars, red. cool: cool climate red cultivars; red.warm: warm climate red cultivars. Note that the primary difference lies in the intercept. *(Data from Jones and Ough, 1985; Ough, 1992).*

solution. Turbidity produces a blurred, often colored, boundary between the shadow and the light in the image projected on to the graticule visible in the eyepiece. If necessary, centrifuge or filter the solution before reading. This is not a problem with optical fiber-based meters as they read at the glass/liquid interface only.

7. Take care not to scratch the glass prism with coarse paper toweling or tissue paper—rinse it under a tap and dab dry, preferably with a fine cotton cloth (optical quality).
8. The ° Brix value is sometimes referred to as representing the total soluble solids (TSS) because compounds other than sugars contribute (Figure 10.3). Sugars, however, dominate, comprising 90–94% of TSS.
9. Electronic refractive index meters are readily available. These increase reliability and reduce person-to-person variation. These are usually temperature compensating. However, they must be treated as a fine camera lens: the surface should not be wiped with abrasive cloths (this includes personal tissues and wipes!). They should be calibrated at least annually.

1.8 Equipment and Materials

- Hand or bench refractometer (analogue or digital) reading to 1 decimal place if graduated in ° Brix, or 4 decimal places if in refractive index (OIV, 2006a).
- Air thermometer
- Distilled water
- 18% w/w sucrose solution (18.0 g sucrose dissolved in 82.0 g distilled water at 20°C)

TABLE 10.3 Temperature Adjustment Values for Baumé

Temp. (°C)	Add to Baumé
15	−0.25
16	−0.20
17	−0.15
18	−0.10
19	−0.05
20	0.00
21	+0.05
22	+0.10
23	+0.15
24	+0.20
25	+0.25

Note: $adj = -1.00 + 0.05°C$.

- The temperature correction table is for a 20° Brix solution (not required for temperature compensated instruments).
- Optional: a certified standard, sucrose or organic.

1.8.1 Storage

Prepare standard freshly, or refrigerate for up to 6 weeks with the addition of a crystal of thymol as an antiseptic (take care as thymol is toxic—check chemical safety data sheet).

1.9 Benchmark Values

- 18–25° Brix depending on wine style, giving an alcohol content of between 11% and 14% depending on yeast efficiency and level of non-fermentable solids
- Dry white wine: 18.0–22.5° Brix
- Fruity white wine: 21.5–24.5° Brix
- Dry red wine: 18.0–23° Brix
- Full-bodied red wine: 23.0–25.0° Brix.

2. BAUMÉ

2.1 Purpose

A ° Baumé hydrometer is used to estimate fruit maturity in the laboratory and to follow the course of fermentation. An alternative is to use specific gravity (e.g. a portable density meter [pycnometer], DMA 35 Anton Paar GmbH). The hydrometer is an informal instrument with no legal standing. The formal method is based on specific gravity using either a graded series of precision hydrometers or a pycnometer (OIV, 2006a). It was devised to represent an estimate of alcohol potential (Table 10.3) but is at best an approximation (Figure 10.3) (see also Blouin and Guimberteau, 2000). Most commonly it is used to follow the progress of fermentation, since density declines as sugar is metabolized and alcohol content rises.

2.2 Occupational Health and Safety

No particular safety issues are associated with this technique, although it should be noted that hydrometers are fragile instruments made of glass. The instrument must not include mercury as ballast. Likewise, the thermometer must be food safe (alcohol type or electronic), not mercury.

2.3 Quality Assurance Records

- Person, date
- Instrument, reference calibration, calibrating solution(s).

2.4 Physics/Chemistry

° Baumé is a measure of the specific gravity of a solution (Boulton et al., 1996) (originally of salt, mass per unit volume; see Table 10.3). Specific gravity is the formal method. ° Baumé is a convenient approximation used to convert specific gravity to potential alcohol. Dissolving sugar in water increases the specific gravity of that solution, while mixing water and alcohol reduces specific gravity.

Specific gravity is measured using a hydrometer that floats higher the greater the specific gravity of the solution. The graph illustrates the relationship between ° Baumé and potential alcohol (assumes 16.83 g of sugar for every 1% of alcohol). An 18° Brix (=18% w/w) solution is approximately equivalent to a 10° Baumé solution and may be used for checking the calibration of both instruments. For a discussion of calibration see Wright et al. (2008).

2.5 Procedure

- Check accuracy of the hydrometer with distilled water (0) and 18% sucrose (m/m)—10° Bé or certified standard.
- Draw off or crush and sieve about 250 mL of juice or must and pour into a tall flask (measuring cylinder). Allow any froth to subside and read off the value on the scale at the bottom of the meniscus where the solution meets the upright tube. If necessary, twirl to remove attached air bubbles. Allow to settle for at least 1 min.

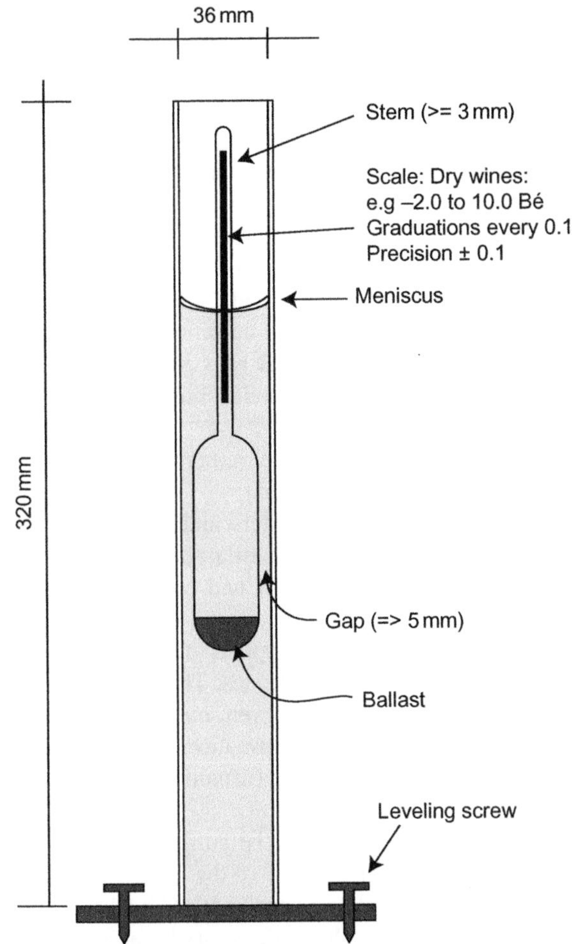

FIGURE 10.4 Illustration of a standard hydrometry/density apparatus. Commonly a 250-mL measuring cylinder is substituted for the standard cylinder.

- Read off the value at the top of the meniscus (Figure 10.4).
- Measure temperature (allow 1 min to equilibrate) and adjust recorded value if necessary (Table 10.3).

2.6 Sources of Error

1. Use of uncertified hydrometer or thermometer.
2. Changing hydrometers or thermometers with different calibrations.
3. Failure to allow time for thermometer to equilibrate with solution.
4. Air bubbles attached to hydrometer.
5. Vessel too small or not held vertical.
6. Unrepresentative sampling of fruit or ferment.
7. Other dissolved substances, even sulfur dioxide (SO_2), can affect the apparent value (Ough, 1992) and careful analyses correct for [SO_2] (OIV, 2006a).

2.7 Notes

1. This technique is useful only as a relative measure for individual musts and ferments (see Figure 10.3).
2. Portable electronic instruments have replaced this technique in research laboratories and large wineries (e.g. Anton Paar® density meter, pycnometer).

2.8 Equipment and Materials

- Baumé hydrometer (− 2° to 15°, non-mercury ballast, calibrated at 20° C) or preferably two, one for juice/must and one for wine nearing the end of ferment (e.g. one −2 to 10° Bé with 0.1° Bé scale graduations, and one 10–20 with 0.1° Bé scale graduations)
- 250-mL measuring cylinder or equivalent: preferably a simple tube, 320 mm × 36 mm (OIV specification) or according to hydrometer dimensions, requires a 5-mm gap
- Kitchen sieve
- Flask for collecting sample
- Thermometer, food safe (e.g. alcohol, not mercury)
- Means of crushing fruit (e.g. cone press, if appropriate)
- Certified reference hydrometer.

2.8.1 Storage

Keep the certified reference hydrometer apart from routine instruments.

2.9 Benchmark Values

- Full-bodied table wine: 13–14.5° Bé
- Examples of minimum values: Pinot noir 10.0° Bé; Merlot 13.5° Bé, Cabernet Sauvignon 13.5° Bé
- End of ferment (dry) $< -1°$ Bé.

3. pH

3.1 Purpose

pH is a key element in managing the wine-making process. It affects SO_2 activity, the growth of yeasts and lactic acid bacteria, sensory aspects of the wine including the release of aromas from precursors, the color of red wine and risk of growth of spoilage organisms.

3.2 Occupational Health and Safety

No particular issues aside from general laboratory safety.

3.3 Quality Assurance Records

- Person, date
- Electrode type, instrument, calibration standards
- Sample.

3.4 Chemistry

See Chapter 9.3.4 for further background information.

3.5 Procedure

- Measure temperature of sample and ensure that the pH standards are near to that temperature. If using an older type pH meter, set the temperature setting to that of the solution.
- Standardize the pH meter at pH 7, then pH 4, and check again at pH 7.

  ■ Take a representative sample of fruit or must or wine and place in a beaker (enough to cover the bulb and junction incorporated in the electrode).
- Read and record the pH, adjust for temperature if necessary. Repeat the measurement, average, and then record to two decimal places.
- Rinse electrode with distilled or deionized water.
- Continue with additional samples from ■.

3.6 Sources of Error

1. Use of a damaged electrode or uncalibrated pH meter.
2. Faulty meter or temperature calibration.
3. Failure to allow solution to come to room temperature (RT).
4. Inappropriate use of apparatus (failing to vent a vented electrode, failing to maintain the electrolytes in the electrode, failing to store the electrode correctly or to activate it appropriately before use).
5. Presence of dissolved carbon dioxide (CO_2) in the diluent or sample: boil or evacuate under a gentle vacuum to remove excess CO_2 in a fermenting must or wine. Not necessary for measurements on juice.

3.7 Notes

1. It is important to calibrate pH meters each time they are used with standards close to the pH of interest. A saturated potassium hydrogen tartrate (KHT) is often used because it is in the mid-range of values found in wine. The two common buffers B and C are more appropriate if also titrating for total acid content. The OIV method uses buffers A and C, then B as a check (OIV, 2006a). Discard after use. Commercial buffer standards are acceptable.
2. If using a vented electrode remember to uncap the vent and recap when finished.
3. The temperature of the buffers and the wine or juice being analyzed should be similar. Modern pH meters include a temperature sensor and correction circuitry. Older pH meters usually possess a dial adjustment or a correction factor can be applied (Table 10.4).
4. Take good care of pH electrodes. They are constructed from very thin glass and even minor scratches will seriously affect their performance. Maintain as per manufacturer's instructions (usually in 4 mol/L KCl, NOT distilled water).
5. The sample volume should be sufficient to cover the bulb and junction at the base of the glass electrode.
6. Frozen samples must be heated to 50° C for 1 h. Cover to prevent evaporation. This is necessary because potassium bitartrate (KHC_4O_6) is present in fruit at supersaturating levels and will not redissolve readily; dilution will also help but may alter the pH slightly.

TABLE 10.4 Temperature Correction Factors for Particular Buffer Solutions

°C	Saturated KHT	0.05 Molal Phthalate (mol/kg)	0.025M KH_2PO_4 + 0.025M Na_2HPO_4	0.01 Molal Borax (mol/kg)
	A	B	C	D
0		4.006	6.982	9.463
5		4.001	6.949	9.395
10		3.999	6.921	9.333
15		3.999	6.898	9.277
20		4.001	6.878	9.226
25	3.557	4.006	6.863	9.180
30	3.552	4.012	6.851	9.139
35	3.549	4.021	6.842	9.102

Note: KHT = potassium hydrogen tartrate.
(Data from Wu et al., 1988, with permission).

3.8 Equipment and Materials

- pH meter
- pH standards (20° C):
  - A—pH 3.57: 3.56: saturated KHT (ca 16 g/L, prepare by adding an excess weight of KHT, heating and then allowing to cool to RT) (plus 0.1 g thymol per 200 mL as a preservative; TAKE CARE)
  - B—pH 4.01: 0.050 mol/L potassium hydrogen phthalate buffer
  - C—pH 6.88: equal parts of 0.025 mol/L potassium dihydrogen phosphate and 0.025 mol/L disodium hydrogen phosphate
  - D—pH 9.81: 0.01M borax (optional)
- 10.0 mL pipette
- Pipetting aid
- Magnetic stirrer and stirrer bar.

3.8.1 Storage

- Electrode in 4M KCl solution
- pH standards, RT, shelf-life is ca 2 months.

3.9 Benchmark Values

- White table wine: pH 3.2−3.3 at start of ferment; pH 3.3−3.4 at bottling
- Red table wine: pH 3.3−3.5 at start of ferment; pH 3.4−3.6 at bottling.

4. TITRATABLE ACIDITY

4.1 Purpose

Titratable acidity (TA) is an important measure of the buffering capacity of a must or wine, i.e. of its ability to maintain pH when acids or alkali are added or removed (e.g. metabolized). It is used as a rough estimate of the amount of malic acid remaining at harvest; the higher the value, the higher the malic acid content.

4.2 Occupational Health and Safety

- Alkali
- Vacuum (risk of implosion)
- Fire or burns if using heat to degas
- If using phenolthalein as an endpoint indicator dye, note that it is a potent laxative!

4.3 Quality Assurance Records

- Person, date
- Standardized values for NaOH against an HCl primary standard.

4.4 Chemistry

Grapes and their products contain two principal organic acids: tartaric (HOOC.CHOH.CHOH.COOH, 150 g/mol, pK_a 3.01, 4.05) and malic (HOOC.CHOH.CHH.COOH, 143 g/mol, pK_a 4.36, 5.05), and many minor acids, e.g. citric. We generally measure the tartaric acid equivalents in grams per liter, although some countries use sulfuric acid (98 g/mol, France) or % (g/100 mL, USA).

4.5 Procedure

- Calibrate the pH meter using the pH 6.88 and pH 4 standards as for the pH determination.
- Add about 80 mL of distilled water to a clean beaker. Check its pH and adjust (only if necessary) to about pH 8.2 with dilute sodium hydroxide.
- Degas a sample of wine by vacuum or heating (fresh juice need not be degassed). Either place in a stoppered Büchner flask under vacuum and shake gently for 3 min (until frothing ceases) or bring to the boil for a few seconds and then cool before adding to the 80 mL of water.
- Pipette in 10.0 mL of juice or must or wine (filtered or centrifuged to clarify and reduce contaminating cell wall fragments).
- Fill burette with 0.100 mol/L NaOH and take a zero reading. Open the burette tap and allow a steady flow until the reading is about pH 5. Reduce the rate of flow to drop-wise as the pH approaches 8.2 (or 7.0, Europe; Figure 10.5). A drop is about 0.025 mL. Smaller additions can be made by touching the tip of the burette with a clean glass rod and mixing that into the solution, but this is rarely necessary.
- Calculate tartaric acid equivalents (g/L) by multiplying the volume (mL) by 0.75; or by 0.49 to give g H_2SO_4/L equivalent [substitute 98 for 150 as molar mass values in the equation; or by 0.075 to give TA g/100 mL (USA)]—the equations below assume moles/L in numerator and denominator which cancel out (1000/1000 = 1, but in the USA use 100/1000 = 0.1):

$$\begin{aligned} TA &= \left(\frac{M(\text{Tartaric acid})}{2}\right) \times c_{\text{NaOH}} \times \frac{v_{\text{NaOH}}}{v_{\text{sample}}} \\ &= \frac{150}{2} \times 0.1 \times \frac{v_{\text{NaOH}}}{10.0} \\ &= 0.75 \times v_{\text{NaOH}} \text{ g/L} \end{aligned}$$

4.6 Sources of Error

1. As for pH.
2. CO_2 dissolved in the wine or in water used to prepare solutions (boil or evacuate to remove).

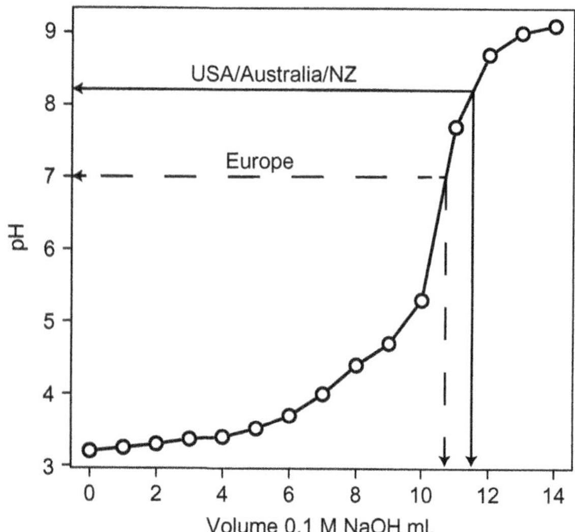

FIGURE 10.5 Generalized titration curve for fresh grape juice. Note the difference between the USA and Australian endpoint definition of pH 8.2 and the European of pH 7.0.

3. Failure to allow solution to come to RT.
4. Failure to store the NaOH in a sealed container or in one protected from CO_2 entry with a Carbosorb® airlock.
5. Failure to prepare an endpoint standard if using an indicator dye.

4.7 Notes

It has been common practice to use an indicator instead of a pH meter. This is not satisfactory for unclarified solutions and is difficult for colored solutions. If using an indicator solution prepare a standard for comparison purposes.

OIV recommend bromothymol blue with an endpoint of pH 7.0 and the preparation of an endpoint standard using sample, indicator and a small volume of pH 7.0 buffer solution to stabilize it. However, the standard practice varies between countries and it is usual in Australia and the USA to report a pH 8.2 endpoint (phenolthalein).

1. In unclarified solutions, suspended galacturonic acids (pectins) may contribute to the acid content and give a misleadingly high value.
2. If working with small samples, use a smaller volume of juice and a lower concentration of NaOH, e.g. 0.010 mol/L, to retain accuracy—a higher concentration will lead to small volumes and thus lower precision; alternatively use a small-bore (precision) burette.

4.8 Equipment and Materials

- 50 mL burettes and stands
- Optional: autotitrator
- Distilled water
- 100-mL beakers
- pH meter and double-junction electrode
- Electrolyte for electrode if refillable type (check manufacturer's specifications)
- pH standards (4 and 7)
- Options: (a) Phenolphthalein, endpoint pH 8.2 or (b) bromothymol blue, endpoint pH 7.0, indicator solution (OIV, 2006a).
    - **a.** Prepare phenolphthalein by dissolving 0.5 g of substance in 100 mL of 1:1 aqueous ethanol. Store in a stoppered flask/bottle at RT.
    - **b.** Prepare bromothymol blue by dissolving 2 g of dye in 100 mL of technical grade ethanol (96% v/v), then add 100 mL of CO_2-free distilled/deionized water. Adjust to pH 7 with dilute NaOH, then take a 7.5-mL sample and make to 1 L with CO_2-free distilled/deionized water. A color standard is prepared by taking 10 mL of CO_2-free wine/juice/must, adding 1 mL indicator, add dilute NaOH until color is blue–green, then adding 5 mL of pH 7 phosphate buffer to stabilize.
- Magnetic stirrers and Teflon bars
- 0.100 mol/L NaOH (freshly prepared or stored with a soda lime filled vent to absorb CO_2); 0.33 mol/L NaOH if using an autotitrator
- Stoppered 100-mL Büchner flasks (for wine)
- Vacuum (preferably) or heat source (for wine).

5. SULFUR DIOXIDE BY ASPIRATION

5.1 Purpose

Management of SO_2 is critical to all but the most extreme and risky forms of organic wine making. SO_2 is used to select for commercially important strains of yeast, to suppress other microbes that may damage the wine, and to prevent excessive oxidation and browning, especially in white table wines.

5.2 Occupational Health and Safety

- Handling, diluting and storing strong acids (H_3PO_4)
- Handling and storing strong oxidants (H_2O_2)
- Open flames in the presence of flammable solvents (if using an alcohol burner).

5.3 Quality Assurance Records

- Person, date
- Source of concentrates and stock solutions, dates of preparation
- Flow rate of gas, air temperature

- NaOH concentration as determined by titration with standard HCl (in duplicate)
- Standard SO_2 (in duplicate)
- Sample temperature.

5.4 Chemistry

This assay is based on:

a. Acid to drive the release of SO_2 (pK_a 1.81, 6.91):

$$SO_3^{2-} + H^+ \Leftrightarrow HSO_3^{1-}$$

$$HSO_3^{1-} + H^+ \Leftrightarrow SO_2 + H_2O$$

b. Excess peroxide (H_2O_2) to convert SO_2 to sulfuric acid (H_2SO_4) and thus drive the reaction to completion to the right.

c. NaOH to estimate the sulfuric acid formed from the SO_2.

d. Heat to dissociate bound SO_2 (Rankine, 1970; Ribéreau-Gayon et al., 2006a).

Thus:

$$H_2O_2 + SO_2 \rightarrow H_2SO_4$$

$$2NaOH + H_2SO_4 \rightarrow Na_2SO_4 + 2H_2O$$

That is, 2 moles of NaOH are consumed for every mole of SO_2 present. The molecular mass of SO_2 is 64 g/L. Therefore, the mass of SO_2 per liter:

$$\begin{aligned} &= v_{NaOH}/v_{sample} \times c_{NaOH} \times 64 \\ &= v_{NaOH}/20 \times 0.01/2 \times 64 \\ &= v_{NaOH} \times 0.016\ g/L \\ &= v_{NaOH} \times 16\ mg/L \end{aligned}$$

5.5 Procedure

- Clean apparatus thoroughly before beginning.
- Connect to vacuum and check that the flow rate is ca 1 L/min.
- Fill burette with 0.010 mol/L NaOH.
- Add 10 mL 0.3% hydrogen peroxide and 4 drops of mixed indicator to the flask through the side arm.
- Using a dropping pipette or the burette, add a few drops of 0.010 mol/L NaOH until the solution changes to green and stays green for at least 30 s. Reconnect the flask to the apparatus.

  ■ Add about 10 mL of 25% v/v phosphoric acid to the round flask and then 20.0 mL of the sample to be tested (do not degas this sample).
- Connect the flask and allow air to bubble through for 15 min. Remove the pear-shaped flask and titrate with 0.01 mol/L NaOH until an emerald green color returns and persists. Record the volume (v_1, Free SO_2).
- NOTE: An excess of NaOH will affect the next reading. If you overshoot the endpoint, then readjust with dilute acid.
- Replace the flask on the apparatus, apply heat to bring the liquid to the boil and continue boiling gently for another 15 min. Retitrate the solution and record the volume (v_2, Bound SO_2).
- Continue with additional samples from ■.

5.6 Calculations

$$\text{Free } SO_2 = v_1 \times 16\ mg/L$$

$$\text{Bound } SO_2 = v_2 \times 16\ mg/L$$

$$\text{Total } SO_2 = fSO_2 + bSO_2$$

5.7 Notes and Sources of Error

1. Take great care not to exceed the balance point when titrating: practice adjusting with alternately dilute acid and NaOH until familiar with the color change—apple green indicates excess NaOH.
2. It is risky to use air pressure rather than a vacuum because the joints in the apparatus may leak and reduce accuracy.
3. It is good practice to disconnect the flask at the ball joint and to use an empty wash bottle to blow into the inlet side of the flask to drive any SO_2 absorbed on to the glass down into the solution SO_2. Then remove the gas bubbler and titrate.
4. Take care when pipetting new solutions not to contaminate the standards with other liquids. Ensure that pipettes are cleaned between uses. Pour off a volume of the peroxide from its container into a small flask and pipette from that.
5. If severe frothing occurs, you may reduce the sample volume to 10 mL or add a non-ionic wetting agent to disperse the froth. (NB. If you change the sample volume, change the calculations.)
6. It is a good idea to test a standard regularly to check the procedure and for contamination.
7. Use cooled water for the condenser and maintain a constant flow of air through the system (1.0 L/min): the condenser acts to eliminate volatile acids from the gas stream.
8. The proportion of SO_2 present in a sample is temperature dependent; therefore, the sample flask may need to be placed in a water bath for the first reading.

5.8 Equipment and Materials

- Rankine aspiration apparatus (Figure 10.6)
- Microburette (readable to 2 decimal places)

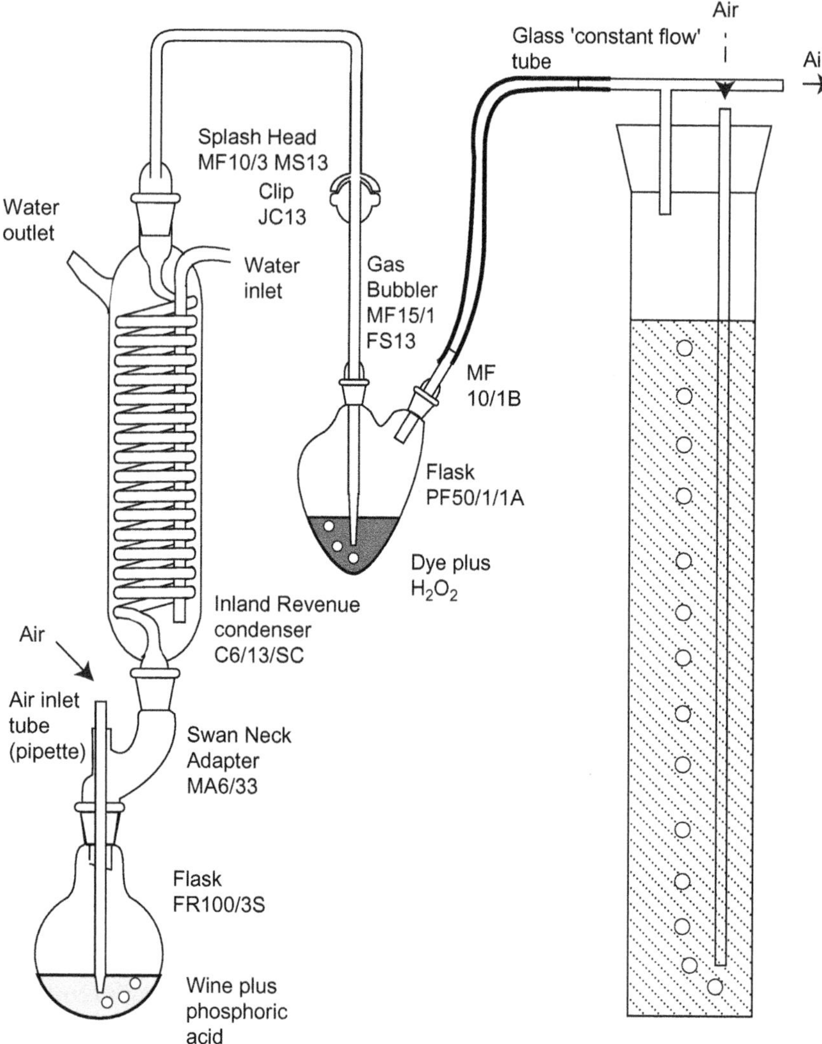

FIGURE 10.6 Illustration of apparatus commonly used for the determination of sulfur dioxide (SO_2) by aspiration and of the individual components used in its construction.

- 10-mL dispenser or pipette and aid
- Empty wash bottle with tip removed (air puffer)
- 20-mL pipette
- Vacuum source, or a small air pump (e.g. an aquarium pump)
- Standard SO_2 solution (10.0 g/L, ca 15.80 g/L of $Na_2S_2O_5$ in 0.1M NaOH, dilute as required, e.g. 1:100 to give 100 mg/L)
- 0.010 mol/L NaOH—standardize against 0.010M HCl
- 0.3% hydrogen peroxide (H_2O_2)
- 25% orthophosphoric acid (dilute concentrated acid, 85% m/m, using an ice bath, proportions approximately 30 vols H_3PO_4 to 70 vols distilled water—ADD the acid TO the water)
- Mixed indicator in a dropping bottle [100 mg methyl red (sodium salt) plus 50 mg of methylene blue dissolved in 100 mL of 50% aqueous ethanol]
- Bunsen or methanol burner.

5.8.1 Storage

- H_2O_2: refrigerate concentrate—stable for 2 months, dilute fresh daily
- NaOH: RT, store using a Carbosorb airlock or prepare fresh daily
- H_3PO_4: RT, stable in a sealed container
- Indicator: RT—stable.

5.9 Benchmark Values

- Red wine: $t\mathrm{SO_2}$ ca 50 mg/L initially and <15 mg/L at the start of the malolactic fermentation
- White wine: $t\mathrm{SO_2}$ ca 50 mg/L initially, <15 mg/L if using a malolactic acid fermentation and 20 + 80 mg/L free and bound SO_2 at clarification
- The legal limit for SO_2 in dry wines (those with <35 mg/L residual sugar) is 250 mg/L and for

sweeter wines is 380 mg/L. (Note: this may vary from nation to nation.)

6. SULFUR DIOXIDE BY TITRATION

6.1 Purpose

This method is included because the apparatus is less expensive than for the aspiration method and the procedure is rapid. However, it may demand handling dangerous concentrations of acids and alkalis, requiring care and expertise. Also, it is prone to errors caused by reacting with other substrates in wine, especially in red wine. The particular variant described here was selected because the addition of bicarbonate resolves some of the problems identified by Joslyn (1955), and it is very simple and rapid (Iland et al. 2004). Other variations are reviewed by Ough and Amerine (1988).

6.2 Occupational Health and Safety

- Handling and diluting strong acids (H_2SO_4) and alkali (NaOH)
- Handling toxic substances: iodine/iodate.

6.3 Quality Assurance Records

- Person, date
- Source of concentrates and stock solutions; date of preparation and name of person who did the preparation
- Standard SO_2 (in duplicate)
- Sample temperature.

6.4 Chemistry

This method is based on the reaction of free SO_2 with iodine, using starch to indicate the presence of excess iodine once the equivalence point has been reached. See also Section 5 on the aspiration method for further details which apply also to this method.

$$I_3^- + SO_2 + H_2O \rightarrow SO_3 + 3I^- + 2H^+$$

6.5 Procedure

6.5.1 Free Sulfur Dioxide

- Add 50.0 mL wine or juice/must to a 250 mL Erlenmeyer flask.
- Add starch indicator (either Vitex® powder, a proprietary solution or prepared).
- Acidify with 5 mL 25% aqueous H_2SO_4, swirl to mix (CARE).
- Immediately before titrating, add ca 1 g Na^- or K-HCO_3.
- Titrate with 10 mM iodine (I_2) to a starch endpoint (dark blue that persists for 30 s), swirl or mix constantly.
- Record titer.
- Calculate Free SO_2 $= v \times 12.8$ (mg/L).
- Or if different values are used, then: Free $SO_2 = \frac{v_{I_2} \times M \times 1000 \times 64.4}{v_{wine}}$, where v_{I_2} is the volume of iodine solution, M is the concentration of the iodine solution (mol/L) and v_{wine} is the volume of wine or must added to the vessel.

6.5.2 Total Sulfur Dioxide

- Pipette 20.0 mL wine or must into a stoppered or screw-cap 250-mL bottle or flask.
- Add 25 mL 1 mol/L NaOH, cap, mix thoroughly and stand for 10 min.
- Add starch indicator (either Vitex powder, a proprietary solution or prepared).
- Acidify with 10 mL 25% H_2SO_4 (v/v/) and then swirl to mix (CARE).
- Immediately before titrating, add ca 1 g Na^- or K-HCO_3.
- Titrate with 10 mmol/L I_2 to a starch endpoint (dark blue that persists for 30 s), swirl or mix constantly.
- Record titer.
- Calculate Total $SO_2 = v \times 32$ (mg/L).
- Or if different values are used, then apply them to the equation above as for free SO_2.
- Calculate Bound SO_2: $= tSO_2 - fSO_2$.

6.6 Notes

1. Why I_3^-? Iodine (I_2) is sparingly soluble in water. So solutions of iodine are prepared by adding KI, which forms a complex that aids solution, $I_2 + KI \rightarrow I_3^- + K^+$. The iodide ion ($I^-$) does not react with starch.
2. Methods using iodine are not appropriate for wines that have had ascorbic acid or erythorbic acids added as these acids react with iodine (unless the amounts are measured and can be deducted).
3. The method tends to overestimate the concentration of fSO_2 because other components of wine also react with iodine (Josyln, 1955).
4. It is a good idea to test a standard regularly to check procedure, for contamination and degree of overestimation. Ideally, conduct a series of standard additions, hold for several days to equilibrate, and then measure the control wine/must and at least two additions to check for linearity.

5. The proportion of SO_2 present in a sample is temperature dependent; therefore, the sample flask may need to be placed in a water bath.
6. A yellow lamp at the side of the flask may help to detect the endpoint, especially for red wines (or decolorize with charcoal).

6.7 Equipment and Materials

- 0.01 mol/L iodine [prepare from a commercial solution, e.g. dilute 1 part 0.05 mol/L I_2 (12.69 g/L I_2 + 20 g/L KI) with 4 parts distilled/deionized water, accurately, e.g. make to volume in a volumetric flask]
- 1 mol/L NaOH
- 25% sulfuric acid [dilute concentrated acid (98% m/m) using an ice bath, proportions approximately 25 vols H_2SO_4 to 75 vols distilled water—ADD the acid TO the water].
- Starch indicator—one of powdered indicator (e.g. Vitex), a commercial solution, or prepared from soluble starch (potato or corn): disperse 1 g of starch in 10 mL distilled water, add to 100 mL boiling distilled water, mix thoroughly for about 1 min. Allow to cool, decant from any precipitate and add a broad-spectrum antimicrobial agent such as a crystal of thymol or a small quantity of salicylic acid or mercury iodide
- 50-mL burette, preferably auto-zero type
- 10-mL dispensing bottle filled with 25% H_2SO_4 (adjustable volume or one 5 mL and one 10 mL)
- 25-mL dispensing bottle filled with 1M NaOH
- 20-mL bulb pipette (for wine, juice/must)
- 50-mL bulb pipette (for wine, juice/must)
- 250-mL beakers or Erlenmeyer flasks for fSO_2 determinations, and screw-cap or stoppered flasks for tSO_2 determination.

6.7.1 Storage

- Actinic bottle for the storage of iodine standard (or wrap in aluminum foil to exclude light). Store at RT.
- NaOH: RT—best to use a plastic screw-cap bottle as glass-stoppered bottles tend to freeze, or use a Carbosorb airlock.
- Indicator: RT—stable, refrigerate.

6.8 Benchmark Values

- Red wine: tSO_2 ca 50 mg/L initially and <15 mg/L at the start of the malolactic fermentation
- White wine: tSO_2 ca 50 mg/L initially, <15 mg/L if using a malolactic acid fermentation and 20 + 80 mg/L free and bound SO_2 at clarification
- The legal limit for SO_2 in dry wines (those with <35 mg/L residual sugar) is 250 mg/L and for sweeter wines is 380 mg/L. (Note that these limits vary from nation to nation)

7. REDUCING SUGARS BY CLINITEST®

7.1 Purpose

The purpose of this test is to provide a quick and cheap method for confirming the end of fermentation. It is used in conjunction with a hydrometer (Baumé) and palate and should be confirmed by a quantitative test (enzyme or colorimetric). Note that fructose, which is the common residual sugar, is 1.7 times as sweet as glucose and sucrose. Hence, even small residual quantities may have a large impact on the apparent sugar/acid balance of the finished wine. This is mainly a problem with white wines owing to lower oxygen levels, poor yeast nutrition and toxic levels of alcohol near the end of fermentation.

7.2 Occupational Health and Safety

- Strong alkali; exothermic reaction: WEAR SAFETY GLASSES and GLOVES.

7.3 Quality Assurance Records

- Person, date
- Reagents (and source), amount per volume.

7.4 Chemistry

The reactions involved in this test are those for an organic reducing agent and were devised originally for estimating sugar in blood and urine (e.g. Benedict, 1928). Many

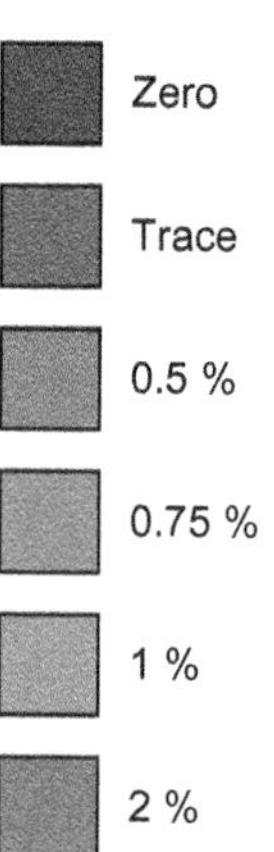

FIGURE 10.7 Guide to colors produced by solutions containing reducing sugars from 0 to 2 g/100 mL.

other variants have been devised as quantitative tests (e.g. Somogyi assay).

Glucose contains an aldehyde group (RHC=O) which can reduce copper salts (cupric sulfate, blue) to produce insoluble red/orange cuprous oxide as a precipitate. Alkali causes a rearrangement of the ketose group in fructose to aldose as occurs in glucose and mannose. The alkali, NaOH, also provides the heat as it dissolves and reacts with the other reagents (the quantitative method uses a heating block or hot water bath). Other components of the tablet are citric acid and sodium carbonate, which assist in dissolving the other reagents and in promoting an exothermic reaction.

7.5 Procedure

- Add five drops of wine to a small test-tube or vial.
- Dilute with a further 10 drops of distilled water.
- Drop one tablet into the tube and wait while the reaction takes place (vigorous bubbling)—do not mix or shake during this phase.
- Allow to stand for a further 15 s, then shake gently.
- Compare with the color chart provided with the kit (dark blue, zero; green to orange, trace to 2% reducing sugar; glucose and fructose) (Figure 10.7).

7.6 Notes and Sources of Error

1. This is a semi-quantitative test only and a quantitative procedure should be used to confirm the analysis once a level of trace or zero is reached. Quantitative versions of the reagent can be prepared using Benedict's reagent or alternatives.
2. This is a test for reducing sugars only: monosaccharides and maltose but not sucrose or lactose. It will react with ascorbic acid if present in a sufficient quantity (unlikely).
3. The contents of the tube are highly toxic and corrosive and the reaction is strongly exothermic, i.e. produces sufficient heat to burn unprotected skin: handle test-tubes by the top! If spilt, the reagent mix must be washed off with copious water.
4. Varying the volume varies the sensitivity of the test and affects the interpretation; doubling the volume of water reduces the sensitivity and expands the range. If you do this you should run your own standards and map these to the color chart. Using an autopipette and specific volumes will improve the reliability of the results. A 2-drop variant changes the range by 2.5 (i.e. upper level is 5%). See the manufacturer's notes.
5. Note: Bayer (e.g. Clinistix) or other company test strips based on the glucose oxidase reaction are unsuitable because the main reducing sugar near the end of a fermentation is fructose, which is unreactive to such tests.

7.7 Equipment and Materials

- Clinitest tablet test kit or equivalent (e.g. EnolTech Clinitest). Note that this item is no longer manufactured by Bayer AG.
- Color chart as provided with kit (or available online from supplier)
- Test-tube and stand
- Safety glasses and gloves.

7.8 Benchmark Values

Trace or <1 mg/L.

8. TARTRATE STABILITY

8.1 Purpose

Tartaric acid exists in wines as a mixture of salts that vary in their solubility and which may precipitate over a period of days, weeks and years, forming unsightly hazes and precipitates, especially in white table wines. Furthermore, during the course of precipitation, pH changes may occur that will affect the sensory balance of the wine. The grape juice/must used to produce wine contains tartaric acid and its salts at concentrations that usually exceed their solubility in water, being held in solution by glycoproteins, tannins and complex salts.

Precipitation is promoted by the presence of alcohol, which lowers the solubility of bitartrate in the solution, by lowering the temperature and by the presence of other ions such as calcium. A wider range of tests is presented in Boulton et al. (1996) and Ribéreau-Gayon et al. (2006b). The methodology was reviewed recently by Lasanta and Gómez (2012).

8.2 Occupational Health and Safety

No particular issues.

8.3 Quality Assurance Records

- Person, date, method, reagents, source, preparation date and name of person who prepared them.

8.4 Chemistry

The simple assays rely on lowering the temperature of the wine to reduce the solubility of the tartrate salts and

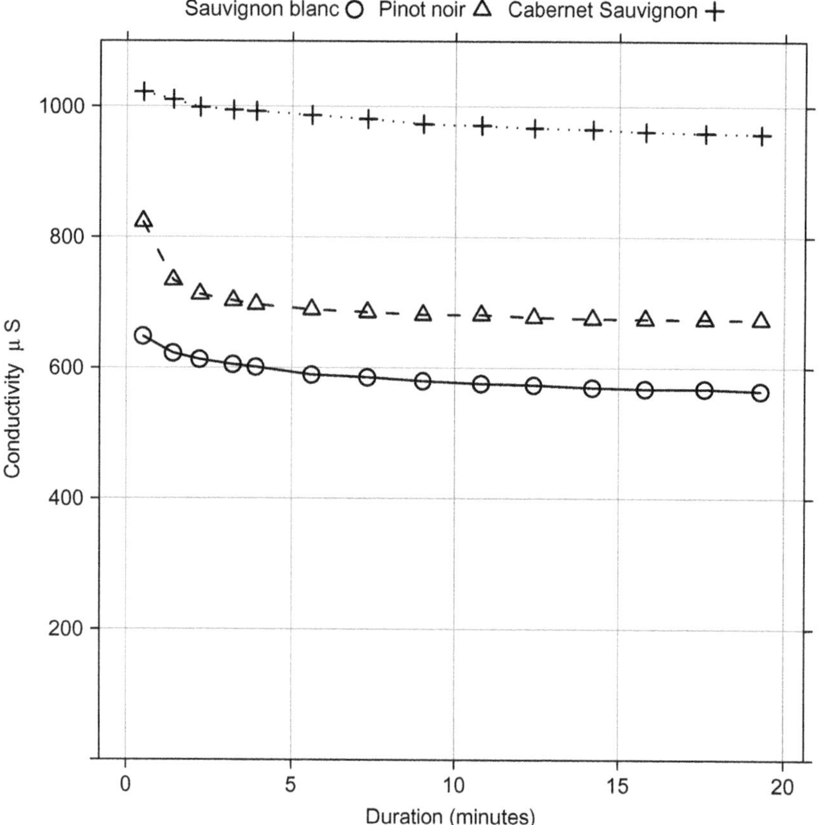

FIGURE 10.8 Change in conductivity of wine of three cultivars seeded with 20 g/L potassium hydrogen tartrate (KHT) and held at 0°C and stirred. *(Redrawn from Boulton et al., 1996, with permission).*

promote their crystallization, usually aided by the addition of excess solid potassium bitartrate, which serves as a nucleating agent, hastens the process and ensures that there is an excess. The outcome is assessed either by observing whether a haze of crystalline bitartrate has developed, or weighing the filtered salts to determine whether there has been an increase or a decrease (dissolved salts have precipitated or the added salts have dissolved) or the conductivity of the wine has increased or decreased owing to the concentration of K^+ either increasing or decreasing. The latter is the method recommended.

The least soluble salt is calcium tartromalate, then calcium tartrate, followed by potassium hydrogen tartrate (KHT), otherwise known as potassium bitartrate or cream of tartar. The relative stability of the supersaturated solutions is due to complexes with other juice components and salts (e.g. amino acids, NH_4^+, tannins and glycoproteins). Calcium is a normal component but in tiny concentrations only, unless used to deacidify wine or the wine was made in an unsealed concrete tank.

Each salt has a physical property known as its solubility product (k_{sp}) which is related to but not the same as solubility (mol or g/L). Solubility is influenced by solute (presence of alcohol), other ions including H^+, and stabilizing solutes and colloids.

8.5 Procedure

8.5.1 *Simple Method*

- Filter a 250-mL sample of wine through a 0.45-μm filter and separate into 50- and 200-mL subsamples.
- Label and hold the 50-mL sample at RT and place the 200-mL sample at −4° C for 3 days.
- Invert both samples and examine for turbidity using a focused light source (see Bentonite Protein Stability Test, Chapter 8.6) or measure turbidity m (Section 9).

8.5.2 *Rapid Conductivity Method (Boulton, 1983)*

- Pour approximately 100 mL of wine into a 250-mL beaker and place in a water bath at a chosen temperature (e.g. 2° C for white and rosé and 12° C for red wines).
- Grind 1.5 g of KHT in a mortar and pestle (note that size is important, <100 μm).
- Insert the conductivity probe into the wine and stir continuously. Check that no bubbles adhere to the electrodes.
- Add ground KHT, note the time, and record conductivity at 1-min intervals until the reading is stable (or after 20 min).

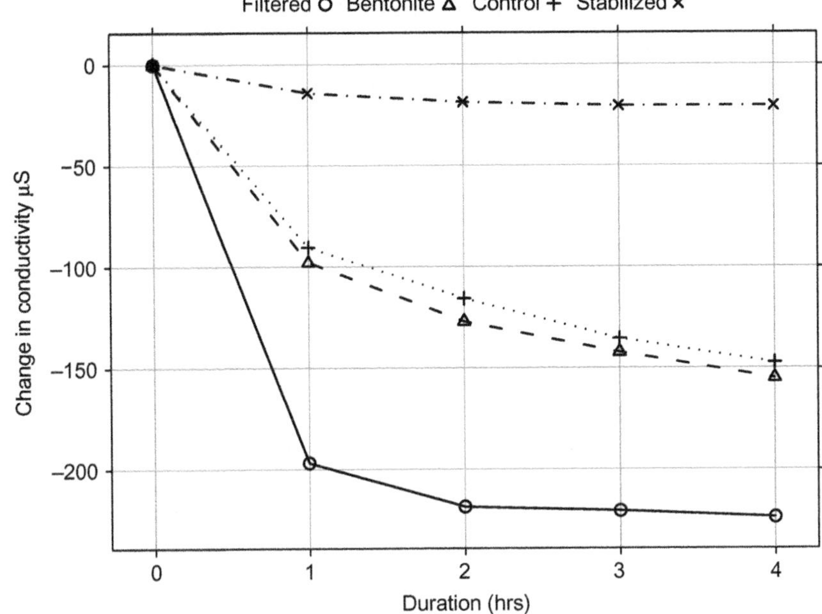

FIGURE 10.9 **Change in conductivity for wine treated by one of the following pretreatments: no treatment (control), filtering with 10^3 Da filter, bentonite or *meta*-tartaric acid (a stabilizing agent) and then held at 2°C and seeded with 5 g/L of potassium hydrogen tartrate (KHT).** *(Redrawn from Maujean et al., 1986 with permission).*

8.6 Notes and Interpretation

1. The conductivity of stabilized wines should not change or may even rise if the saturation temperature is greater than the test temperature.
2. Glycoproteins such as mannoproteins are sometimes added to wine to enhance low-temperature stability, as may *meta*-tartaric acid (especially to sparkling wines). The formation of crystalline deposits as wines age is probably a result of the breakdown of the stabilizing complexes (e.g. tannins). Red wines are more stable than white wines and take longer to cold-stabilize than white wines (Figure 10.8). The method is also useful for evaluating stabilizing agents and bentonite treatment (Figure 10.9). For additional information see Berg and Keefer (1958), Boulton et al. (1996) or Ribéreau-Gayon et al. (2006b).
3. The proportion of tartaric acid in the insoluble form is about 65% in wine but is subject to pH, H^+, K^+ and tartaric acid concentration, and percent alcohol (Berg and Keefer, 1958). KHT ionizes, with the direction of the equilibrium being determined by the relative concentration of each component and the solubility product of KHT.
4. The rapid conductivity method provides a means of estimating the temperature at which the wine is fully saturated with KHT. This may then be used to estimate the lower limit of temperature stability of the wine assuming normal levels of stabilizing colloids and substances and an alcohol level of 11%. The method should be validated for each product using the extended method (Ribéreau-Gayon et al., 2006b).

8.7 Equipment and Materials

8.7.1 Both Methods

- Mortar and pestle
- KHT
- Balance.

8.7.2 Simple Method

- A water bath set at −4° C (glycerol solution) or a modified freezer or other selected temperature
- A focused light source (or spectrophotometer)
- Vacuum Büchner filter and filter discs 0.45 µm.

8.7.3 Conductivity Method

- pH-conductivity meter and conductivity probe (ranges: 100 µS to 1.0 mS, 1.0–10.0 mS
- A water bath set at the desired temperature (e.g. 0–5° C for white and rosé wines, 10–15° C for red wines). A glycol solution will be required for the low-temperature settings. Alternatively, a refrigerator may suffice.
- 250-mL beaker
- 100-mL volumetric cylinder
- Magnetic stirrer and bar (air-driven if in water)
- Clock.

8.8 Benchmark Values

The saturation temperature for red wines generally should be between 10 and 15° C, while white wines require 2–5° C and sparkling wines may require −4° C.

9. TURBIDITY

9.1 Purpose

Turbidity measurements are used to determine the minimum level of bentonite required to ensure heat stability of white wines and, prior to bottling, to monitor clarity after settling and filtration. Wines that appear cellar bright following settling and filtering or centrifugation will usually contain particulate matter, colloids, which are invisible in normal light but which nonetheless will serve as nuclei for crystallization of tartrates and other salts or proteins, risking instability; this is not just about appearance.

Turbidity may also be used in quality control during a bottling run to ensure that the filtering process has not failed.

9.2 Occupational Health and Safety

Issues only arise if using a standard, e.g. formazine, hexamethylene–tetramine and hydazinium sulfate are hazardous to extremely hazardous reagents. Check the MSDS before purchasing and using.

9.3 Quality Assurance Records

- Person, date, method, standard and source, including batch and date of manufacture.

9.4 Chemistry

This measure depends on the difference in the refractive index of particles suspended in a solution. High refractive index particles reflect or divert the light causing lower transmission, or in a nephelometer, which measures light incident at right angles to the incident light, the light that is scattered. This is compared with the light that is transmitted.

9.5 Procedure

- Zero instrument using ultrafiltered, distilled water.
- If using standards, measure their transmission.
- Measure samples and plot against bentonite addition (see also Chapter 8.6).

9.6 Sources of Error

1. Dirty cuvette.
2. Leakage in membrane filter apparatus or other cause of residual particles within the zero standard.
3. If using a nephelometer, then instrumental drift or aging.

9.7 Notes

1. Few wineries will possess a nephelometer and thus the values recorded using a spectrophotometer are quite arbitrary and useful only for comparison purposes. Even with a specialized nephelometer it is difficult to obtain values th at may be compared from one laboratory to another. Thus, when recording data, details of the instruments and circumstances should be fully specified.
2. As this is usually just a comparison, no benchmark values can be provided. It is simply a matter of following the trend when undertaking a bentonite fining trial and comparing the impact of filtering treatments prebottling (see Bentonite Protein Stability Test in Chapter 8.6).

9.8 Equipment and Materials

- Nephelometer (OIV recommends 620 nm; OIV, 2006a) or a spectrophotometer (700 nm) or a collimated light source
- Ultrapure water twice filtered through a 0.1-μm membrane filter
- Formazine standard (optional).

9.8.1 Turbidity Standard 400 NTU (OIV, 2006a)

- Dissolve 10.0 g hexamethylene–tetramine ($C_2H_4N_2$) in filtered (0.1 μm) distilled water, bring to 100.0 mL with filtered distilled water.
- Dissolve 1.0 g hydrazinium sulfate in filtered distilled water and bring to 100.0 mL with filtered distilled water.
- Mix 5.0 mL of each and hold at 25 ± 3° C for 24 h. Make to 100.0 mL with filtered distilled water. Store at RT in the dark or in an actinic vessel for up to 4 weeks.
- Dilute 1.0 mL to 400 mL with filtered distilled water (1 NTU std). Keep for up to 1 week at RT in the dark or in an actinic vessel.

Chapter 11

Quality Assurance, Teaching and Research

Chapter Outline

1. FREE AMINO NITROGEN

1.1 Purpose

Nitrogen is a key element in the nutrition of yeast and lactic acid bacteria. Normally it comprises both ammonia (NH_3) and amino acids other than proline (also termed yeast-assimilable nitrogen, YAN). Proline cannot be metabolized under anaerobic conditions. Proline is, however, a major component of the free amino acids in grapes. The most prevalent metabolizable amino acid is arginine, and this assay is based primarily on that compound. Certain amino acids are also critically important aroma precursors formed largely as yeast cells die (autolyze) and during the malolactic fermentation. Ammonium

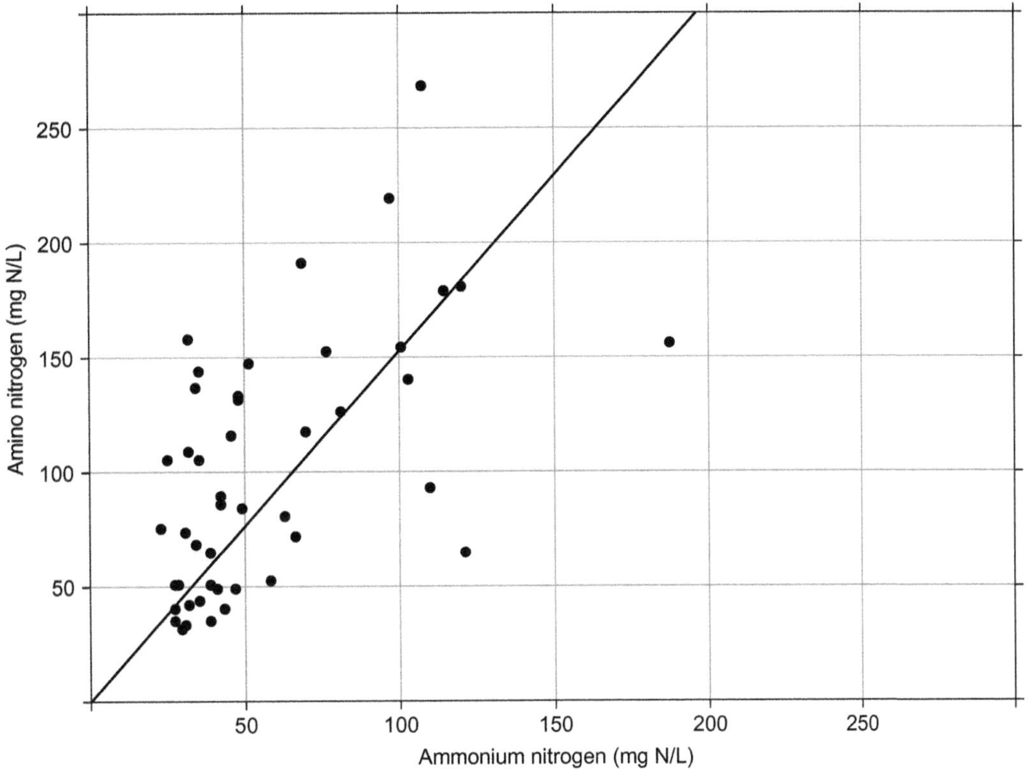

FIGURE 11.1 **Scatterplot of values for ammonium and α-amino acids for 46 samples of must from Californian vineyards.** $x = 1.53 + 0.115y$, $r^2 = 0.79$. *(Data reanalyzed and replotted from Dukes and Butzke, 1998, with permission).*

ions are the other half of this equation but this assay is insensitive to ammonium (Dukes and Butzke, 1998). While a high level of free amino nitrogen (FAN) may seem desirable, a contraindication is the relationship between arginine and the risk of excessive levels of ethyl carbamate, a carcinogen, in aged wine in particular (Butzke and Bisson, 1997). Therefore, this measurement also serves as a risk management element in the wine-making process.

There is a trend for ammonium and amino acids to increase in tandem (Figure 11.1) but the relationship is not predictive. One can simply add diammonium phosphate (DAP) to minimize the risk of a stuck or sluggish ferment, but measuring both FAN and NH_3 will enable the wine-maker to manage nutrition more exactly and to develop an understanding of the nutritional and quality potential of a given source of fruit in the long term.

This measurement is not just about the current vintage, it is also knowledge that should be used to develop long-term management practices in the vineyard, the place where the production of fine wines begins.

1.2 Occupational Health and Safety

- Corrosive alkali
- Hazardous substance: *o*-phthalialdehyde (OPA).

1.3 Quality Assurance Records

- Person, date, reagent sources, product manufacturer, purchase date and batch number/code.

1.4 Chemistry

The chemistry of this reaction is considered to involve the reaction of OPA under alkaline conditions, reversibly with both a thiol (R-SH), in this instance *N*-acetyl cysteine, and a primary amine, which in turn react to produce a stable end-product, 1-(alkythio)-2-alkisoindole) (Wong et al., 1985).

1.5 Calculations (Single-Beam Spectrophotometer)

$$A_{test} = A_{obs} - A_{blank}$$

Read concentration off a standard curve using the absorbance value or enter into the equation.

$$A = a + b.FAN$$

where A is the spectrophotometer reading at a particular nitrogen concentration (FAN), and a and b are constants. Then rearrange to the form:

$$FAN = (A - a)/b$$

TABLE 11.1 Standard Curve for Amino Nitrogen

	Amino N as L-isoleucine[a]				
Solution (Volume)	28 mg/L	56 mg/L	84 mg/L	112 mg/L	140 mg/L
Distilled water (μL)	40	30	20	10	0
10 mM L-isoleucine standard (μL)	10	20	30	40	50
OPA (mL)	3.00	3.00	3.00	3.00	3.00
OD					

Note: Include two blanks as per the assay (see Table 11.2).
[a]*The molecular mass of isoleucine is 131.17 g/mol while the atomic mass of nitrogen (N) is 14.007 g/mol: 10 mmol isoleucine contains 1.312 g/L of substance but only 0.139 g/L of that is N (ca 140 mg/L).*

TABLE 11.2 Tube Contents for Samples and Their Blanks

Solution	Blank – Buffer	Blank – NOPA	Sample – Blank	Sample – Test	Sample – Blank	Sample – Test
Distilled water (μL)	50.0	50.0	–	–	–	–
Sample (μL)	–	–	50.0	50.0	50.0	50.0
Buffer (mL)	3.00	–	3.00	–	3.00	–
NOPA (mL)	–	3.00	–	3.00	–	3.00
OD						

Note: NOPA = nitrogen by *o*-phthaldialdehyde; OD = optical density.
(Dukes and Butzke, 1998).

1.6 Procedure

- Filter (0.45 μm), centrifuge or cold-settle sample.
- Prepare the following solutions according to Tables 11.1 and 11.2 (Note: two solutions for each sample).
- Cover the tubes with Parafilm® or a plastic-covered sponge, invert several times to mix thoroughly, and stand for 10 min.
- Read the absorbance of the buffer blank and the standards at 335 nm (use buffer as the zero or reference in a dual-beam spectrophotometer).
- Use the buffer blank as the reference and read the sample blanks and tests at 335 nm. Calculate the adjustment required for each test sample. If using a dual-beam spectrophotometer, place the sample blank in the reference beam.
- Either draw a graph of the standard (absorbance as the *y* axis and nitrogen concentration as the *x* axis) or calculate a linear regression (see Chapter 12 or spreadsheet (The spreadsheet can be found on the companion website to this book.)).

1.7 Notes

1. Ammonium and free amino acids serve as nutritional sources for yeast. If inadequate levels occur then fermentation ceases or becomes stuck.
2. Ammonium and primary amino acids are not well correlated and thus their content should be measured independently (Figure 11.1). This method is insensitive to ammonium (3.5% of total only indicated), which has an absorption maximum at 335 nm (also the excitation maximum as this reagent fluoresces at an emission maximum of 450 nm).
3. The method may be readily modified as a highly sensitive assay for a plate reader using absorption or fluorescence.
4. This method does not measure proline (or hydroxyproline), which is an amino (imino) acid that is often high in berries but which cannot be metabolized by yeast under anaerobic conditions.
5. The need for correction comes from the absorbance of catechins (a class of phenolics) at the wavelength used in this assay. Alternatively, these could be removed by the use of polyvinylpolypyrrolidone (PVPP) (or use activated charcoal, which would also remove anthocyanins). Therefore, a juice blank must be assessed.
6. Many primary amino acids serve as yeast-derived aroma precursors, adding to their importance in a 'must'.

1.8 Equipment and Materials

- Ultraviolet (UV)-grade disposable cuvettes (10 mm pathlength)
- Parafilm®
- 0–50-μL micropipettes

- 0–3-mL micropipettes
- Analytical balance to 0.001 g
- 10-mL syringe with cannula and 0.45-μm membrane filters
- Centrifuge with 15-mL centrifuge tubes
- Two 1-L volumetric flasks
- One 250-mL volumetric flask
- Two 1-L glass storage bottles
- One 250-mL glass storage bottle
- Reagent buffer (pH 9.5):
  - 3.837 g NaOH [analytical grade reagent (AR)]
  - 8.468 g boric acid (H_3BO_3) (AR)
  - 100 mL ethanol (99.5% AR)
  - 0.816 g *N*-acetyl-L-cysteine
  - Dissolve reagents in ethanol and make to 1 L with distilled water.
- OPA (pH 9.5):
  - As for reagent buffer, but add:
  - 0.671 g OPA at the beginning.
- 10 mM L-isoleucine standard solution:
  - Dissolve 0.328 g of L-isoleucine in 250 mL of distilled water.

1.8.1 Storage

- Store buffer and OPA reagents at 4°C for no longer than 3 weeks.
- Standard is stable for 1 week at 4°C (better to freeze aliquots for future use).

1.9 Benchmark Values

- Red wine must: 110 mg/L
- White wine must: 165 mg/L.

2. AMMONIUM BY ENZYMATIC ASSAY

2.1 Purpose

Ammonium is the other half of the berry-derived nitrogen, the companion for the assimilable amino acids. Together, these assays provide the winemaker with knowledge regarding the starting point for nitrogen nutrition for the fermentations. The minimal amount of supplement can then be added. Keeping to the minimum not only saves costs but also reduces the risk of having an excessive amount of NH_3 that may interfere with lactic acid bacteria and which may also exceed legislative limits. As with the FAN assay, this knowledge helps to determine long-term vineyard management strategies (Figure 11.2).

2.2 Occupational Health and Safety

Standard laboratory requirements only.

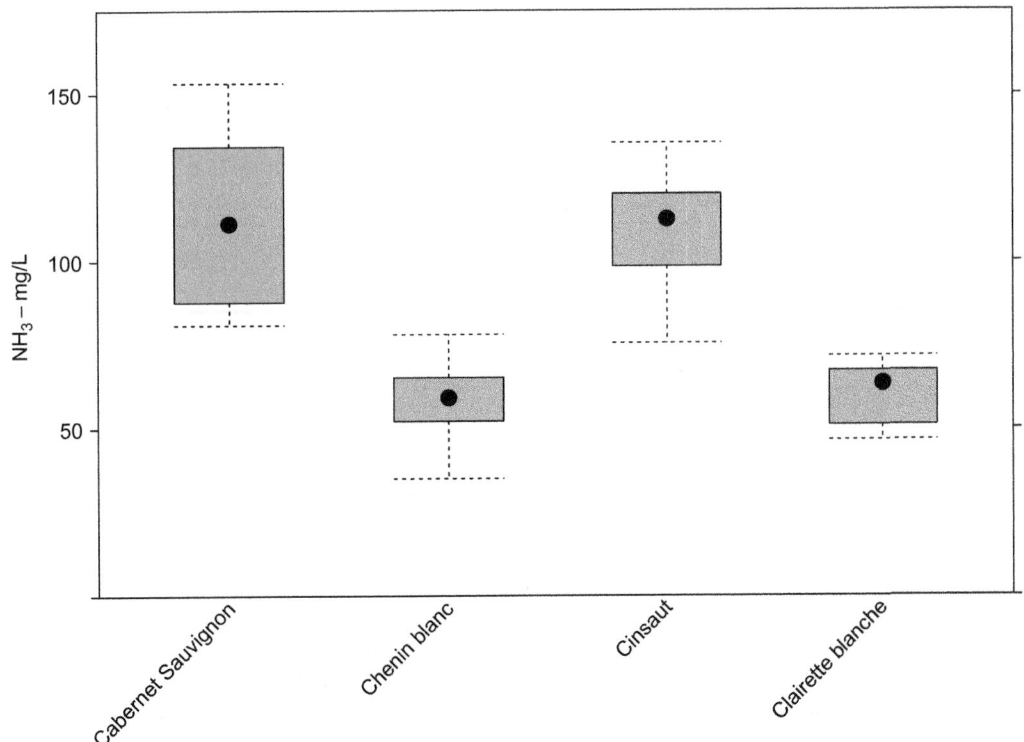

FIGURE 11.2 Comparison of the ammonia (NH_3) content (mg/L) in fruit at delivery from vineyards in Stellenbosch, South Africa. The rectangles include 50% of the observed values; the central dot, the mean; and the dotted horizontal lines, the quartiles. *(Data from Ough and Kriel, 1985. The article describes the relationship between vineyard management and NH_3 concentration).*

2.3 Quality Assurance Records

- Enzyme kit code, manufacturing date, batch number, date of purchase, storage location and conditions, date and name of person preparing working solutions, standard analysis value.

2.4 Chemistry

The kit contains four reagents: NADH (reduced form of nicotinamide adenine dinucleotide), 2-oxoglutarate, glutamate dehydrogenase and an ammonium standard.

The reaction is:

$$\text{2-Oxyglutarate} + \text{NADH} + NH_4^+ \rightarrow \text{L-Glutamate} + \text{NAD}^+ + H_2O$$

Thus, one mole of NADH is consumed for every mole of ammonium consumed.

2.5 Calculations

Calculate net absorbance (A) as:

$\Delta A = (A_1 - A_2)_{\text{sample}} - (A_1 - A_2)_{\text{blank}}$

Calculate the concentration (c) of ammonia as:

$$c = 151.3 \times \Delta A \times DF \text{ mg/L}$$

2.6 Procedure

- Dilute the grape or must sample by 10-fold, i.e. 1 vol. + 9 vols redistilled water.
- Pipette 1.00 mL of solution #1 into a cuvette for each sample to be assayed plus one for a standard (2-oxoglutarate and buffer).
- Dissolve one tablet of NADH (#2) in each cuvette including the blank (do not handle directly, use forceps).
- Pipette 1.900 mL distilled water into each cuvette except the blank, into which you place 2.000 mL.
- Pipette 100 μL of sample or standard into all tubes except the blank (this may be varied from about 20 to 200 μL, but if you vary, remember to adjust the calculations).
- Cover with Parafilm, mix by inversion, and stand for 5 min.
- Record the absorbance (A_1) at 365 nm using distilled water to zero the instrument (Table 11.3).
- Pipette in 20 μL of enzyme (#3) to each cuvette, mix as before. Stand for 20 min. Take a second reading of the absorbance (A_2) at the wavelength used earlier (glutamate dehydrogenase) (Table 11.4).

TABLE 11.3 Example of Cuvette Contents for Ammonium by Enzyme Assay (Using R-Biopharm Kit)

	#1 (ml)	#2 Tablet	DI Water (mL)	Sample or Standard (μL)	OD A_1 at 5 min	#3 (μL)	OD A_2 +20 min
Blank	1.000	1	2.000	–		20	
Standard	1.000	1	1.900	100		20	
Sample 1	1.000	1	1.900	100		20	
Sample 2	1.000	1	1.900	100		20	
Sample *n*	1.000	1	1.900	100		20	

Note: Recommended volumes may vary for other manufacturers.
DI = distilled/deoinized; OD = optical density.

TABLE 11.4 Example Worksheet for Ammonium by Enzyme Assay

	OD 1 (A_1)	OD 2 (A_2)	$\Delta = (A_1 - A_2)$	$\Delta A_{\text{Sample}} - \Delta A_{\text{Blank}}$
Blank	0.458	0.453	0.005	
Standard	0.459	0.321	0.138	0.133
Example	0.634	0.544	0.090	0.085
Sample 1				
Sample *n*				

Note: OD = optical density.

2.7 Notes

1. The quantity of NADH remaining is measured by the absorbance of the solution at 365 nm in a UV spectrophotometer space (other substances in wine interfere at the wavelengths recommended by the manufacturers of the kit).
2. If necessary, remove phenolics with PVPP or charcoal. It is a good idea to run an additional blank with sample but no enzyme and/or to do a set of standard additions using aliquots of the standard solution.

Derivation of Equation

The concentration (c) of NH_4^+ in the sample solution is calculated as:

$$c = \left[\frac{\text{vol} \times 1.01 \times M(\text{NH}_4^+)}{\varepsilon \times d \times v_s \times 1000}\right] \times \Delta A \times DF$$

$$c = \left[\frac{0.5143}{3.4}\right] \times \Delta A \times DF \times 1000\ \text{mg/L}$$

$$c = 151.3 \times \Delta A \times DF\ \text{mg/L}$$

where vol = total solution volume (times 1.01 to allow for the volume of the tablet), $NH_4^+ = 17.03$ g/mol; d is the pathlength (1 cm); v_s is the sample volume (0.1 mL); ε is the extinction coefficient of NADH, which is 3.4 mmol cm^{-1} at 365 nm; and DF is the dilution factor (10) in these notes but may be varied if the solution concentration is too low ($\Delta A < 0.1$) or too high ($\Delta A > 0.5$ at 365 nm).

2.8 Equipment and Materials

- R-Biopharm® or equivalent ammonium kit—note that the details of the assay may vary between manufacturers
- Forceps
- Parafilm
- 0–2.000-mL autopipette and tips
- 100-μL autopipette and tips
- 20-μL autopipette and tips
- Disposable UV-grade cuvettes, 10-mm pathlength (Kartell® 1939)
- UV–visible spectrophotometer (Hg 365 nm)
- Redistilled water.

2.8.1 Storage

Store all reagents at 4°C and bring to room temperature (RT) before use.

2.9 Benchmark Values

- Minimum NH_3 in red wine must before fermentation: 50 mg/L
- Minimum NH_3 in white wine must before fermentation: 140 mg/L
- Maximum DAP addition: 960 mg/L (USA), 604 mg/L (EU) and <850 mg/L (Australia on the basis of [P]).

3. REDUCING SUGARS BY ENZYMATIC ASSAY

3.1 Purpose

Residual sugar concentration is a critically important value for dry table wines. A value of less than 2 mg/L is necessary to ensure a clean palate and to guard against spoilage. Enzyme kits are the most common way of conducting these assays, although many other methods are available. In the larger winery or commercial laboratory, near-infrared spectrophotometry (NIR) may be the method of choice. A strength of the enzyme method is that it is highly specific and capable of distinguishing between glucose and fructose or measuring both simultaneously. Small quantities of other sugars are present but these are not measured; they may be included in other, less specific 'wet' chemistry analyses such as the Somogyi–Nelson, described later. See Possner and Kliewer (1985) for a list of other sugars (but note that their methods were relatively insensitive compared with modern methods and thus their list will be far from exhaustive).

3.2 Occupational Health and Safety

No particular issues.

3.3 Quality Assurance Records

- Person, date, reagents, source and batch number, date prepared.

3.4 Chemistry

$$\text{Glucose} + \text{ATP} \xrightarrow{\text{hexokinase}} \text{Glucose-6P} + \text{ADP}$$
$$\text{Fructose} + \text{ATP} \xrightarrow{\text{hexokinase}} \text{Fructose-6P} + \text{ADP}$$
$$\text{Fructose-6P} \xrightarrow{\text{phosphoglucose isomerase}} \text{Glucose-6P}$$
$$\text{Glucose-6P} + \text{NADP}^+ \xrightarrow{\text{glucose-6P dehydrogenase}} \text{Gluconate-6P} + \text{NADPH} + \text{H}^+$$

3.5 Procedure

- Prepare the reagents as per the manufacturer's instructions and bring to RT.
- Estimate the level of sugar using the Clinitest® method and then dilute the sample to give an estimated concentration of between 0.15 and 1 g/L in the sample (e.g. if ca 5 g/L, dilute 1:9). Record the

TABLE 11.5 Example of Cuvette Contents for Reducing Sugars Assay (Using R-Biopharm Kit)

Pipette into Cuvettes	Blank (mL)	Standard (mL)	Standard Sample Volume (mL)	Alternative Sample Volume (mL)
DI water	2.000	1.900	1.900	2.000
Sample or standard solution	–	0.100	0.100	0.010
Solution 1	1.000	1.000	1.000	1.000
Mix and read absorbance (A_1) after 3 min at 20–25°C				
Suspension 2	0.020	0.020	0.020	0.020
Mix and read absorbance (A_2) after 10–15 min at 20–25°C				
Suspension 3	0.020	0.020	0.020	0.020
Mix and read absorbance (A_3) after 10–15 min				

Note: Recommended volumes may vary for other manufacturers.
DI = distilled/deoinized.

dilution factor. Alternatively, use a smaller volume of sample (see e.g. Table 11.5).

- If necessary (e.g. if using undiluted red wine), decolorize using the method described in the Notes section of the Malic Acid by Enzymatic Assay (Section 6.7).
- Set the spectrophotometer to 365 nm and zero using water as the blank.
- Pipette the requisite volumes as per Table 11.5 and ensure that they are at RT (20–25°C). Mix by inversion and set the timer to 3 min.
- Read absorbance A_1 at 365 nm—do not rezero!
- Pipette 20 μL of suspension #2, mix by inversion, and stand for 10–15 min (until readings are stable).
- Read absorbance A_2 at 365 nm—do not rezero.
- Pipette 20 μL of suspension #3, mix by inversion, and stand for 10–15 min (until readings are stable).
- Read absorbance A_3 at 365 nm.
- Calculate glucose and fructose concentrations:

$$[c_g] = (A_2 - A_1) \times 1.55457 \times \text{DF}$$

$$[c_f] = (A_3 - A_2) \times 1.56482 \times \text{DF}$$

where DF is the dilution factor (10 in this example).

3.6 Notes

1. This method is highly specific and sensitive and requires the use of less dangerous chemicals and procedures than the chemical assays (Somogyi–Nelson, Fehlings, Rebelin, Lane and Eynon). The latter assays also react with pentose sugars, which are not readily fermentable and which account for the bulk of the residual sugar in wine. Grapes contain approximately equal quantities of glucose and fructose but yeasts metabolize glucose preferentially.
2. This assay is similar in principle to the malic acid and ammonium assays (see Sections 6 and 2) and measures the change in $NADP^+$ to NADH.
3. An alternative method may be provided in the manufacturer's brochure, which can provide for longer storage times for the reagents (up to 8 weeks).
4. Glucose and fructose may be determined together by adding solutions #2 and #3 together.
5. If the readings after addition of each enzyme are not stable, repeat the reading at intervals until the reading stabilizes. Note the time.
6. Other manufacturers' kits may vary so check if using another kit and modify the calculations in the spreadsheet accordingly. (The spreadsheet can be found on the companion website to this book http://booksite.elsevier.com/9780124080812.)

3.7 Equipment and Materials

- D-Glucose/D-fructose enzyme kit (e.g. R-Biopharm or equivalent)
- 1-mL adjustable autopipette and tips
- 100-mL autopipette and tips
- 10-mL autopipette and tips
- PVPP
- Cotton wool
- Low-speed centrifuge (4000 rpm)
- Centrifuge tubes
- 10-mm pathlength, UV-grade cuvettes
- Spectrophotometer (365 nm)
- 0.45-μm filters or high-speed bench-top centrifuge
- Parafilm
- Laboratory timer.

3.7.1 Reagents

The kit contains (R-Biopharm example, other manufacturers' kits may vary):

#1: 5 g powder, including NADP+, adenosine triphosphate (ATP), buffer, $MgSO_4$—dissolve in 80 mL redistilled water. Stable for 3 days 4°C
#2: 0.7 mL suspension of hexokinase (HK, 200 units), glucose-6-phosphate dehydrogenase (G6P-DH, 100 units)—add to solution #1
#3: 0.7 mL phosphoglucose isomerase (PGI, 490 units)—add to solution #1.

3.7.2 Storage

- Store all reagents at <−10°C before use.
- Prepare reagents at 4°C and bring to RT before use. Stable for 3 days or more depending on procedure (see manufacturer's notes).

3.8 Benchmark Values

- Dry wine styles: <2 mg/L.

4. REDUCING SUGARS (SOMOGYI–NELSON)

4.1 Purpose

Sugar assays are primarily of interest to the winemaker when assessing the end of fermentation and as an aid in diagnosing problems related to stuck or sluggish ferments. Normally it is best to use an enzymatic analysis that can distinguish between glucose and fructose because yeasts preferentially absorb and metabolize glucose, leaving a relative excess of the more slowly utilized fructose. This is expensive if using macroassays such as normally available to the small-scale winemaker (a microtiter plate reader reduces the costs markedly as the amounts per assay are greatly reduced). This assay does not distinguish between glucose and fructose but is suited to large-scale analyses at either a macro- or micro-level (as may be required for experimental purposes). It is not suited to use in the winery because it involves the use of a class 1 poison—arsenic.

4.2 Occupational Health and Safety

- S1 Poison—arsenate: must be stored under lock and key and disposed of according to local regulations.
- Boiling water.

4.3 Quality Assurance Records

- Person, date, acid (and source), amount per volume.

4.4 Chemistry

Reducing sugars possess a terminal aldehyde group (–HC = O) which is capable of being oxidized to a carboxylate ($-COO^-$) while the oxidizing agent (cupric ions, Cu^{2+}, in alkali) is in turn reduced to a cuprous ion, Cu^+, in this instance insoluble cuprous oxide. Such sugars are termed aldoses. While fructose is strictly a ketose (–CO–CHOH), it is converted under alkaline conditions (and heat) to an aldose, glucose and mannose. Sucrose is a non-reducing sugar and must first be hydrolyzed to its components, glucose and fructose, before it can be measured in this assay.

The partial equation below shows the steps:

$$\text{Glucose} + 2Cu^{2+}5OH^- \rightarrow \text{Gluconate} + Cu_2O + 3H_2O \quad (1)$$

The cuprous oxide is red and insoluble, which drives the equation to the right in the presence of excess reagents.

The second step is to oxidize the cuprous oxide in acidic conditions to produce a blue copper arsenotungstate chromophore with $\lambda_{max} = 870$ nm (not 520 nm), the common value used.

4.5 Procedure

- Clarify an aliquot of the sample to be analyzed by centrifugation or filtration.
- If necessary, remove interfering substances by the addition of charcoal to decolorize and/or PVPP to remove phenolics, or cation-exchange resin to remove metals (not tested for wine).
- Dispense 0.5 mL of distilled or deionized water into each test-tube.
- Use duplicates for all samples and standards.
- Dispense 10 μL of clarified sample(s) into a glass test-tube (may need to adjust volume or dilute depending on sugar content as estimated, e.g. by Clinitest).
- Dispense 10 μL of each standard: 0–10 g/L of glucose (or fructose or 1:1 mix), which gives 0–100 μg per tube.
- Dispense 0.5 mL of reagent D to each test-tube and mix with a vortex mixer.
- Cover tube with a glass marble (or an aluminum foil cap, dimpled) and place in a boiling water bath for 15 min, exactly.
- Remove and place in a cold water bath at RT, or stand on the bench for 5 min.
- Dispense 3.0 mL of reagent E and mix well.
- Stand for 15 min and then mix again.
- Read optical density at 520 nm or, better, at 870 nm (Farnet et al., 2010).
- Calculate mg/L reducing sugar using a regression equation based on standards or read off a graph.

4.6 Notes

1. This method is not suited to use in wineries because the reagent is toxic and its presence is incompatible with a food-producing facility. The method is suited to laboratory use where large numbers of samples need to be processed and where expense can preclude the use of enzyme-based methods (although microplate techniques substantially resolve the cost issue). The Schinner method may be suitable in that circumstance in wineries but has not been tested (Schinner and von Mersi, 1990).
2. Dry wine styles normally have reducing sugars of less than 2 g/L; higher levels than this are noticeable on the palate and may lead to microbial instability.
3. This method is highly sensitive but reacts with all reducing sugars, not just glucose and fructose (Fielding et al., 1986). It therefore is not as specific as the glucose oxidase enzymatic method but is more economical, reliable and well suited to wine.
4. The method can be adapted for microtiter plates if very large numbers of samples need to be analyzed. One variant is given here but, as Somogyi and Nelson observe (Somogyi, 1952), the method is highly flexible and can be modified to suit the end-use.
5. If interference of other substances is an issue then a potassium ferric hexacyanide method may be used (Schinner and von Mersi, 1990). This is not only much less toxic but is claimed to be more sensitive than the Somogyi–Nelson and therefore less sample needs to be added (5–10 μL).

4.7 Equipment and Materials

- Boiling water bath and test-tube racks
- 12-mm Pyrex test-tubes
- Washed glass marbles
- Visible light spectrophotometer (520 nm)
- Disposable 10-mm pathlength cuvettes
- 3-mL autopipette or dispensing bottle
- 0.5-mL autopipette or dispensing bottle
- 10-μL autopipette
- Timer
- Glucose standards: 10, 5, 2.5, 1, 0.5 g/L
- Vortex mixer.

4.7.1 Reagents

A. 25 g anhydrous sodium carbonate (Na_2CO_3), plus 25 g sodium potassium tartrate, and 200 g of anhydrous sodium sulfate (Na_2SO_4) are dissolved in 800 mL of demineralized water and the volume is adjusted to 1 L (see Farnet et al., 2010 for a modified version with less Na_2SO_4). The solution is filtered if necessary.

B. 30 g of copper sulfate pentahydrate ($CuSO_4.5H_2O$) is dissolved in 200 mL of demineralized water containing 4 drops of concentrated sulfuric acid.

C. 50 g of ammonium molybdate is dissolved in 900 mL of demineralized water and 42 mL of concentrated sulfuric acid is added carefully. 6 g of sodium arsenate heptahydrate is dissolved separately in 50 mL of water, and this is added to the above solution. The volume of the solution is adjusted to 1 L. If necessary, warm the solution to 55°C to give complete dissolution of the components.

D. Add 1 mL of reagent B to 25 mL of reagent A.

E. Dilute solution C five-fold with demineralized water just before use (this is stable at 4°C for about 4 weeks).

4.7.2 Storage

- Store all reagents at 4°C and bring to RT before use.

4.8 Benchmark Values

- Dry table wine: <2 mg/L.

5. PROTEINS

5.1 Purpose

The ability of certain proteins to form a haze when heated is a problem principally of white wines but possibly also of rosé-style wines. Grape proteins largely comprise thaumatins (sweet proteins) and chitinases or pathogenesis-related proteins (Robinson et al., 1997; Tattersall et al., 1997; Waters et al., 1996). They also fall into two classes: stable and unstable. The unstable proteins are thought to be bound to low molecular weight polyphenols and together react with bentonite (Somers and Zeimelis, 1973). Protein content is not a faithful measure of protein stability but is a useful measure of the progress of fining treatments (Mesquita et al., 2001; Weiss and Bisson, 2001). For an example see Figure 8.6 in Chapter 8). Factors affecting protein stability are also discussed by Sarmento et al. (2000).

Measurement of proteins, reliably in the laboratory, has been problematic owing to the presence of reactive peptides (small amino acid polymers) and interfering substances including sugars, phenolics and pigments (Boyes et al., 1997; Upreti et al., 2012). The method chosen here is the modified Bradford protocol (Coomassie brilliant blue) because of its suitability for use in a minimally resourced wine laboratory (Boyes et al., 1997). However, there are many other well-characterized assays to choose from, e.g. Amido black (fewer interfering conditions and greater linear range) (Weiss and Bisson, 2001) and the more traditional Lowry assay (adapted for a microplate reader) (Upreti et al., 2012).

These assays should be regarded as in-house and suited only for comparisons within a particular cultivar and perhaps season, owing to the diversity of proteins and polypeptides in grapes and wine and the range of potential interfering substances and conditions.

5.2 Occupational Health and Safety

- Corrosive alkali and concentrated mineral acids.

5.3 Quality Assurance Records

- Person, date, sources, manufacturer, date of manufacture, codes
- Source of reagents and their batch number
- Ratio of absorption of the three states: anionic, neutral, cationic; 595, 650, 470 (see below).

5.4 Chemistry

Coomassie brilliant blue is an amphoteric dye with two sulfonic acid groups and three basic nitrogen groups: color is therefore pH dependent, from red to blue to green as pH rises. Color is also dependent on whether the molecule exists freely in solution or bound. The dye binds preferentially to arginine, giving rise to the blue color of the bound form (Compton and Jones, 1985).

5.5 Calculations

Data may be reported simply in terms of absorbance, corrected for any dilution, or preferably as bovine serum albumen (BSA) equivalents using a standard curve prepared from BSA, in which case the values may be read off from a graph or a regression equation calculated as per Chapter 12.

5.6 Procedure

- Prepare standard BSA samples and make to 100 μL.
- Dispense 100 μL of sample (dilute with distilled water as necessary).
- Dispense 50 μL of 1M sodium hydroxide (NaOH).
- Mix gently but completely and stand for 5 min.
- Dispense 3.0 mL Coomassie brilliant blue solution.
- Read absorbance at 595 nm.
- Check for interference by reading at 470 and 650 nm.

5.6.1 Recipes

Coomassie Brilliant Blue (Stable for 1 month at RT)

- Dissolve 100 mg of dye in 50 mL aq. ethanol (90% v/v).
- Add 100 mL conc. H_3PO_4 [CARE].
- Bring to 1 L by adding to ca 500 mL distilled/deionized (DI) water, mixing, making up to 1 L and then mixing again.
- Filter twice through Whatman no. 542 or equivalent.
- Check absorbance at 470, 595 and 650 nm (reference values: 1.528, 0.55 ± 0.005, 0.716) (Boyes et al., 1997).

BSA Standards (Stable for 1 day)

- Prepare standards in duplicate at 2.5-μg intervals from 0 to 25 μg by pipetting the appropriate volume of the stock (if 250 μg/mL then 0, 10, 20 ... 100 μL); bring to 100 μL with DI water (100 to 0 μL).
- Then add NaOH and DI water as per Procedure, above.

5.7 Notes

1. The authors state that the relationship between concentration and absorbance is non-linear (cubic), but the deviance in the range is small and not unreasonable for the purpose of this assay (which, strictly speaking, is not analytical or quantitative).
2. Browning reactions can interfere with the binding of Coomassie brilliant blue and therefore it is best to work with fresh samples or to ensure protection from oxidative browning by adding ascorbic acid.
3. The reaction should be related to turbidity as a measure of haze-forming potential but does not replace that assessment.
4. If haze formation remains a problem but the protein assay is satisfactory then some other source of haze formation should be examined (bacterial or metallic, iron and/or copper) (Iland et al., 2004; Zoecklein, 1995).

5.8 Equipment and Materials

- Coomassie brilliant blue—G250
- Ethanol, technical grade (90% v/v)
- Phosphoric acid, technical grade ($\rho^{25°C} = 1.685$, 850 g/L)
- Whatman no. 542 (or equivalent)
- Spectrophotometer suitable for assay at 595 nm (supplementary at 470 and 650 nm)
- Disposable cuvettes, 10-mm pathlength, 4.5-mL capacity
- Autopipettes, 25, 50 and 100 μL
- Dispensing pipette, 3.00 mL
- BSA 250 μg/mL (or other known concentration).

6. MALIC ACID BY ENZYMATIC ASSAY

6.1 Purpose

Malic acid forms the basis of the malolactic acid fermentation. This assay enables a winemaker to assess whether there is sufficient acid to support that fermentation and to assess whether it has finished. If too little, then the wine may need to be blended with another wine with a higher level or the level may need to be supplemented with natural, L-malic acid; or accept and proceed to stabilize and age/bottle (<0.1 g/L malic acid is regarded as finished and stable) (Ribéreau-Gayon et al., 2006a).

6.2 Occupational Health and Safety

No particular issues.

6.3 Quality Assurance Records

- Enzyme kit code, manufacturing date, batch number, date of purchase, storage location and conditions, date and name of person preparing working solutions, standard analysis value.

6.4 Chemistry

Malate is measured by a system based on the production of NADH and its measurement (cf. NH_3 assay). The system requires two enzymes: L-malate dehydrogenase (MDH) to produce NADH from NAD, and glutamate-oxaloacetate transaminase (GOT) to ensure that the first reaction goes to completion by removing one of the products, oxaloacetate:

$$\text{L-Malate} + \text{NAD}^+ \overset{\text{L-MDH}}{\Leftrightarrow} \text{Oxaloacetate} + \text{NADH} + \text{H}^+$$

$$\text{Oxaloacetate} + \text{L-Glutamate} \overset{\text{GOT}}{\Leftrightarrow} \text{L-Aspartate} + \text{2-Oxoglutarate}$$

The system measures free malic acid. If you wish to measure esterified malates then these must first be converted to the free acid by hydrolysis with NaOH (see manufacturer's instructions). The kits usually contain five components: #1, buffer and glutamate; #2, NAD (dissolve in distilled water according to manufacturer's instructions); #3, GOT; #4, MDH; and #5, L-malate standard.

Derivation of Equation

The concentration is calculated as (see Ammonium assay, Section 2, for definitions of terms):

$$c = \left[\frac{v \times \text{MW}}{\varepsilon \times d \times sv \times 1000}\right] \times \Delta A \times DF$$

$$c = \left[\frac{2.220 \times 134.09}{3.4 \times 1.0 \times 0.1 \times 1000}\right] \times \Delta A \times DF$$

$$c = 0.8775 \times \Delta A \times DF \text{ g/L}$$

6.5 Calculations

Calculate net absorbance (ΔA) as:

$$\Delta A = (A_2 - A_1)_{\text{sample}} - (A_2 - A_1)_{\text{blank}}$$

Calculate the concentration (c) of malic acid as:

$$c = 0.8755 \times A \times DF$$

6.6 Procedure

- Dilute the grape or must sample by 10-fold, i.e. 1 vol. + 9 vols redistilled water.
- Clarify by ultrafiltration (0.45-μm filter) or by centrifugation.
- Pipette 1.000 mL of solution #1 (buffer, stabilizers and L-glutamic acid) into a cuvette for each sample to be assayed plus one for a standard and one for a blank (Table 11.6).
- Pipette 0.900 mL distilled water into each cuvette except the blank, into which you place 1.000 mL.
- Pipette 200 μL of solution #2 (NAD^+) into each cuvette.
- Pipette 100 μL of sample or standard (take care not to contaminate this solution) into all tubes except the blank.

TABLE 11.6 Example of Tube Contents and Record Sheet for Malic Acid Assay (Using R-Biopharm Kit)

Solution	Blank	Standard	Sample 1	Sample 2	Sample *n*
#1 (mL)	1.00	1.00	1.00	1.00	1.00
#2 (μL)	200	200	200	200	200
#3 (μL)	10	10	10	10	10
Distilled water (mL)	1.00	0.90	0.90	0.90	0.90
Sample or standard #5 (μL)	–	100	100	100	100
Read absorbance A_1 at 3 min					
#4 (μL)	10	10	10	10	10
Read absorbance A_2 + 5–10 min					

Note: May need to be altered for kits supplied by other manufacturers.

- Pipette in 10 mL of suspension #3 (GOT).
- Cover with Parafilm, mix by inversion and stand for 3 min.
- Record the absorbance (A_2) at 365 nm using distilled water to zero the instrument.
- Pipette 10 μL of #4 (L-MDH) into each cuvette, mix as before and stand for 5–10 min. Take a second reading of the absorbance (A_2) at the wavelength used earlier.

6.7 Notes

1. Calculated as malic acid, not malate ion.
2. Concentration should be between 0.35 and 3.5 g/L. If higher, dilute 10-fold; if lower use undiluted.
3. If necessary (blank or A_1 readings too high), decolorize by sealing an autopipette tip with cotton wool, add in PVPP and then sample, place in a centrifuge tube and centrifuge at about 4000 rpm for 5 min. Remove the tip and sample the cleared solution. If a centrifuge is not available, filter.
4. When dispensing samples with an autopipette, rinse the tip with each new solution.
5. The standard usually contains an exact amount of about 0.2 g/L malic acid.

6.8 Equipment and Materials

- Malic acid enzyme kit (e.g. R-Biopharm or Macrozyme®)
- 1-mL adjustable autopipette and tips
- 200-μL adjustable autopipette and tips
- 10-μL autopipette and tips
- PVPP
- Cotton wool
- Low-speed centrifuge (4000 rpm)
- Centrifuge tubes
- 10-mm pathlength, UV-grade cuvettes (Kartell PMMA code 1939 or 1961)
- Spectrophotometer (365 nm)
- 0.45-μm filters or high-speed bench-top microfuge
- Parafilm.

6.8.1 *Storage*

- Store solutions #1–#4 at 4°C.
- Solution #2, when made up, is stable for 3 weeks at 4°C and 8 weeks at −20°C. Bring all solutions to RT before use.

7. ACETIC ACID BY ENZYMATIC ASSAY

7.1 Purpose

Acetic acid is the principal component of volatile acidity. Its presence indicates a serious fault due usually to inadequate hygiene and/or control of oxygen status during fermentation or barrel aging, or conducting a malolactic fermentation in the presence of residual glucose or the use of moldy fruit. Often the implications are not only sensory but also regulatory, as many administrations set upper limits. The enzyme method has been selected because it is quick and accurate. The steam distillation method is not only much slower but also prone to serious errors due to incomplete extraction and the confounding influence of other acids including lactic, proprionic and sorbic, and sulfur dioxide (SO_2) (McCloskey, 1976).

7.2 Occupational Health and Safety

No particular issues.

7.3 Quality Assurance Records

- Enzyme kit code, manufacturer, manufacturing date, batch number, date of purchase, storage location and conditions, date and name of person preparing working solutions, standard analysis value.

7.4 Chemistry

The kit contains five reagents (based on Megazyme®, but others use similar chemistry, although volumes vary):

#1: 30 mL buffer and L-malic acid
#2: NAD^+, ATP and coenzyme A (CoA) (dissolve in 5.5 mL distilled water)
#3: 1.1 mL MDH and citrate synthase (CS) suspension
#4: Acetyl coenzyme A synthase (ACS), 1.1 mL
#5: Acetic acid standard 5 mL, 0.10 mg/mL—check label for exact value.

Acetic acid (vinegar acid) is determined by a three-step enzymatic reaction resulting in the reduction of NAD^+ to NADH:

$$\text{Acetate}^- + \text{CoA} \xrightarrow{\text{ACS}} \text{AcetylCoA} + \text{AMP} + \text{PP}$$

$$\text{L-Malate} + \text{NAD}^+ \xleftrightarrow{\text{L-MDH}} \text{Oxaloacetate} + \text{NADH} + \text{H}^+$$

$$\text{AcetylCoA} + \text{Oxaloacetate} + \text{H}_2\text{O} \xrightarrow{\text{CS}} \text{Citrate}^- + \text{CoA}$$

Acetate is the ionized form of acetic acid. The third step ensures that the reaction goes to completion. ATP provides the energy for the reaction, producing adenosine monophosphate (AMP) and pyrophosphate (PP). CoA facilitates the reaction and is regenerated, while the enzymes are ACS, L-MDH and CS.

7.5 Calculations

The concentration is calculated as acetic acid (see Ammonium assay, Section 2, for definitions):

$$c = \left[\frac{\text{Vol} + M(\text{Acetic acid})}{\varepsilon \times d \times sv \times 1000}\right] \times \Delta A \times df$$

$$c = \left[\frac{3.840 \times 60.05}{3.5 \times 1.0 \times 0.1 \times 1000}\right] \times \Delta A \times df$$

$$c = 0.4873 \times \Delta A \times df \text{ g/L}$$

7.6 Procedure

- If necessary, clarify by ultrafiltration (0.45-μm filter) or by centrifugation.
- Pipette 1.000 mL of solution #1 into a cuvette for each sample to be assayed plus one for a standard and one for a blank (Table 11.7).
- Pipette 200 μL solution #2 into each cuvette.
- Pipette 100 μL of sample #3 or standard into each cuvette, except the blank.
- Pipette 1.000 mL distilled water into the blank and 0.900 mL into each sample or standard cuvette.
- Cover with Parafilm, mix by inversion and stand until reaction ceases (5–15 min).
- Record the absorbance (A_0) at 365 nm using distilled water to zero the instrument.
- Pipette 10 μL of MDH and CS (#3) into each cuvette, mix as before and stand for 3 min.
- Take a second reading of the absorbance (A_1) as previously.
- Pipette in 20 μL of #4 and acetyl CoA, and stand for 20 min.
- Take a third reading, A_2, as previously.
- Calculate the concentration (c) of acetic acid as:

$$\Delta A = \left[(A_2 - A_0) - \left[\frac{(A_1 - A_0)^2}{(A_2 - A_0)}\right]\right]_{\text{sample}} - \left[(A_2 - A_0) - \left[\frac{(A_1 - A_0)^2}{(A_2 - A_0)}\right]\right]_{\text{blank}}$$

and

$$c = 0.4873 \times \Delta A \times df$$

7.7 Notes

1. This procedure is based on the Megazyme enzyme kit. Other manufacturers' kits (e.g. R-Biopharm) may vary in detail and the method may require adjusting. The chemistry used in these kits differs from that used originally by McCloskey (1976).
2. The equations and the table used here differ from those in the Megazyme protocol in that an additional blank is recorded, A_0. If tests show that this is not useful for your samples, then simplify the assay by removing this step (the spreadsheet calculation assumes that the assay has been simplified by removing this step: the spreadsheet calculation assumes a zero value if a value is not entered (The spreadsheet can be found on the companion website of this book)).
3. Calculated as acetic acid, not acetate ion.
4. The method follows the reduction of NAD^+ to NADH and depends on its well-characterized extinction characteristic. The standard is for checking and may best be added to a sample to check for linearity and for problem solving.

TABLE 11.7 Example of Cuvette Contents for Acetic Acid Assay by Enzymatic Protocol (Megazyme)

Solution	Blank	Standard	Samples
#1 (μL)	500	500	500
#2 (μL)	200	200	200
Redistilled water (mL)	2.10	2.00	2.00
Sample (or standard: #5) (μL)	0	100	100
Mix, stand for 3 min, read A_0			
#3 (μL)	20	20	20
Mix, stand for 4 min, read A_1			
#4 (μL)	20	20	20
Mix, stand for 12 min, read A_2; if unstable, wait 4 min and reread			

Note: Recommended volumes may vary for other manufacturers.

5. Concentration should be between 15 and 300 mg/L. If higher, dilute 10-fold or add a smaller volume; if lower, use undiluted.
6. If necessary (blank or A_1 readings too high), decolorize by sealing a pipette tip with cotton wool and add PVPP (0.2 g/10 mL). Then sample, place in a centrifuge tube and centrifuge at about 4000 rpm for 5 min. Remove the tip and sample the cleared solution. Note that the Megazyme kit contains PVPP so this should not be required in that case.
7. When dispensing samples rinse the tip with each new solution.
8. The standard contains about 0.10 g/L acetic acid (check bottle for exact amount).
9. For problem solving see the manufacturer's pamphlet.

7.8 Equipment and Materials

- Acetic acid enzyme kit (Megazyme, R-Biopharm or equivalent)
- 1-mL adjustable autopipette and tips
- 200-μL adjustable autopipette and tips
- 10-μL autopipette and tips
- PVPP
- Cotton wool
- Low-speed centrifuge (4000 rpm)
- Centrifuge tubes
- 10-mm pathlength, UV-grade cuvettes
- Spectrophotometer (365 nm)
- 0.45-μm filters or high-speed bench-top centrifuge
- Parafilm.

7.8.1 Storage

- Store #s 1–4 at 4°C.
- #2, when made up, is stable for 3 weeks at 4°C and >2 years at −20°C. Make up to volume, dispense aliquots into vials, cap and freeze until required.
- Solution #4 is stable for 5 days at 4°C (>2 years if Megazyme kit).
- Bring all solutions to RT before use.

7.9 Benchmark Values

Commonly, wine contains ca 400 mg/L acetic acid. The sensory threshold is about twice this level and the legal limits are about 1.1 g/L (dry white) to 1.2 g/L (dry red) of wine in California but a little lower in Europe (1.08 and 1.2 mg/L, respectively).

8. VOLATILE ACIDITY BY DISTILLATION

8.1 Purpose

As for the enzymatic method, this procedure provides for the estimation of acetic acid in wine. Despite having recognized problems, it remains a legal method in most jurisdictions. The official Organisation Internationale de la Vigne et du Vin (OIV) method provides for measuring contaminants, especially SO_2, sorbic and salicylic acids (if present) (OIV, 2006a). In this protocol, SO_2 is eliminated prior to steam distillation by conversation to sulfuric acid, which is not volatile (Iland et al., 2004). The method is valid for comparative purposes. Confidence in this method may be strengthened if an analysis is conducted together with a standard addition of a small, known quantity of acetic acid to a paired analysis to check for recovery. This method is, as are many others, sensitive to the matrix, i.e. the composition and pH of the solution being assayed, which means that the result may differ depending on the wine.

8.2 Occupational Health and Safety

- Steam poses a serious risk of burning and great care should be exercised to avoid such burns. Safety glasses should be worn in the event of an exploding flask.
- Risk of superheating and an explosion is greatly reduced by the addition of porcelain chips or beads into the boiling flask used as the source of steam. See Notes (Section 8.7).

8.3 Quality Assurance Records

- Person, date, sample, standards, source and concentration, air temperature.

8.4 Chemistry/Physics

The boiling point of acetic acid is higher than that of water, being 118.1°C. Even though it does not form an azeotrope with water it is not practical to purify it by distillation: the water would distill first but would take with it some acetic acid. Steam distillation overcomes this problem, especially when dealing with small concentrations. Here, you are relying on acetic acid vapor codistilling, i.e. being entrained by the steam. This takes many volumes to go toward completion. The OIV method calls for the collection of a distillate volume of at least 12–13 times the sample volume (250 mL for a 20 mL sample).

8.5 Calculations

$$CH_3COOH + NaOH \leftrightarrow CH_3COONa + H_2O$$

$$[\text{Acetic acid}] = \frac{v_{NaOH}}{1000} \times M_{NaOH} \times \frac{1000}{v_{wine}} \text{ moles/L}$$

$$[\text{Acetic acid}] = v_{\text{NaOH}} \text{ moles/L}$$

$$\text{Acetic acid} = v_{\text{NaOH}} \times 60 \text{ mg/L}$$

8.6 Procedure

8.6.1 Distillation

- Assemble still and receival flask (Figure 11.3).
  ■ Clean still by adding about 10 mL of distilled water to the stoppered inlet on the upper side of the Markham vessel. Turn on the steam inlet valve and open the drain tap at the bottom of the Markham vessel. Run steam until the vessel is hot, open the stopper to admit the water, close the drain and continue for a few minutes. Open the vent tap on the steam vessel, turn off the steam tap, open the drain tap and allow the water to automatically siphon out as the internal pressures change.
- Adjust the pH of about 50 mL of distilled water to pH 8.2 with 0.01M NaOH using a pH meter or phenolphthalein indicator (a few drops). Pour into receival flask. Ensure that the bottom of the delivery tube is immersed in the water.
- Degas a sample of wine either under vacuum or by heating, i.e. eliminate carbon dioxide (CO_2).
- Turn ON steam tap and turn vent tap to OFF. Pipette 10.0 mL of this wine into the stoppered funnel inlet on the Markham still, add 0.2–0.3 g tartaric acid, then add 0.5 mL of 0.3 mL/100 mL aqueous hydrogen peroxide (H_2O_2).
- Remove the stopper and allow the wine to enter the still. Rinse with a small volume of water and restopper. Close the drain tap to begin the distillation. Collect about 100 mL of distillate (ca 30 min).
- Open the vent tap on the steam source and close the steam tap. Allow any remaining distillate to empty into the receival flask and then remove it for analysis.
- Repeat the washing step (■) before each use.

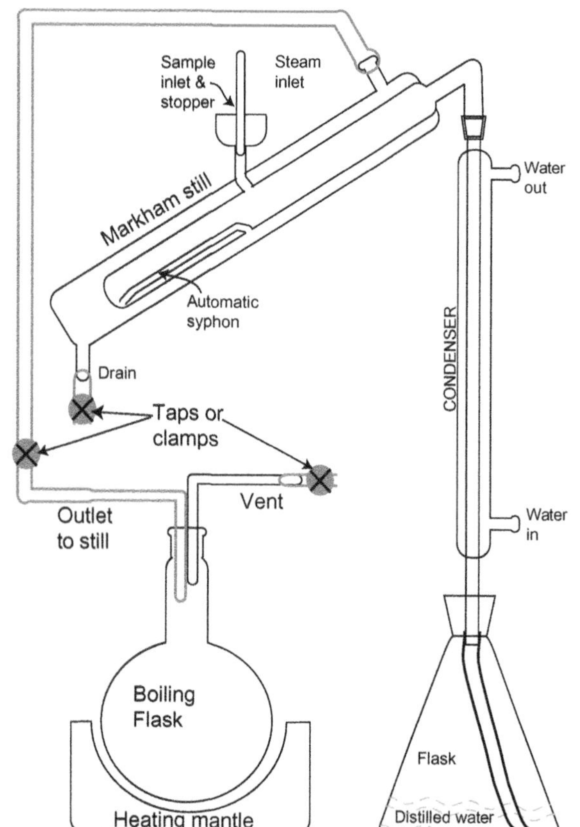

FIGURE 11.3 Diagram of a steam distillation apparatus suited to preparing volatile acids for analysis.

8.6.2 Analysis

- Titrate the entire distillate with 0.01 mol/L NaOH to a pH 8.2 (phenolphthalein) endpoint.
- Multiply the volume by 60 to give acetic acid in mg/L (provided volumes and concentrations are as given).

8.7 Notes

1. When beginning a sample or cleaning run, ALWAYS close the vent tap last.
2. When ending a sample or cleaning run, ALWAYS open the vent tap first. These steps will prevent a buildup in pressure and reduce the risk of an explosion.
3. CO_2 and SO_2 potentially interfere with the assay and hence the wine must be degassed and treated with H_2O_2 before testing.
4. The tartaric acid is to acidify the sample and thus maximize the proportion of acetic acid that is in the associated, hydrogen, form.
5. Test for carry-over of volatiles by running regular blanks of distilled water (at the beginning and then, say, every five to 10 assays). If necessary, subtract this value (Iland et al., 2004).
6. Ensure that no CO_2 remains in the system by running steam through for at least 10 min before running the first sample.
7. This test measures volatile organic acids only (e.g. acetic, proprionic and butyric) and not other volatiles such as ethyl acetate that may be more apparent. If necessary, such compounds may be measured using gas–liquid chromatography (GLC).
8. Because the method is sensitive to the matrix, recovery will vary from wine to wine. If conducting a

standard addition ensure that the same wine is measured plus and minus the addition.

8.8 Equipment and Materials

- Burette and stand
- Magnetic stirrer and bar
- 0.01 mol/L NaOH (freshly prepared or stored under Carbosorb®)
- Phenolphthalein indicator (or pH meter)
- 0.3% w/v H_2O_2
- 10 mL pipette
- 250-mL conical flask
- Source of steam
- Porcelain chips
- Steam distillation apparatus (Figure 11.3)
- Safety glasses
- Source of vacuum or heat (for degassing).

8.9 Benchmark Values

As for enzyme method (see Section 7.9).

9. POTASSIUM (SODIUM) BY FLAME PHOTOMETRY

9.1 Purpose

Potassium (K^+) is a major cation present in fruit. It is transported coincidently with sugar in the phloem and is unavoidable in that sense. However, the total quantity is also a reflection of the nutritional status of the vine and of its thermal and water-deficit history (Mpelasoka et al., 2003). High concentrations of potassium are associated with high pH in the juice and may impart an earthy flavor to the wine and have a strong impact on tartrate stability. Red wine from warm regions may contain more than 1 g/L K^+, while white and rosé wines usually have lower values. The highest concentration of potassium is in the skin of the berry, hence the difference between red and white wines (Possner and Kliewer, 1985).

Sodium, while an essential element and therefore always present, is required in tiny amounts only. Regulators set upper limits for health reasons and winemakers set upper limits for palate reasons. Wines from coastal or arid regions may accumulate excessive levels through foliar or root uptake. Rootstock selection can alleviate this problem in some instances.

9.2 Occupational Health and Safety

- Strong acids
- Toxic metals (cesium)
- Explosive gases (acetylene).

9.3 Quality Assurance Records

- Person, date, acid (and source), amount per volume, standards and sources.

9.4 Calculations

Grape Berries

$$K = \frac{c}{1000} \times \frac{v}{n} \times DF \text{ mg/berry}$$

where n is number of berries in the sample (5 here) and v is the volume of the homegenate (50 mL).

$$K = \frac{c}{1000} \times \frac{1}{W} \times DF \text{ mg/g fwt}$$

where v is the volume of the sample (50 mL), W is the combined weight of the berries (g) and c is the concentration of potassium in mg/L.

Wine

$$K = c \times 39.1 \times DF \text{ mmol/L}$$

9.5 Procedure

9.5.1 *Fruit: General Procedure*

Weigh five berries and macerate in a mortar and pestle (you may add some acid-washed sand as an aid). Add about 10 mL of HCl/CsCl, mix thoroughly and transfer to a centrifuge tube. Rinse mortar and pestle with a further aliquot of HCl/CsCl, add to centrifuge tube and then centrifuge at 3000 rpm for 5 min. Carefully pour the supernatant into a 50-mL volumetric flask using a small glass funnel and make up to volume with the dilute HCl/CsCl. Make a further dilution for measurement (e.g. 1.0 mL made to 50 mL with HCl/CsCl, $DF = 2500$).

Note: if measuring sodium then you may need less dilution (e.g. 1:10): remember to change the calculation equation.

9.5.2 *Wine*

Pipette 1.0 mL of wine and make up to 100.0 mL with the HCl/CsCl solution ($DF = 100$).

9.6 Measurement

- Prepare a standard curve using the standards provided (see Equipment and Materials, Section 9.8).
- Aspirate 10 mL of unknown into a syringe, fit a 0.45 μm filter and filter the solution into a sample tube. Label the

tube. The filter may be used a number of times but must be rinsed with the next solution before use.

- Aspirate a small amount of solution into the photometer to ensure that it is clean of previous solutions. Aspirate the remainder and take a reading.
- Calculate the potassium level either by eye from a graph or by fitting a curve (line) or read directly from instrument if it has an internal calibration.

9.7 Notes

1. The alkaline earth metals emit light of a particular wavelength when heated. In flame photometry the solution is evaporated in a flame and the finely dispersed ions are heated at the same time. The method is linear over a narrow range only (0–10 mg/L) and grape and wine samples must be diluted before measurement (they may contain about 1000 mg/L). In a complex mix such as the juice of a grape, other elements may interfere. This can be resolved by the standard additions approach or by using an ion suppressant (2–5 g/L cesium chloride in the case of potassium or sodium). We use the ion suppression method; for details of the standard addition method see Iland et al. (2004).
2. Hydrochloric acid is included to aid the extraction of salts from the skin and pulp and to ensure that the salts are not precipitated as tartrates.

9.8 Equipment and Materials

- Flame photometer with detectors tuned to potassium (sodium) ionization
- KCl standards, 0, 2.5, 5, 7.5, 10 mg/L in 2 g/L CsCl in 0.5M HCl
- 2 g/L CsCl in 0.5M HCl (for diluting unknowns) in a dispensing bottle
- 10-mL disposable syringes
- 0.45-μm filters to suit syringe
- Disposable 15–20-mL screw-cap sample tubes
- Stands to suit
- Permanent label pen
- Mortar and pestle
- Acid-washed sand
- 50-mL volumetric flask
- 10-mL adjustable dispenser (for the diluent)
- 1-mL transfer pipette
- Small, glass funnel
- Talcum-free latex gloves
- 15-mL Falcon® centrifuge tubes.

9.9 Benchmark Values

- Dry red wines: Australia—27–77 mmol/L, ca 1–3 g/L; Bordeaux—22–32 mmol/L, ca 0.9–1.3 g/L
- Berries: Australia—seeds 2.2–3.3 mg/g fresh weight; pericarp 1.3–2.9 mg/g fresh weight; skin 4.8–8.8 mg/g fresh weight.

10. ANTHOCYANINS AND TOTAL PHENOLICS IN RED GRAPES

10.1 Purpose

Phenolics and their derivatives are a good part of the components that distinguish wines made from fruit of the grapevine from those beverages made from fermentation of other fruits (a non-fermentable acid, tartaric acid, is another). In red wines, depth of color, color stability and mouth feel without undue bitterness are vital elements in the quality equation. It is not possible yet to conduct a simple analysis that will reliably predict wine quality, but it is possible to make comparative analyses for particular cultivars and within particular seasons.

While analyses to measure maturity from the standpoint of sugars and acids are well established, those for the equally important phenolic flavor components are not and the winemaker tends to rely on sensory analysis (see Chapter 7). This set of analyses was developed by Yves Glories at the Faculté d'Oenologie de Bordeaux, Université Bordeaux Seglen (reviewed by Ribéreau-Gayon et al., 2006b, but see also www.bordeauxraisins.fr). The data are used as a guide to maturity in red wine grape cultivars and to seed maturity. Seed phenolics are estimated on the assumption that flesh and skin phenolics, measured optically at 280 nm, are about 40 times that of the anthocyanins measured in this protocol, and thus any balance when this value is subtracted from the total is due to seed-derived phenols. This fraction declines as the berry and seed mature, although the total content continues to rise until maturity. Fruit is considered to be mature, phenol-wise, when this fraction reaches a maximum value and begins to decline (Ribéreau-Gayon et al., 2006b, p. 188 ff.). This procedure is designed to emulate somewhat processes that may occur during extraction and is not designed as a standard measure for anthocyanins (Lee et al., 2005).

This is but one approach to the assessment of phenolics but it is one that is publicly available and which has a long history. A simpler approach has been developed by the Australian Wine Research Institute which simply requires a range of optical density values to be entered. However, such an approach requires a very broad database with a long history. A recent review of phenol measurement in the winery supports the use of optical methods but notes their limitations (Harbertson and Spayd, 2006).

10.2 Occupational Health and Safety

- Strong acids and oxidants.

10.3 Quality Assurance Records

- Person, date, acid (and source), amount per volume.

10.4 Chemistry

The chemistry underlying this assay is set out in Chapter 3. This is a complex assay which attempts to estimate the impact of maturity on wine color and palate after fermentation. Monomeric anthocyanins are sensitive to pH, absorbing most intensely at pH 1 (flavylium form) and almost not at all at pH 4.5 (hemiketal form). The accepted standard method is to measure at both pH values and subtract (Lee et al., 2005). Degraded and polymerized anthocyanins absorb approximately equally at both. In this assay, however, pH 3.2 is chosen to estimate absorptivity in wine and the impact of sulfite is also determined to remove that as a factor (Ribéreau-Gayon et al., 2006b).

All unsaturated, cyclic ring structures such as those prevalent in phenols and polyphenols (including anthocyanins) absorb strongly in UV light at 280 nm. The observation that in all cultivars assessed the content of flesh polyphenols is 40 times (range 35–45) the absorbance due to anthocyanins enables the phenol content of seeds to be estimated by subtraction. This is a comparative assay, not one that is analytically robust. It has, however, proven useful in that context (see e.g. www.bordeauxraisins.fr).

10.5 Calculations

Malvadin monoglucoside ($\varepsilon = 28{,}000\ \text{L} \times \text{mol}^{-1} \times \text{cm}^{-1}$; molecular mass = 493.2 g/mol).

A is the absorbance value, A' is absorbance of SO_2-treated sample, DF is the dilution factor (61.2), l is the cuvette pathlength (1 cm), convert to mg (× 1000), and Δ is difference (here between the treated and untreated samples):

$$\Delta A_{520} = A_{520} - A'_{520} \quad (1)$$

(i.e. the difference between the control and the SO_2-treated samples).

Anthocyanin potential (P_{An}, pH 1):

$$P_A = \frac{(A_{520} - A'_{520}) \times 493.2 \times DF \times 1000}{\varepsilon \times l} \quad (2)$$

$$= \frac{\Delta A_{520} \times 493.2 \times 61.2 \times 1000}{28{,}000} \quad (3)$$

$$= \Delta A_{520} \times 1078\ \text{mg/L Malvidin equivalents} \quad (4)$$

Fraction extractable anthocyanin content (%):

$$E_{As} = \frac{\Delta A_{520}^{pH1} - \Delta A_{520}^{pH3.2}}{\Delta A_{520}^{pH1}} \times 100 \quad (5)$$

Fraction seed-derived phenolics (%):

$$P_s = B_{280} - \frac{\Delta B_{520} \times 40}{1000} \times 100 \quad (6)$$

10.6 Procedure

- Sample harvest unit as per Chapter 7 and record weight and bunch number.
- Prepare and weigh two subsamples of 200 berries.
- Press sample 1 using a standardized protocol. Record either the weight of the expressed juice or the weight of the retained marc. Calculate juice fraction (yield). Measure and record °Brix and calculate or measure juice density (g/mL).
- Homogenize sample 2 to a complete macerate, again using a standardized protocol, instrument, speed setting and duration. Using juice density from sample 1, calculate the weight of 50 mL of macerate; collect two 50-mL samples into tared 250-mL Erlenmyer flasks. From the balance take samples for pH and titratable acids (see separate protocols, above).
- To subsample A, add 50.0 mL of the pH 1 diluent. To subsample B, add 50.0 mL of the pH 3.2 diluent. Place on a rocker or swirl at 10-min intervals by hand over a period of 4 h or place in a fridge overnight.
- Filter or centrifuge (3500 rpm for 5 min) a portion of each and pipette 1.0 mL of each into separate flasks (50 mL). Dilute and acidify each by adding first 1.0 mL of acidified ethanol, then 20.0 mL of dilute HCl (CARE—strong acid). Mix each thoroughly (cap and invert), then stand at RT for at least 5 min.
- As per Figure 11.4, take a 10.0-mL subsample of each and add to two fresh flasks (control and sulfited), i.e. four flasks/tube in total. To the control of each add 4.0 mL of DI water. To the other, add 2.0 mL DI water and 2.0 mL sodium bisulfite (CARE—strong irritant).
- Record optical density for each of the four preparations at 520 nm (using DI water as a blank) and 280 nm for the pH 3.2 control (but first dilute 1:100 in DI water; Figure 11.4).

10.7 Notes

1. Various authors apply different values in the calculation (Equation 3), e.g. some round the molecular weight to 500 and others use a different extinction coefficient (20,000, 28,000 or even 5000). We have chosen to use Malvidin as the base because, commonly, it is the most

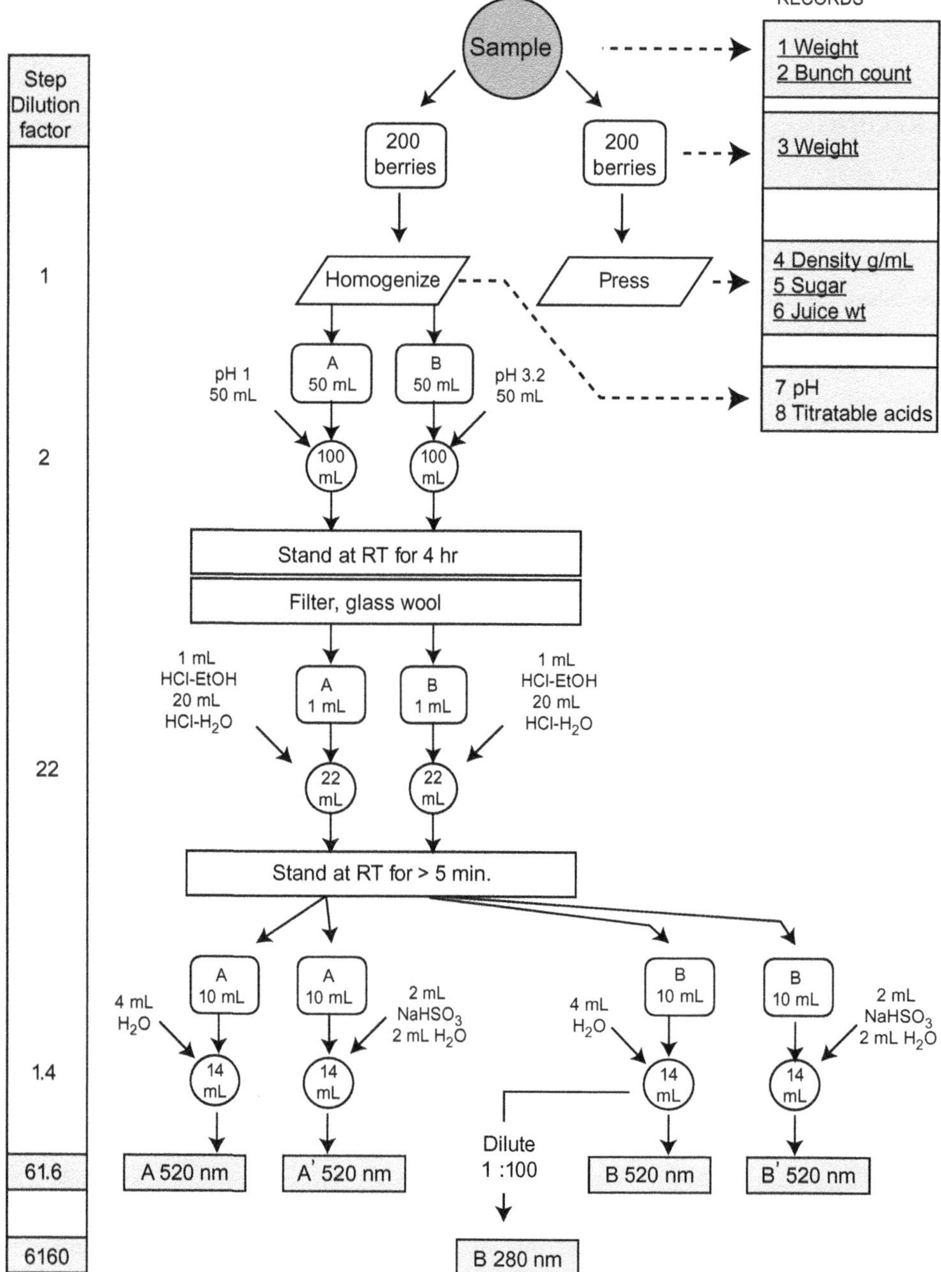

FIGURE 11.4 Flow diagram illustrating the processes and stages in the measurement of anthocyanins and total phenolics in red grapes using the protocol developed by Yves Glories. *(Ribéreau-Gayon et al., 2006b, p. 199 ff.)*

abundant form in *Vitis vinifera*. This is, however, of little consequence because assays of this type are for comparison purposes only: it is the trend, not the value, that is important. The French use a factor of 875, so their values will be a little lower than those produced by the equation used here. Aqueous ethanol solutions reportedly have a lower extinction value (Lee et al., 2005).

2. The 1000 denominator is to convert the units back to g/L.
3. The precision of the optical density value declines with density. You may need to dilute the sample to ensure that the values are between 0.1 and 1.0 optical units. If the density is routinely too high, then adjust the volumes used in the procedure (and the equation).
4. Some authors recommend that an additional reading be taken at 700 nm and the value subtracted. This is intended to correct for turbidity. It is better to ensure that all solutions are filtered or centrifuged before reading absorbance.
5. Some variants press both samples to improve consistency but this may reduce extraction efficiency. Also, we have opted to measure the pH and TA on the homogenate, rather than the pressings, because in our view this will be closer to the values achieved in

fermentation. This means that the values obtained will differ from those published on the basis of measurements made on pressings.

6. The sample for the phenolics measure at 280 nm is taken off after the addition of the HCl–ethanol, whereas other variants take the subsample before that point but still from the pH 3.2 sample after standing. If this is done, then the factors in the equations will need to be changed.

10.8 Equipment and Materials

- Top-loading balance (precision ± 0.01 g)
- Spectrophotometer (preferably UV–visible 280–800 nm; if visible only then can still determine anthocyanin content)
- Plastic or glass cuvettes (10-mm path, UV grade for the 280-nm reading)
- Low-speed bench-top centrifuge (or high-speed if only centrifuging before reading absorbance)
- Centrifuge tubes
- Blender or homogenizer (suited to larger volumes, e.g. 1–2 L)
- pH 1 diluent: 0.1 mol/L HCl
- pH 3.2 diluent: 5 g tartaric acid in 800 mL DI water; add 22.2 mL 1 mol/L NaOH, bring to 1 L with DI water; check pH and adjust if necessary
- Ethanol (technical grade, 95% v/v); 100 mL + 0.1 mL conc. HCl (37%) [CARE]
- Dilute HCl (5.4 mL, add to DI water and make to 100 mL) [CARE]
- Sodium bisulfite, 30 g/100 mL (or equivalent as metabisulfite).

10.9 Benchmark Values

See, for example, www.bordeauxraisins.fr and Figure 7.7, Chapter 7.

11. COLOR AND PHENOLICS IN RED WINE

11.1 Purpose

Color is important in all wine because it presents either a visually appealing prospect or not. Intensity and hue are important measures because they are related to quality: intensity to the quality of the fruit, small berries, low yield and exposed fruit; and hue as an index of age and oxidation state—the older and the more oxidized the higher the intensity of browning (Oliveira et al., 2011). The method presented here was developed at the Australian Wine Research Institute (AWRI) (Somers and Evans, 1977). This is a routine protocol but is not designed to measure all aspects of color in a technically correct manner (chromatic characteristics) (OIV, 2006a); see also Harbertson and Spayd (2006) for a review of progress and alternatives.

11.2 Occupational Health and Safety

- Strong acids and bases, volatile aldehydes.

11.3 Quality Assurance Records

- Person, date, sample, reagent details and sources.

11.4 Chemistry

This assay measures the principal components of color (anthocyanins, red; oxidized and condensed phenolics, brown) and bitterness/astringency/browning potential (phenols and catechols). It is a crude assay but useful for comparative purposes. Some key values are as follows:

Anthocyanins:

$$\lambda_{max} = 510 - 520\ \text{nm};\ \varepsilon = 28,000\ \text{L mol}^{-1}\ \text{cm}^{-1}$$

Diphenylquinone (a product of browning):

$$\lambda_{max} = 400\ \text{nm};\ \varepsilon = 69,000\ \text{L mol}^{-1}\ \text{cm}^{-1}$$

Catechol:

$$\lambda = 280\ \text{nm};\ \varepsilon = 2300\ \text{L mol}^{-1}\ \text{cm}^{-1}$$

Thus, oxidized quinones and their many condensation products absorb light in the visible spectrum and may have a molar absorption coefficient (ε) about three times higher than that of anthocyanins. Hence, even small quantities will be noticeable. In general, oxidative browning is not dependent on enzymes but often proceeds as follows (red and white wines) (reviewed by Oliveira et al., 2011; Singleton, 1987):

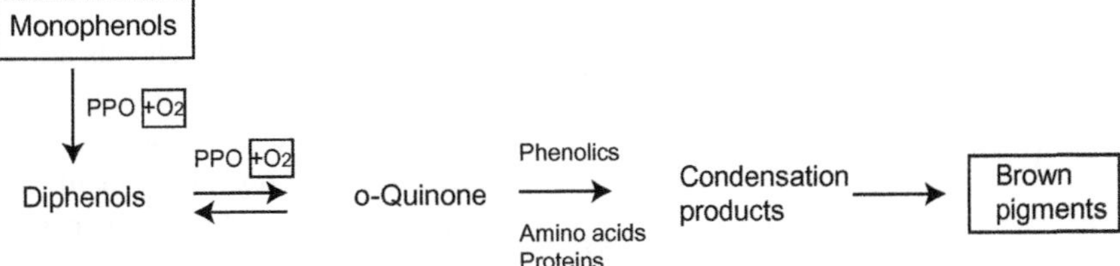

where PPO is polyphenyl oxidase.

Acetaldehyde is an important product of autoxidation of catechol and its derivatives in the presence of ethanol; first, to produce a reactive quinine and hydrogen peroxide which, in turn, oxidizes ethanol to acetaldehyde:

$$\text{Catechol} + O_2 \rightarrow \text{Diquinone} + H_2O_2$$

$$H_2O_2 + CH_3CH_2OH \rightarrow CH_3CHO + 2H_2O$$

The impact of pH, SO_2 and acetaldehyde (CH_3CHO) are shown here. Sulfur dioxide and high pH drive the equilibrium toward colorless forms while low pH and the removal of SO_2 by the addition of an excess of acetaldehyde drive the equilibrium toward the colored flavilium form.

Colorless pseudo base

pH 1 (H+)

pH 4.5 (OH-)

Colored flavilium ion

SO_2

CH_3CHO

Colorless bisulfite form

11.5 Calculations

$$\text{Color} = A_{420} + A_{520} \quad (1)$$

$$\text{Hue} = \frac{A_{420}}{A_{520}} \quad (2)$$

$$\text{Red} = A_{520}^{\text{HCl}} \times DF \quad (3)$$

$$\%\text{Red} = \frac{A_{520}}{A_{520}^{\text{HCl}} \times DF} \times 100 \quad (4)$$

$$SO_2 \text{ impact} = A_{520}^{CH_3CHO} \quad (5)$$

$$SO_2 \text{ stable} = A_{520}^{K_2S_2O_5} \quad (6)$$

$$\text{Phenolics} = (A_{280}^{\text{HCl}} - 4) \times DF \quad (7)$$

11.6 Procedure

- Adjust a portion of the wine (ca 100 mL) to pH 3.5 with dilute NaOH or HCl as appropriate (or to another pH, e.g. Glories recommends 3.2 if you wish to have consistency, but remember to record the value used).
- Prepare following solutions in Table 11.8 for both raw wine and wine adjusted to pH 3.5 (alternatively 3.2) (directly in cuvette if using 10-mm pathlength cells).
- Cover and invert to mix.

TABLE 11.8 Cuvette Contents for Red Wine Color and Hue Analysis (Age/Oxidation)

Measure	Volume of Wine (mL)	Volume of Reagent	Holding Time at RT
Color and hue of raw wine	2.00	–	
Color and hue of adjusted wine (pH 3.5)	2.00	–	
Color and hue with SO_2 removed	2.00	20 μL CH_3CHO	45 min
Color and hue—resistance to SO_2	2.00	30 μL $K_2S_2O_5$	1 min
Color and hue (pH 3.5) with SO_2 removed	2.00	20 μL CH_3CHO	45 min
Color and hue (pH 3.5)—resistance to SO_2	2.00	30 μL $K_2S_2O_5$	1 min
Red color and phenolics	0.100	10.0 mL 1 mol/L HCl	4 h

RT = room temperature; SO_2 = sulfur dioxide.
Note: The original article recommended undiluted wine and narrow, quartz cuvettes, whereas these values assume standard, disposable cuvettes. This should have little impact on ratio values but may affect direct measures because non-linearity in the dilution/absorbance curve will affect the application of a dilution factor, which assumes linearity (Ribéreau-Gayon, 1974; Somers and Evans, 1977). This set may be abbreviated by omitting the unadjusted wine. Holding time is the waiting period between addition and measurement.

- Switch spectrophotometer on, turning on both the tungsten and the mercury (Hg) vapor lamps, and allow to warm up and stabilize.
- If available, read all but the last mixture in a 1-mm pathlength quartz cuvette using water as a blank. DO NOT DILUTE THESE SAMPLES.
  - Alternatively, dilute ca 10-fold and use a 10 standard pathlength cuvette.
  - Be sure to rezero when changing wavelength; record A_{280}, A_{420} and A_{520}.
- Read the last mixture using a 10-mm pathlength cuvette. This mixture may be further diluted. Remember to allow for the dilution factor (DF = 101) when calculating % red and phenolics.

11.7 Notes

1. A standard 10-mm cuvette is frequently used instead of a short pathlength cuvette, but be aware that red wines may not meet the assumptions of the Beer–Lambert law when diluted (Ribéreau-Gayon, 1974).
2. Color intensity of simple anthocyanins is strongly influenced by pH and SO_2 concentration [SO_2]. Anthocyanins may be measured exclusively, and simply, using the high/low pH method (Lee et al., 2005).
3. The absorbance of polymerized anthocyanins is relatively stable to pH and to [SO_2], reportedly increasing only by 5/3 at pH 1 cf. pH 3.5 (Somers and Evans, 1977).
4. Acetaldehyde binds far more strongly with SO_2 than do anthocyanins, so it is added in excess to effectively remove all free SO_2.
5. As for white wines, Equation (7) contains an 'average' constant. This is fine for in-house comparisons of like with like, but risky if wider comparisons are intended. If that is the intention then charcoal and/or PVPP should be used as a comparison when estimating the phenolics.

11.8 Equipment and Materials

- UV–visible spectrophotometer
- 1-mm pathlength quartz cell [CARE—expensive and fragile]: not usually justified
- 10-mm pathlength disposable cells (Kartell PMMA 1939 or 1961)
- 1-mol/L HCl
- 1 mol/L NaOH
- 25% w/v $K_2S_2O_5$
- 10% w/v acetaldehyde (CH_3CHO)
- pH meter and standards
- Magnetic stirrer and bar (small).

12. ANTHOCYANINS BY CELLULOSE CHROMATOGRAPHY

12.1 Purpose

Anthocyanins are a signal component for all red wine and table grapes. In *V. vinifera*, they are present as a glycosylated derivative (usually the 3 position of the C ring). Anthocyanins from other species (e.g. species native to America) are frequently diglycosides and their presence can be readily distinguished by partial hydrolysis, releasing a mixture of aglycone, monoglycone and diglycone. Thus, for diglycosides, one spot becomes three and for monoglycosides, one spot becomes two. Tests for adulteration may be conducted in this manner or using high-pressure liquid chromatography.

12.2 Occupational Health and Safety

- Concentrated acids and volatile solvents.

12.3 Quality Assurance Records

- Person, date, acid (and source), amount per volume.

12.4 Chemistry

Many forms of chromatography separation comprise a stationary phase and a mobile phase. Separation depends on the relative solubility of each substance in the two phases. Those more soluble in the mobile phase will tend to move with that phase and remain close to the advancing solvent front. Conversely, those substances that are sparely soluble in the mobile phase but more soluble in the stationary phase move slowly as they tend to spend more time in the stationary phase than the mobile phase. In paper or cellulose chromatography, the mobile phase is the solvent and the stationary phase is water bound to the cellulose, and indeed the cellulose itself, which can absorb some substances and thus impede their progress.

12.5 Calculations

Calculate the R_f as:

$$R_f = \frac{d_s}{d_f}$$

where d_s is the distance from point of application to the midpoint of substance of interest and d_f is the distance the solvent has moved beyond the point of application of the substances.

12.6 Procedure

12.6.1 Extraction

- Peel or chop the tissue coarsely, record its fresh weight, place into methanol–HCl (99:1, v/v) and leave covered in an explosion-proof fridge at 4°C overnight. Use 100 mL solvent per 10 g tissue.
- Filter and rinse the tissue in methanol and then dry under vacuum (or under a stream of dry nitrogen). Redissolve and make up to volume with methanol (5 mL).

12.6.2 Hydrolysis

- Pipette about 1 mL of extract into a test-tube and add 1 mL of 4M HCl.
- Place in a water bath or heating block at 80–90°C and cover with a glass marble.
- At 0, 15, 30, 45 and 60 min, remove 100 μL and dispense into Eppendorf® tubes for spotting on paper (usually backed thin-layer cellulose or glass slides coated with a thin layer of ultrafine cellulose).

12.6.3 Chromatography

- Cut a 7 × 10–15-mm rectangle of cellulose thin-layer chromatography (TLC) paper; very lightly mark an origin with a pencil at about 2.5 cm from the base. Label five lanes about 1 cm apart.
- Drop by drop, spot the extracts on to the origin, allowing them to dry between drops (keep the spot size as small as possible).
- Place the paper into a chromatography tank containing the solvent mixture (HCl–formic acid–water, 19.0: 39.6:41.4); BE CAREFUL. Ensure that the line at which the substances were spotted sits about 5–10 mm above the surface of the solvent.
- When the solvent has run about 100 mm, remove the chromatogram from the tank and allow to dry in a fume hood or well-ventilated area. Do not allow the solvent to move to the top of the chromatogram sheet.
- Measure the distance of the spots from the base and the solvent front and calculate the R_f for each spot.

12.7 Notes

1. Examine and record the impact of the hydrolysis of the sugar on the R_f of the spots. Present in *V. vinifera* are malvidin >> petunidin, peonidin, cyanidin, delphindin.

12.8 Equipment and Materials

- Fruit (extract)
- Red cabbage as a comparison (extract)
- Aluminum-backed cellulose TLC plates
- Chromatography tank (TLC)
- Solvent: conc. HCl–formic acid–water, 19.0:39.6: 41.4 (v/v/v)
- Methanol–HCl, 99:1 (v/v)
- Nitrogen gas for concentrating the extract
- Water bath (80–90°C)
- Test-tubes (12 mm diameter), marbles and racks (submersible)
- 100-μL transfer pipette and tips.

13. PHENOLICS IN WHITE GRAPES AND WINE

13.1 Purpose

Most white wines are prepared from pressings with limited contact with the primary sources of phenolics: seeds and skin. However, this is not always the case, as in some Sauvignon blanc styles that involve some skin and seed contact in strainers and which may involve minimally mature fruit. Also, higher than desirable levels of phenolics may arise from hard pressing or from juice recovered from lees. Techniques such as those presented here provide a rapid and economical means of determining the phenol content. If analyzing wines, then it is important to include a measure at 420 nm as an estimate of oxidative browning.

13.2 Occupational Health and Safety

No particular issues.

13.3 Quality Assurance Records

- Person, date, samples and sampling protocol.

13.4 Chemistry

See previous sections on anthocyanins, color and phenolics.

13.5 Calculations

Abbreviations: CAE, caffeic acid equivalents; TP, total phenolics.

Caffeic Acid

Molar mass = 180.15 g/mol

ε (10 mg/L), $\lambda_{280} = 0.90$ and $\lambda_{320} = 0.60$ au (10-mm pathlength);

$$\text{CAE} = (2) \times \frac{10}{0.9}\ \text{mg/L}$$

or

$$= (A_{320} - A_{320\text{pvp}}) \times \frac{10}{0.9}\ \text{mg/L}$$

and

Catechin

Molar mass = 290.28 g/mol

ε (10 mg/L) $\lambda_{280} = 0.14$ and $\lambda_{320}\ \text{nm} \cong 0.0$

$$\text{TF} = (3) \times 10/0.14\ \text{mg/L}$$

or

$$= (A_{280} - A_{280\text{pvp}}) - (A_{320} - A_{320\text{pvp}}) \times \frac{0.6}{0.9} \times \frac{10}{0.14}\ \text{mg/L}$$

13.6 Procedure

13.6.1 Standard Method

- Prepare sample by pressing or drawing off a sample (blending or homogenizing may be appropriate for research or teaching purposes). See Figure 7.6 in Chapter 7 for sample processing.
- Clarify juice samples and homogenates by centrifugation at 4000 g for 10 min or by coarse then fine filtering (0.45 μm). Simply fine filter wine.
- Measure and record the absorbance at 280 and 320 nm (and 420 nm for wines) against a DI water blank. Dilute if necessary to bring within an absorbance range of < 1.0 au (absorbance units).
- Make the following calculations. The constants are average values derived from a survey of over 400 white wines of Australian origin (Somers and Ziemilis, 1985):

$$\text{Total phenolics (TP)} = (A_{280} \times DF) - 4 \qquad (1)$$

$$\text{Total CAE} = (A_{320} \times DF) - 1.4 \qquad (2)$$

$$\text{Total flavonoids (TF)} = (1) - \frac{0.6}{0.9} \times (2) \qquad (3)$$

$$\text{Brown pigments} = A_{420} \times DF\ \text{(wine only)} \qquad (4)$$

where A is measured absorption.

13.6.2 Alternative Method

- As above, except prepare about 10 mL; split into two; pour 5 mL into a clean centrifuge tube and add 0.5 g PVPP (important to add an excess unless evaluating fining protocols). Mix thoroughly and leave for 30 min. Add a pellet of dry ice to prevent browning and cap or cover both samples (or add 1 mg ascorbic acid to the original 10 mL).
- Centrifuge for 10 min at 4000 g or filter using 0.45-μm membrane.
- Either: read both the plus and the minus PVPP against water or use the PVPP as the blank. The difference between the two readings represents the true level of phenolics in your sample (Figure 11.5).

$$\text{Total phenolics (TP)} = (A_{280} - A_{280\text{PVPP}}) \times DF \qquad (5)$$

$$\text{Total CAE} = (A_{320} \times DF) - 1.4 \qquad (6)$$

$$\text{Total flavonoids (TF)} = (4) - \frac{0.6}{0.9} \times (5) \qquad (7)$$

$$\text{Brown pigments} = A_{420} - A_{420\text{PVPP}} \qquad (8)$$

13.7 Notes

1. The standard method is a rough method, suitable for comparisons of similar material but subject to many assumptions due to the diversity of phenolic compounds in grapes and wine, the impact of other substances which absorb in similar regions of the

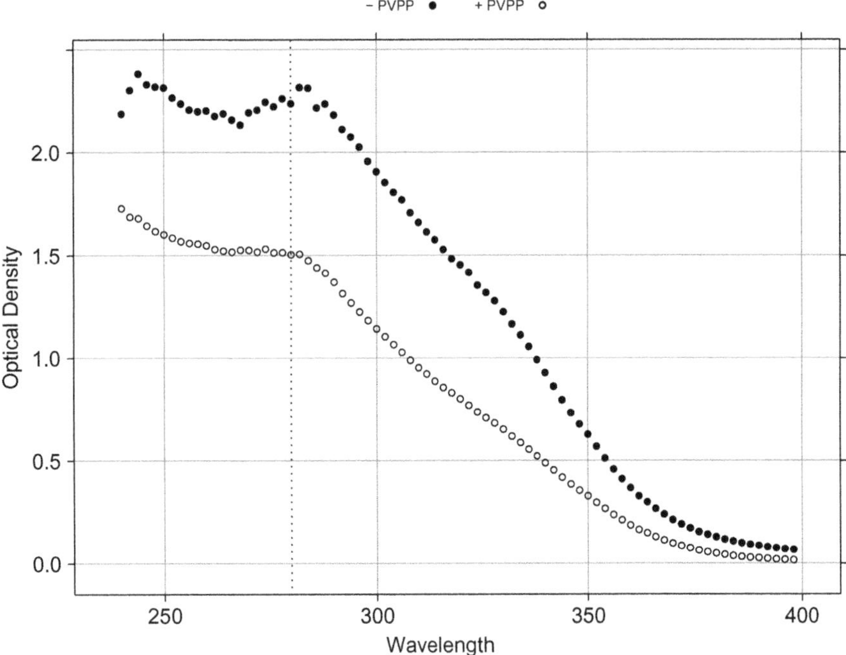

FIGURE 11.5 Graph showing the difference that an addition of polyvinylpolypyrrolidone (PVPP) makes to the absorbance spectrum of a commercial Chardonnay wine. The lower curve (+ PVPP) represents the absorbance due to non-phenolic substances, and the upper (− PVPP), that due to cinnamic acids and related compounds and to catechins and their relatives.

spectrum and incomplete absorption of many phenolics by PVPP (used to fine white wines by removing undesirable phenolics: catechins and related compounds). Phenolics are not normally in high concentration in white wine because they are located principally in the skin and seeds, which are usually discarded immediately. The method may be improved if before and after treatment with PVPP is undertaken, as originally suggested by Glories (Figure 11.4) and Somers and Ziemilis (1985).

2. Sulfur dioxide and sorbic acid (added to sweet white and rosé wines) interfere with the determination in wine. To prevent oxidative browning of must for analysis, add ascorbic acid (ca 1 g/L) or keep under nitrogen or CO_2.
3. Flavonoids (tannins, procyanidins and catechins) absorb strongly at 280 nm while the monocyclic phenols (cinnamates and hydroxycinnamates) absorb strongly at 320 nm. Tryon et al. (1988) report on limitations of the method. Our experience suggests that the standard method is not suitable for juice or berry components.
4. The two methods presented represent one based on Australian experience (AWRI), while the other provides absolute values. Note that the absorbance values reported are >> 1; therefore, be prepared to dilute your samples and adjust the equations accordingly.
5. This assay presents some problems in clarifying the solutions: PVPP filters well but has a density close to water and so is difficult to remove by centrifugation, while blended grape samples may be difficult to filter owing to the presence of pectins and proteins. Thus, a combination may be required: first centrifugation, then filtration. Use non-absorbent filter media.

13.8 Equipment and Materials

- UV–visible spectrophotometer (280 and 320 nm)
- PVPP (e.g. Polyclar AT®)
- 10-mm, UV cuvettes (Kartell PMMA 1939 or 1961)
- 10-mL disposable centrifuge tubes
- Centrifuge ($4000 \times g$)
- Blender or homogenizer
- 0.45-μm filter (or high-speed bench-top centrifuge)
- Dry ice or ascorbic acid.

13.9 Benchmark Values

- Red wines: typically $A_{280} = 5-15$ au (17–130 mg/L).
- White wines: $A_{320} = 3-13$ au (0–3.1 mg/L).
- A_{CAE}/A_{TF} >1.5 in wines from free-run juice. Lower values indicate coarse, bitter wine. Such wines are also strongly susceptible to oxidative browning.

14. TOTAL PHENOLICS BY COLORIMETRY

14.1 Purpose

This an alternative method to the direct optical density procedures described previously. It has one particular and potential advantage over those: it can be conducted with a low-cost visible light or filter spectrophotometer and it can give a more detailed analysis of the composition. The reaction also produces an end-product that is the same irrespective of the nature of the reacting phenol, greatly simplifying standardization. It is possible to remove flavonoids by precipitation with formaldehyde, but formaldehyde is highly noxious and not suitable for use in a general wine laboratory (see review by Harbertson and Spayd, 2006).

14.2 Occupational Health and Safety

- Heavy metals
- Strong acids.

14.3 Quality Assurance Records

- Person, date, sample details and origin, reagents and their details.

14.4 Chemistry

The reaction is based on the reduction of phospomolybdic/phosphotungstic acids to form complex colored compounds (mixed salts of these) which show increasing absorption from about 550 to 750 nm. Most commonly, this method is used to measure protein concentration using the Lowry procedure (in copper/alkali/tartrate).

14.5 Procedure

- Prepare about 2 mL of sample by pressing, sampling, blending or homogenizing.
- Filter (coarse then 0.45 μm) or centrifuge in a bench-top centrifuge to clarify.
- Dilute 1:10 (or take 100 μL and dispense into test-tube and then add 900 μL distilled or deionized water). Prepare a duplicate sample.
- Dispense 2.5 mL of 1:10 Folin–Ciocalteu reagent.
- Mix thoroughly and stand for 30 s (no longer than 8 min).
- Dispense 4 mL Na_2CO_3 reagent and mix.
- Stand for 2 h (can be less if prepare standard under the same conditions).
- Read against a water blank at 765 nm.
- Compare with standard and calculate content.

14.6 Notes

1. This is based on the Association of Official Analytical Chemists (AOAC) method. It is for total phenols and includes both cinnamates and catechins. These can be separated but the method requires the use of formaldehyde—a highly noxious substance—and, as such, is not suitable for general use. This is unfortunate because the method is otherwise simple and robust and uses relatively inexpensive equipment. It gives values for condensed tannins which are approximately equal to the sum of the number of units.

14.7 Equipment and Materials

- Folin–Ciocalteu reagent diluted 1:10 with distilled water in a dispensing bottle set to 2.5 mL (store at RT in a tightly sealed bottle)
- 75 g/L anhydrous Na_2CO_3 in distilled water in a dispensing bottle set to 4.0 mL
- 1-mL pipette for dispensing juice
- Bench-top centrifuge (2 mL) or 0.45-μm filter
- Centrifuge tubes (disposable)
- Spectrophotometer: 765 nm
- 10-mm disposable cuvettes
- 5 g/L gallic acid standard
- Approved fume hood if using the formaldehyde step.

15. TANNINS

15.1 Purpose

Low molecular weight polyphenols (tannins) are responsible for the astringent sensation associated with young red wines in particular. These can be assessed sensorially but the method has many limitations including sample number for individual tasters, as well as being subjective. An objective measure helps to ensure consistency in evaluation and in fining.

15.2 Occupational Health and Safety

- Hot water.

15.3 Quality Assurance Records

- Person, date, sample details, sampler, reagents, batch numbers, manufacturer and date purchased.

15.4 Chemistry

Proteins and certain other polymers react with polymeric phenolics, either condensed or hydrolyzable, to form large complexes which precipitate depending on the matrix (e.g. pH, ionic concentration, solvent) and temperature. It is this reaction that produces the 'drying' sensation when an astringent wine is tasted. Several have been developed based on this reaction for use in wineries: the gelatin index, the ovalbumin test (egg white), the ferric chloride/bisulfite test (Adams–Harbertson UC Davis tannin test, reviewed by Harbertson and Spayd, 2006; Herderich and

Smith, 2008) and a recent methyl cellulose test refined by Sarneckis et al. (2006). The latter method is presented here as it is simple, rapid, tolerant of variation in pH and alcohol levels and closely related to more traditional tests and to wine sensory attributes. Furthermore, it involves reagents that are of low health risk. It and the Adams–Harbertson test have been further modified for high-throughput assays using a multiwell plate reader (Mercurio and Smith, 2008).

15.5 Procedure

15.5.1 Standard Curve—Optional

- Prepare a standard series of (−)epicatechin dissolved in DI water, e.g. 10, 50, 100, 150, 200, 250 mg/L. Read absorbance at 280 nm against a DI water blank. Use this to record results in epicatechin equivalents as a way of standardizing the results and assisting in comparisons across laboratories and seasons.
- Prepare and zero the spectrophotometer.
- Prepare grape extracts if necessary (homogenize in 1:1 v/v DI water/95% ethanol, centrifuge or filter).
- Begin preparation of controls and samples by pipetting matched volumes into tubes, one for each, as per Table 11.9.
- Complete the control preparation:
  - **a.** Add ammonium sulfate.
  - **b.** Make up to volume with DI water and mix (seal with Parafilm and invert 10 times or use a vortex mixer).
  - **c.** Stand for 10 min at RT.
  - **d.** Centrifuge for 5 min (4000 rpm if 10 mL, 10,000 rpm if 1 mL).
  - **e.** Transfer to cuvette and record absorbance at 280 nm.
- Red wine or ferment check beyond day 3 (dilute these if necessary and record dilution) or grape extract:
  - **a.** Dispense 1 vol. of methyl cellulose.
  - **b.** Shake to mix and stand for 2–3 min.
  - **c.** Add ammonium sulfate.
  - **d.** Make up to volume with DI water and mix (seal with Parafilm or foil and invert 10 times or use a vortex mixer).
  - **e.** Stand for 10 min at RT.
  - **f.** Centrifuge for 5 min (4000 rpm if 10 mL, 10,000 rpm if 1 mL).
  - **g.** Transfer to cuvette and record absorbance at 280 nm.
- Calculate tannins in arbitrary units by subtraction.
- Optional: calculate tannins in epicatechin equivalents according to the standard curve.

15.5.2 Calculations

Wines and Musts (Juices)

$$A_{280}^{T} = (A_{280}^{\text{control}} - A_{280}^{\text{sample}}) \times DF$$

where DF = dilution factor = 40 for wine/must × any subsequent dilution to bring sample in range of spectrophotometer reading (<1.0).

TABLE 11.9 Suggested Volumes of Solutions for the Methyl Cellulose (MC) Tannin Assay Using Standard Cuvettes

Part	Units	Sample	MC	$(NH_4)_2SO_4$	DI Water
Wine: 10 mL total volume					
Control	mL	0.25	0.0	2.0	7.75
Sample	mL	0.25	3.0	2.0	4.75
Grape extract: 10 mL total volume					
Control	mL	1.0	0.0	2.0	7.0
Sample	mL	1.0	3.0	2.0	4.0
Wine: 2.0 mL total volume (standard cuvette)					
Control	μL	50	0.0	400	1550
Sample	μL	50	600	400	950
Grape extract: 2.0 mL total volume					
Control	μL	200	0.0	400	1400
Sample	μL	200	600	400	800

Note: DI = distilled/deoinized.
(Sarneckis et al., 2006; AWRI, 2009).

Option—epicatechin equivalents:

$$[\text{Tannin}] = E^T_{280}/b^{\text{std}} \text{ mg/L}$$

where b^{std} = slope of the regression line of A plotted against [epicatechin] in (mg/L) as per the standard curve (this should intersect zero).

Grape Extracts

$$A^T_{280} = (A^{\text{control}}_{280} - A^{\text{sample}}_{280}) \times DF \times \frac{V_e}{W_g}$$

where DF = dilution factor = 10 for wine/must × any subsequent dilution to bring sample in range of spectrophotometer reading (<1), V_e is the final volume of the extract, and W_g is the fresh weight of the grape sample.

Option—epicatechin equivalents, as per wine samples: divide the adjusted absorbance by dividing by the slope of the regression coefficient for the standard curve.

15.6 Notes

1. The A^{control}_{280} should always be much greater than the A^{sample}_{280} because the methyl cellulose should have precipitated most of the polymeric phenolics that also absorb at that wavelength. This method separates the phenolics into those that can be precipitated and those that cannot. The control is a measure of the total phenolics while the sample measures that which remains after the polymeric phenols have been removed by the methyl cellulose.
2. The composition of tannins varies markedly even with *V. vinifera* and through the wine-making process; thus, even if using the epicatechin equivalent value, differences will arise that are due to tannin composition and differences in their extinction coefficient (Sarneckis et al. 2006).

15.7 Equipment and Materials

- UV–visible spectrophotometer (<280 nm)
- Quartz or UV-grade acrylic cuvettes (1-cm pathlength)
- Centrifuge (4000 rpm, 1800 *g*) and 10-mL tubes or microfuge and 1.5-mL tubes
- Autodispenser/pipettes: range 0.25–7.75 mL- or 25–775-μL autopipettes
- Vortex mixer or rolling or orbital table
- Test-tube or other appropriate racks
- (−) Epicatechin (for standard curve)
- Saturated ammonium sulfate solution
- 0.04% w/v methyl cellulose solution
- Hotplate or gas burner
- Fridge/freezer (or an ice bath)
- Ammonium sulfate, $(NH_4)_2SO_4$, reagent grade
- Methyl cellulose, viscosity of 2% w/v in water of 1500 cP at 20°C
- 1-L volumetric flask
- Magnetic stirrer and medium stirrer bar
- Plastic or glass ice bath (that can sit on the stirrer)
- Crushed ice or a volume of cold water (near 0°C).

15.8 Reagent Preparation

Ammonium Sulfate

- Add reagent grade $(NH_4)_2SO_4$ to a volume of distilled or deionized water (e.g. 300 mL), stirring continuously until no further salts dissolve.
- Continue adding until a level of about 10% of the volume of excess, undissolved salts is present.
- Store reagent in a sealed vessel. Reagent is stable at RT.

Methyl Cellulose

- Store at RT for up to 2 weeks; discard if flocculant develops.
- Cool 700 mL of DI water to between 0 and 5°C.
- Heat 300 mL of DI to 80°C.
- Pour the hot water into a 1-L volumetric flask and slowly add the methyl cellulose while stirring vigorously and continuously: avoid allowing lumps to form. Use a Pasteur pipette to wash down the neck of the flask.
- Now add the cold water and stir for a further 30 min while immersed in cold water or ice.
- Remove from the ice and continue stirring until clear (overnight).
- Allow to come to RT and then make to volume with DI water.

15.9 Benchmark Values

See Table 11.10.

16. ALCOHOL BY DISTILLATION

16.1 Purpose

The method given here is no longer commonly used by certified laboratories for the purposes of meeting legal labeling requirements because it is time consuming and thus expensive. The method can, however, serve as the basis for obtaining distillate to meet legal requirements in a certified laboratory. More commonly, alcohol content is determined using one or other forms of near-infrared (NIR) or Fourier transform infrared (FTIR) spectroscopy, or by GLC. These latter methods also require much smaller samples, which is a boon for researchers.

TABLE 11.10 Average Concentration of Tannins in Selected Cultivars Sampled from a Range of Locations in Australia (Epicatechin Equivalents)

Sample Type	Statistic	Shiraz/Syrah	Cabernet Sauvignon	Merlot
Grape extract (mg/g)	Average	4.15	5.5	5.35
	Min.	2.39	2.56	1.93
	Max.	8.0	7.86	8.35
Red wine (g/L)	Average	1.85	1.97	1.54
	Min.	0.36	0.12	0.31
	Max.	4.8	4.17	2.98

Source: (AWRI, 2009, with permission).

However, the distillation and the boiling point methods are suitable for the small wine laboratory.

16.2 Occupational Health and Safety

- Heat: flame or electric mantle, steam; risk of explosion of boiling sample if overheated.

16.3 Quality Assurance Records

- Person, date, sample, adjustments, [SO_2], volatile acidity, water bath temperature, hydrometers and their certified standards.

16.4 Chemistry/Physics

There are two processes in play in this procedure: the separation of volatile and non-volatile components by distillation, and the codistillation of two solvents that interact and alter each other's behavior. The method used to estimate the alcohol content in this procedure relies on the physical properties of ethanol and water. Here, the property of interest is density. The density of the solution is affected by all dissolved substances. Sugars in particular affect density. Thus, the first step is to separate ethanol and water from the other largely non-volatile components.

Ethanol and water form a negative azeotrope and at high proportions of ethanol, a constant boiling mixture of about 96% ethanol will form. The boiling point of this mixture is a little lower than that of pure ethanol—other means are required to remove the remaining water. Here, we ensure an excess of water to force the composition of the vapor phase toward 100% water because, when the proportion of ethanol is less than the azeotropic value, ethanol predominates in the vapor phase, and thus the composition is driven toward pure water as the distillation proceeds.

16.5 Procedure

- Fill a 250-mL volumetric flask with the sample and place it in the water bath for 15 min to equilibrate to 20°C. Adjust volume with a glass Pasteur pipette to precisely 250 mL. If a water bath set to 20°C is not available, then a large vessel of water that has equilibrated to RT will suffice as this will hold a relatively constant temperature over the course of the distillation.
- If necessary, adjust the wine to pH 8.2 with 0.1 mol/L NaOH (i.e. if SO_2 >200 mg/L or volatile acidity >1 g/L).
- Pour the wine into the boiling flask, rinsing the volumetric flask and other containers used subsequently with a volume of about 200 mL of additional water. Add the boiling chips to the flask (essential safety issue to prevent overheating and a consequential explosion!).
- Assemble the still (Figure 11.6), reusing the volumetric flask as the collection vessel. Ensure that all seals are leakproof, and that water is running in the condenser. A minimal volume of water should be placed in the volumetric flask and the flexible delivery tube from the condenser set below the surface of the water. A good precaution is to pack ice around the volumetric flask. This will ensure complete condensation of the condensate.
- Heat the boiling flask to a vigorous boil, adding a few drops of an anti-foaming agent if necessary. Collect about 200–220 mL. This will take about 30 min.
- Turn off the heat and remove the volumetric flask and place it in the water bath at 20°C (until the temperature has equilibrated). Adjust the volume again to precisely 250 mL with a glass Pasteur pipette using water equilibrated in the water bath.
- Rinse a hydrometer tube and appropriate hydrometer with a small amount of the distillate. Check the reading (read off just below the meniscus). Check

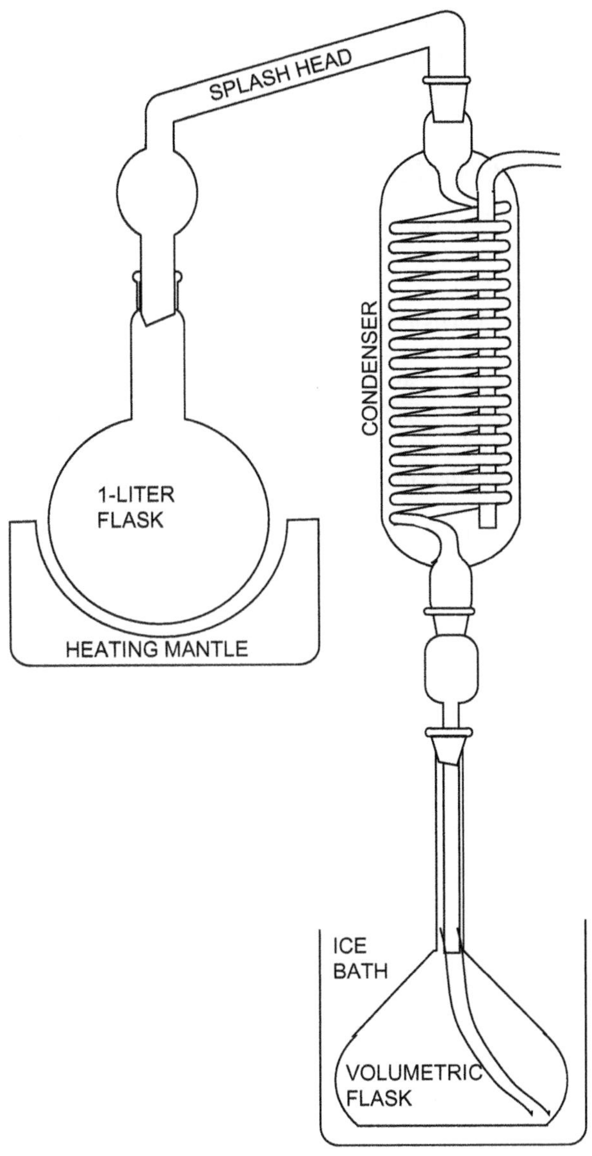

FIGURE 11.6 Diagram of an example of a simple distillation apparatus used to prepare alcohol for analysis by density (physically or electronically).

the temperature and adjust if necessary using Figure 11.7.

16.6 Notes

1. The purpose of this exercise is to separate the alcohol from all other wine components except water and then to measure its concentration by hydrometry (alcohol reduces the density of water, whereas sugars increase it). It is important to ensure that the final volume exactly equals the original volume; hence the use in this procedure of volumetric flasks and a constant temperature.

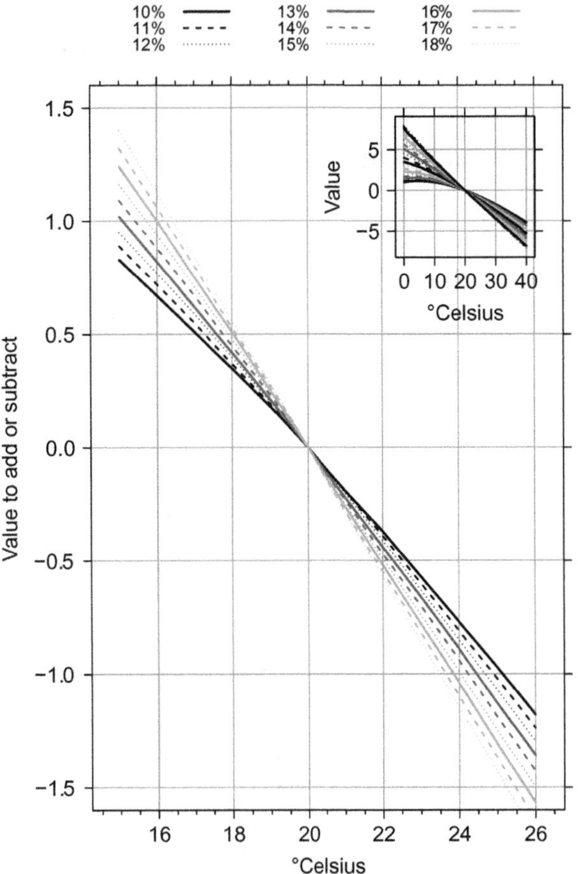

FIGURE 11.7 Graph of correction values to add to or subtract from the observed density at a limited range of temperatures and alcohol proportions. This data range is well fitted by the equation: $y = 0.9931 - 0.05280C + 0.2577alc - 0.01274C \cdot alc$, where C is observation temperature (°C), and alc is the recorded alcohol fraction (%, v/100 v). Adjusted $R^2 = 0.999$ (14.9 > < 25°C, 9.9% > < 18% v/100 v). The fit declines progressively outside this range due to curvilinearity at lower temperatures and percentage alcohol levels. The OIV table encompasses a range from 0 to 40°C and 0 to 30% alcohol, as represented in the inset graph. *(OIV, 2006a).*

2. Wine for sale or export must be accompanied by a certified document stating the level of alcohol (v/100 v at 20°C). Therefore, this is a legally important assay and should be verifiable independently.

16.7 Alternative Methods

Alcohol content may also be measured using either boiling point or freezing point depression, also known as psychrometry. Both methods are affected by the presence of other dissolved substances, especially sugars, but both use smaller amounts of wine and are suited to experimental and routine analysis. Neither method is appropriate to meet legal requirements, although other methods such as NIR or FTIR spectrometry or GLC (not discussed here) may be necessary and are the usual methods conducted for legally binding assessments. An electronic densitometer (pycnometer) may also be used, as may refractometry (OIV, 2006a).

16.8 Equipment and Materials

- 250-mL volumetric flask
- Water bath at 20°C (this will usually be a refrigerated, thermostatic, bath)
- Ice
- Hydrometer flask
- Alcohol hydrometers, e.g. 0–10 and 10–20 v/100 v or smaller range precision instruments
- Alcohol still (Figure 11.6)
- Heating mantle
- Porcelain chips (anti-bumping granules)
- Precision thermometer (± 0.1°C)
- Anti-foaming agent.

16.9 Benchmark Values

A minimum of 10% v/100 v is required to ensure microbial stability. Some regulators also set upper limits, e.g. 14% v/100 v. Levels higher than about 13% have an appreciable impact on the palate.

17. ALCOHOL BY EBULLIOMETRY

17.1 Purpose

This method fulfills a similar function to that described in Section 16. The difference is that this is not as exacting, is executed more rapidly and does not require a preliminary distillation step. It relies on another physical property of liquids, one in which the interaction between molecules of the solvent and the solutes, whether volatile or not, affects the temperature at which the solvent(s) boil or freeze.

17.2 Occupational Health and Safety

- Open flame, flammable liquid.

17.3 Quality Assurance Records

- Person, date, acid (and source), amount per volume, air temperature and air pressure.

17.4 Chemistry/Physics

As discussed briefly in Section 16, the distillation method, the presence of alcohol lowers the boiling point of water when the two are combined, and the degree of reduction is related to the proportions. The equipment used was developed in France by Dujardin-Sallerin in 1870 and is suited only to dry table wines, but many versions exist (Lefco, Mallignard, Braun).

17.5 Procedure

- Rinse the chamber with a little wine and fill to mark with wine (usually it will be necessary to dilute with water, 1:1).
- Fill the condenser with cold water.
- Fill the methanol burner and light.
- Follow the rise of the mercury in the thermometer and, when it has stabilized for 30 s, take an accurate reading: T_1 (°C).
- Remove the burner; empty the condenser and the chamber.
- Add distilled water to the mark (20 mL).
- Reapply the burner (do not add water to the condenser).
- Take a reading when the thermometer stabilizes and steam issues continually—this is the boiling point of water: T_2 (°C).
- Calculate $\Delta T = T_2 - T_1$ (°C).
- Read the percentage alcohol from a table and apply any dilution factor used.
- Or calculate as:

$$\text{Alcohol\%} = (0.0105 + 0.9189\Delta T + 0.0502^{\circ}\Delta T^2) \times DF$$

17.6 Notes

1. This is a simple method but not of analytical quality. The accuracy is usually ± 0.5% v/100 v, although some modern versions claim a value of ± 0.15% v/100 v. It relies on the boiling point of water being reduced by the addition of alcohol. However, the boiling point is raised by other substances such as sugars and salts and an adjustment should be made for these. Sugars are the most important and if there is significant residual sugar then the apparent % (v/100 v) alcohol should be corrected using the following formula:

$$\text{Adjusted alcohol\%} = \text{Observed alcohol\%} \times (1 - {}^{\circ}\text{Baumé} \times 0.015)$$

2. A further source of error is the changing air pressure: a 0.4 kPa change will cause an error of about 0.5% v/100 v. Thus, take regular readings of T_2. Note that this method is not suitable for legal purposes and GLC or infrared spectrophotometry is normally used to obtain precise values.
3. Precision may be improved by measuring a dilution series: undiluted, 1:1, 1:4 and 1:9, wine/distilled water,

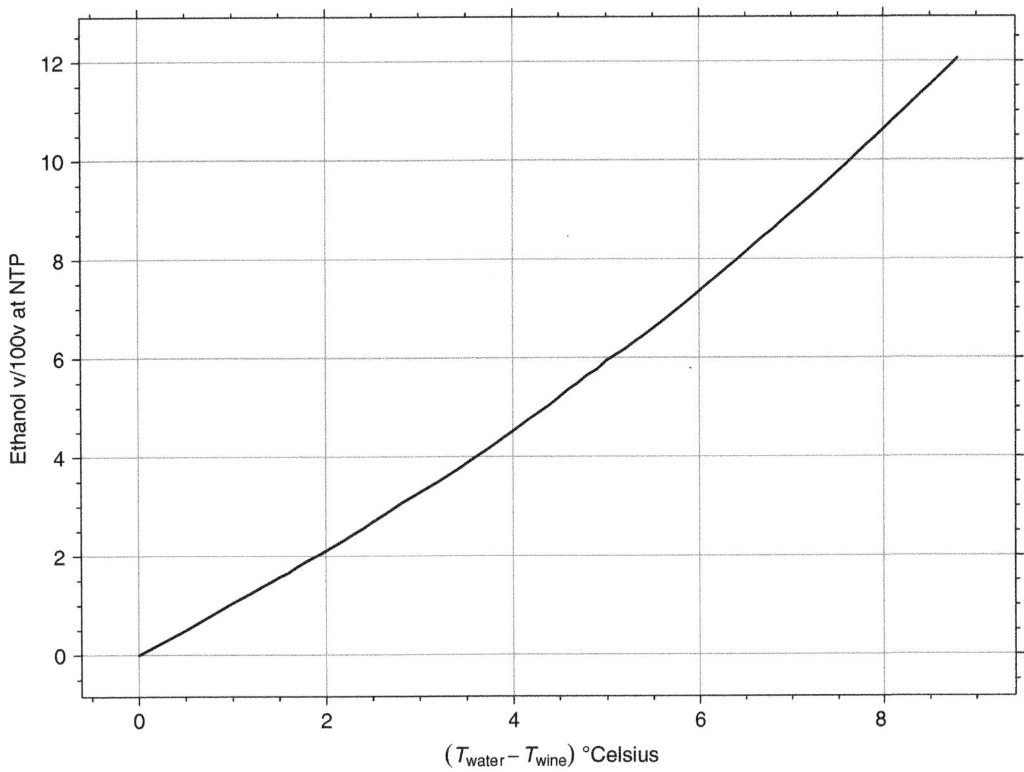

FIGURE 11.8 Plot of proportion of alcohol in wine as a function of observed boiling point reduction. This data is well fitted by the equation: $y = 0.0105 + 0.9189\Delta T + 0.0502\Delta T^2$, adjusted $R^2 = 1.0$, where y is the alcohol content at normal temperature and pressure (NTP) and ΔT is the difference between the boiling point of water and that of the wine. *(Data from IRS, 1955).*

and then adjusting for the dilution. Diluting minimizes the effect of other dissolved substances.

4. Some modern versions use a digital or an electronic thermometer as an aid to precision.
5. The boiling vessel should be washed regularly, with dilute alkali if necessary to remove tartrates. Rinse thoroughly and check for repeatability of the boiling point of distilled water before reusing.

17.7 Equipment and Materials

- Ebulliometer and tables supplied
- A standard wine of known composition as a check
- Conversion tables or calculator (or see Figure 11.8 and read or calculate using the equation provided)
- Methanol
- Spare cotton wick
- Distilled water.

Chapter 12

Sample Statistics

Chapter Outline

1. INTRODUCTION

Confidence in any measured value requires a minimal repetition of two. When we make comparisons that involve wine making we have an additional source of variation that is introduced by the fermentation itself. This is over and above that of any experimental treatment we may wish to assess, whether that treatment be the yeast selection, the vineyard or a management treatment within a vineyard. Ferments can vary substantially from one to another even if the juice and yeast culture are from the same batch and conditions are identical, as they were in Figure 12.1. Three sample ferments is generally the minimum number required to reduce the risk that the difference observed at the final analysis is that of the treatment and not that of a random effect of the ferment itself. While this is an extreme example, one can observe that the average of the three ferments for each plot is quite different. But, if by chance, you ran a single ferment for each plot and that ferment was the fastest of plot 1 and the slowest of plot 2, then you would not observe any significant difference between the two plots. Also, it is apparent that on day 4 in plot 1, an event occurred that led to one ferment becoming sluggish, while each of the three ferments made from plot 2 experienced a delay of 3 days before fermentation began, resulting in at least two that are unlikely to finish to dryness (it also has the highest ° Baumé, possibly a contributing factor). Be vigilant in managing your fermentations.

2. SIMPLE DESCRIPTIVE STATISTICS

In this text we have applied simple statistics of dispersion of values and relationships. These may be calculated manually or with the aid of functions in Excel® spreadsheets or in more specialized software such as 'R', which was used to prepare all figures in this text (R Development Core Team, 2012). Correct use of statistics provides a way of accurately communicating results and the reliability value of those results. The important statistical measures are as follows (Figure 12.2).

2.1 Range

This is the difference between the minimum and maximum values and is an indication of dispersion of the data set.

The range is calculated in Excel as:

$$= MAX(x_1{:}x_n) - MIN(x_1{:}x_n)$$

where $x_{1\ \text{to}\ n}$ represents the range of values in cells in a row(s) or a column(s) in an Excel spreadsheet.

2.2 Mean

Strictly speaking, the mean is the middle point between the minimum and the maximum of any given set of data. While, commonly, the terms 'mean' and 'average' are

A Complete Guide to Quality in Small-Scale Wine Making.

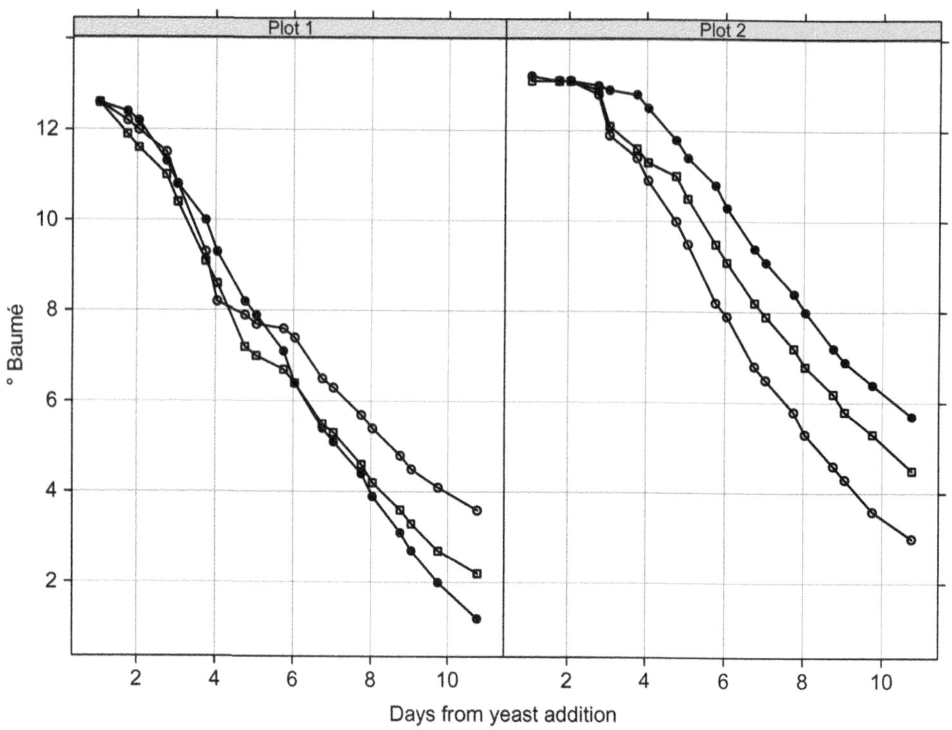

FIGURE 12.1 Fermentation of Chardonnay wines made in triplicate from two plots from one vineyard. Each of the three 10-L fermentations for each plot was made from a sample of the same juice (ca 45 L), fermented under identical conditions and with an equal initial density of yeast of the same yeast culture.

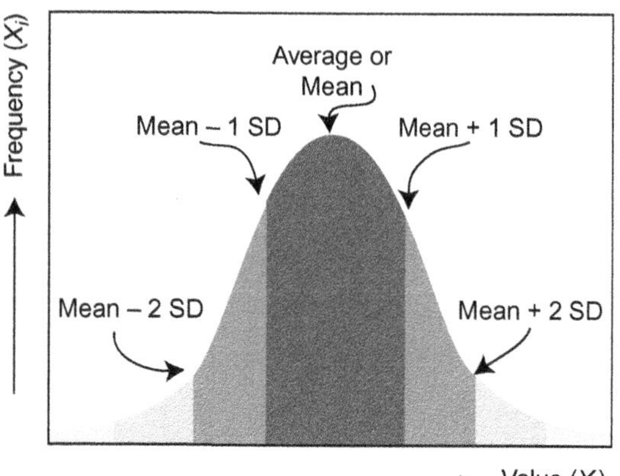

FIGURE 12.2 Frequency distribution of values for a normally distributed population of samples. In a normal population, mean = average = mode.

used interchangeably, here we will make the distinction, i.e. the mean can be calculated as:

$$\text{Mean } (\overline{x}) = \frac{x_{max} - x_{min}}{2}$$

The mean is calculated in Excel as:

$$= \frac{MAX(x_1{:}x_n) - MIN(x_1{:}x_n)}{2}$$

It is not widely used outside meteorology, where the daily mean temperature is the midpoint between the maximum and minimum temperature.

2.3 Average

Average is the sum of the data in a given data set divided by the number of values. This is the commonly applied method of calculation and that used in this text. It may differ from the mean as defined above because all values are used in the calculation, not just the maximum and minimum. It is therefore referred to as a weighted measure.

When choosing to use this as the method we assume that the frequency distribution of the values is approximately normal, i.e. if we plotted the number of times a particular value or small range of values occurs and then plotted this frequency against the value we would obtain a graph like that shown in Figure 12.2. In this case, the mean calculated either way would be the same. However, it is frequently not so and therefore we use this method of calculation because we desire a value that is less subject to the chance occurrence of unusual values. x-bar ($\overline{x}$) is the shorthand term used to represent the average in any measured set of values. Sigma (Σ) is shorthand for 'the sum of x_i^n', where x_i represents any measured value, and n the number of values being summed.

This is the value calculated by the Excel function:

$$\overline{x} = \text{AVERAGE}(x_1{:}x_n)$$

2.4 Mode

The mode is defined as the most common value. In a normally distributed set of data, the mean, average and mode are approximately equal. If they are not, then some other method of calculation may be required to normalize the mean. This is termed transformation, but is beyond the scope of this text.

This is the value calculated by the Excel function:

$$= \text{MODE}(x_1{:}x_n)$$

2.5 Standard Deviation

This is a measure of the spread of values around an average value. It is calculated as the square root of the 'sum of the deviations from the average, squared, divided by the number of values less one'.

$$s = \sqrt{\frac{\sum(x_i - \bar{x})}{(n-1)}}$$

$$s = \sqrt{\frac{\sum x_i^2 - (\sum x)^2}{(n-1)}}$$

In a normally distributed set of data the average ±1 standard deviation (SD) includes about two-thirds of all values, ±2 SD includes 95%, while ±3 SD includes 99%. It is thus a very useful measure of dispersion (variability). In an ideal 'grape factory', all values would be similar and the standard deviation would be low. The higher this value, the less controlled and more variable the product. Likewise, a good analyst produces values with a low standard deviation. A common measure of this is the coefficient of variation (CV, see below).

SD is calculated in Excel as:

$$s = \text{STDEV}(x_1{:}x_n)$$

where x_1 and x_n represent the first and last numbers in the cells of an array (either a column or a row or a set of columns and rows—an array).

2.6 Frequency Distribution

This is a graph or numerical display of the counts (frequencies) of values falling within particular ranges of the total range (termed bins by statisticians). An ideal distribution is bell shaped and symmetrical (Figure 12.2).

The central area in Figure 12.2 represents ±1 SD, the next 2 SD, and the third 3 SD. If you measure a single sample or sample a single bunch or berry, then two out of three times, that sample will lie within 1 SD of the average of all possible samples, i.e. one out of three times it will lie outside that range. Therefore, it is essential to measure more than one independent sample: the more samples, the greater the confidence that the value you get will represent the 'real' value (e.g. the ° Baumé measured after crushing all the fruit from a vineyard block, compared with the samples you took to judge when to harvest).

2.7 Coefficient of Variation

The CV is calculated as the SD divided by the average and multiplied by 100 to express the value as a percentage:

$$C = \frac{s}{\bar{x}} \times 100$$

CV is calculated in Excel as:

$$= (STDEV(x_1{:}x_n) \times 100)/AVERAGE(x_1{:}x_n)$$

2.8 Correlation Coefficient

This value estimates the degree to which two separate measures move in relation to one another. The range of the values varies from +1 to −1. A value of +1 or −1 indicates a perfect correlation and that the parameters measured are tightly associated, either positively or negatively. Lesser values indicate a less perfect relationship (i.e. there is error or natural or systematic variation).

The underlying assumption is that the values being compared are related in a simple arithmetic way, i.e. that the relationship is linear, and an increase or a decrease in the value of one of the correlates implies a corresponding increase or decrease in the other. However, this is not always the case. Thus, a low value may not indicate 'no relationship'. Also, two values may be related because they in turn are related to a common third factor and the apparent correlation has no causal implication. So it is always good practice to visualize a plot of the data as well as conducting the analysis.

A good way to assess relationships is to plot a scatter diagram of one value against the other (e.g. Figure 12.3). Plotting the values from 0 to 60 years would give a near-zero correlation coefficient, 0–30 years a positive (+0.45) and 30–100 years a negative relationship (−0.1, +0.45 and −0.65, respectively). Hence, great care needs to taken when applying correlation analyses. Always plot the data before calculating the correlation coefficient and be aware that you may have sampled only part of the whole, and that your conclusion may not apply generally, just to the data set you have analyzed.

A further statistical value derived from the correlation coefficient (r) is r^2. This value is widely used to indicate the proportion of variance that is explained by the correlation (however, with the loss of the sign + or −). For example, an r value of −0.44 has an r^2 value of 0.19, suggesting that only 19% of the observed variation can be accounted for by a simple linear relationship and 81% is

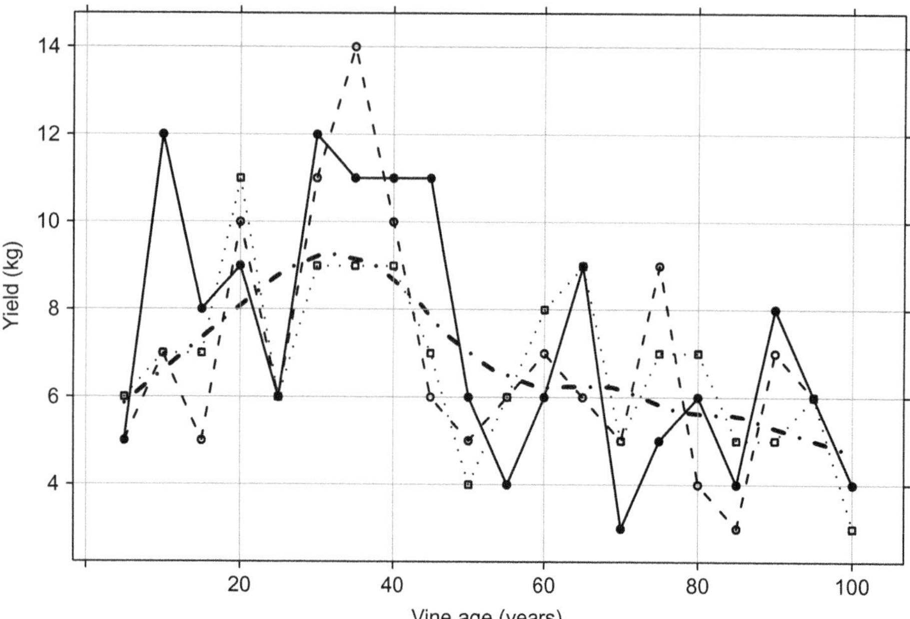

FIGURE 12.3 Scatterplot of the yield of three vines over a life of 100 years (hypothetical). Applying a correlation analysis to the data yields a statistically significant value of −0.44 but, depending on the age range, a range of positive or negative values would have been found, none of which would have had any value in assessing the overall long-term relationship between vine age and yield (e.g. the continuous dashed curve).

not accounted for: statistically significant the value may be, but it is of limited importance.

The correlation coefficient is calculated in Excel as:

$$r = CORREL(x_1{:}x_n,\ y_1{:}y_n),$$

where x_1 and x_n represent the first and last numbers in the cells of an array (either a column or a row) and y_1 and y_n represent the first and last numbers in the cells of a corresponding array (either a column or a row).

2.9 Regression Coefficient

This value represents the slope of a regression line between two related variables, such as optical density and concentration of a substance (e.g. NADH). For simple, linear regressions of the form $y = a + bx$, this may also be calculated using Excel as follows:

$$b = SLOPE(x_1{:}x_n,\ y_1{:}y_n)$$

and

$$a = INTERCEPT(x_1{:}x_n,\ y_1{:}y_n)$$

where x and y are as defined above.

One final issue regarding regression and which is important in the context of fitting standard curves is that it is usually appropriate to omit the intercept, i.e. to assume that at zero concentration of the substance of interest, the optical density should also be zero (Figure 12.4). If it is not, then there may be a problem with the assay or an inappropriate blank has been chosen. The correlation coefficient (or more often r^2) is frequently used as a measure of goodness of fit. For more details see any introductory statistical text.

3. EXPERIMENTATION

All winemakers experiment. It may be with blends, yeasts, barrel toasts or origin, or with sources of fruit, or even the outcomes of adopting particular management practices in the vineyard. The critical issue in experimentation is repeatability: could another person repeat the experiment and, if so, would they be likely to obtain results that are consistent with those of the original? If you do it again next year will you get a similar result? A good part of this objective can be met through meticulous record-keeping, but as Figure 12.1 illustrates, the outcome of any single fermentation can be highly irregular. Therefore, the second issue is one of repetition. Two fermentations is the absolute minimum if one is to ensure that the results have any validity, but three is much better.

The reliability of any mean or average is estimated by the standard error. This is calculated as the standard deviation (s) times the square root of 1 divided by the number of values used to estimate the mean ($s_{\bar{x}} = s \times \sqrt{1/n}$). Thus, the larger n, the smaller the value and the greater the confidence in the mean. To convert this to confidence limits with a probability, multiply the standard error by t

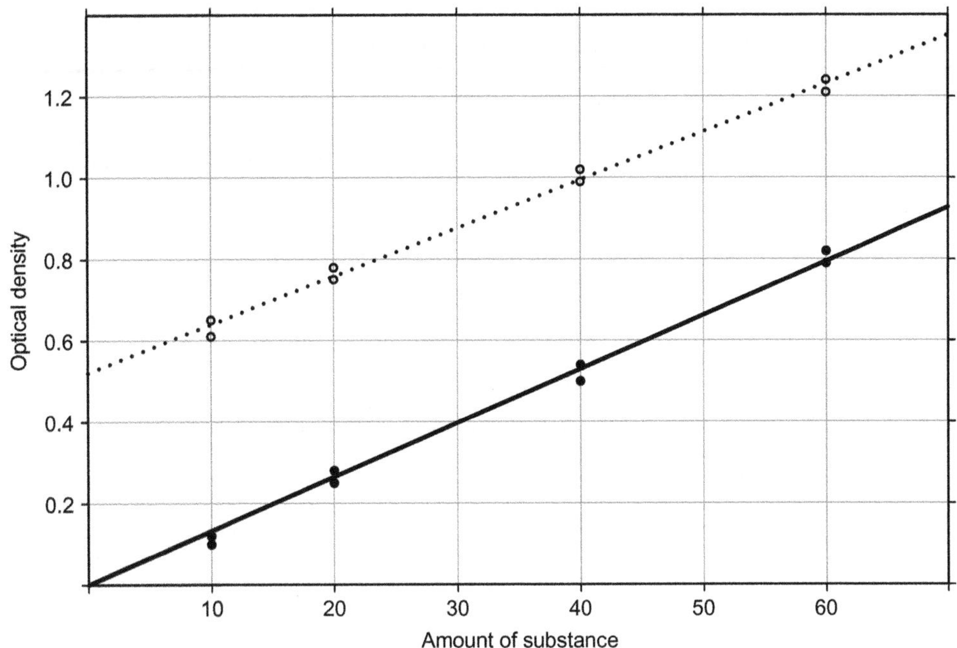

FIGURE 12.4 Examples of two hypothetical standard curves with fitted lines. One has been forced through a zero intercept, as may apply in the case of reducing sugars or potassium (solid line; substance A). The other has a notable background, such as may be encountered in an analysis of phenols in wine, where there is a background value of non-phenolic substances where water is used as the blank (dotted line; substance B). Substance A: $y = bx$, therefore $x = y/0.0132$, $r^2 = 0.998$; substance B: $y = a + bx$, therefore $x = (y - 0.521)/0.0132$, $r^2 = 0.992$.

TABLE 12.1 A Simplified Version of the Student's Distribution of *t* for the Estimation of Confidence Limits (Two-Sided)

Probability of a Larger Value Is = or <	Degrees of Freedom ($n - 1$)										
	1	2	3	4	5	6	7	8	9	10	∞
0.05	12.71	4.30	3.18	2.78	2.57	2.45	2.36	2.31	2.26	2.23	1.96
0.01	63.6	9.93	5.84	4.60	4.03	3.71	3.50	3.36	3.25	3.17	2.58

Note: See any statistical text or a reputable web site address for a complete table of values.

for the probability [e.g. 0.05 and $n - 1$, which is the degrees of freedom (df); see any statistical text]. This will allow you to report the mean value ± the confidence limits with a defined probability. Values in the table also serve to emphasize the importance of replication as a determinant of the reliability of the value of a mean (Table 12.1: compare df 1 with df 2 and 3).

For more complex experiments, it is always important to include a control; commonly this will be standard practice but it may be a zero level of an addition. It is not within the scope of this text to provide specific guidance for experimental design. Experiments are usually very expensive in time if not in resources and, if you intend to rely on the outcome for ongoing practice, then it would be wise to consult an expert. Frequently, simple designs are all that is required, such as a one-way, completely randomized design (Table 12.2). However, even in this case, if levels or time series are involved, then a more sophisticated analysis may yield more information from the same data; consider instead regression or analysis of covariance. The latter enables means to be adjusted for one or more factors that influence the mean, and which can be measured along with the value of interest; for example, temperature or pruning weight (if recording yield). There are many possibilities but only a professional statistician will be able to provide sound advice.

TABLE 12.2 Selected Experimental Protocols for Use in the Laboratory, Winery or Vineyard

Procedure	Description	When to Use
Mean $\pm s_{\bar{x}}$	Single sample with repetition	Description without comparison, e.g. alcohol content. Used in quality control in conjunction with specific boundary values
t Test	Comparison, pairwise of two means, with the hypothesis that they are not different at a given, predetermined, level of probability	Comparison of the stability of a wine, treated vs untreated, or the comparison of two treatments
One-way ANOVA, completely randomized	Comparison of a set of two or more replicated treatments, assuming that the error not explained by the treatments may be used to evaluate all treatments	Comparison of a set of barrel sources or yeasts conducted in a common environment
One-way ANOVA, blocked	As above, but environment not common to all replicates (plots), so a single replicate (usually) of each treatment is allocated at random to each environment (plot)	Enables the effect of the environment to be estimated and subtracted, increasing design efficiency, but only if the environment is important, e.g. research wines in several fermentation rooms
Two-way ANOVA	As for one-way ANOVA, but allows for analysis of factorial combinations of treatments and optimization. Most efficient if treatments do not interact but enables interaction to be assessed (i.e. whether the treatments are additive, the effect of adding one or other is equal across the range tested)	Possibly useful when evaluating combinations of fining agents. Widely used in field research
Regression	A very powerful analytical tool that works best when a wide range of values is being assessed and when a linear model can be fitted. It can be used in conjunction with designed experiments and has the strength that the model may be used predictively to estimate optimal (in the case of multiple linear regression) or intermediate treatments in a simple linear model	Widely used for the estimation of the amounts of unknown substances using values derived from a small set of known values, e.g. a standard curve

Note: See specific texts for details.
ANOVA = analysis of variance.

Bibliography

Alcalde-Eon, C., Escribano-Bailón, M.T., Santos-Buelga, C., Rivas-Gonzalo, J.C., 2006. Changes in the detailed pigment composition of red wine during maturity and ageing: a comprehensive study. Anal. Chim. Acta. 563, 238–254.

Allen, T., Herbst-Johnstone, M., Girault, M., Butler, P., Logan, G., Jouanneau, S., et al., 2011. Influence of grape harvesting steps on varietal thiol aromas in Sauvignon blanc wines. J. Agric. Food Chem. 59, 10641–10650.

Anfang, N., Brajkovich, M., Goddard, M.R., 2009. Co-fermentation with *Pichia kluyveri* increases varietal thiol concentrations in Sauvignon blanc. Aust. J. Grape Wine Res. 15, 1–8.

Arnold, R.A., Noble, A., Singleton, V.L., 1980. Bitterness and astringency of phenolic fractions in wine. J. Agric. Food Chem. 28, 675–678.

ASA (Australian Standards Association), 1999. Plastic Materials for Food Contact Use. Australian Standards Association, Sydney.

ASTM (American Society for Testing and Materials), 2006. Standard Practice for Cleaning, Descaling, and Passivation of Stainless Steel Parts, Equipment, and Systems. ASTM International, West Conshohocken, PA. Available from: <http://www.astm.org/>.

AWRI (Australian Wine Research Institute), 2009. Measuring Tannins in Grapes and Red Wine Using the MCP (Methyl Cellulose Precipitable) Tannin Assay. Australian Wine Research Institute, Adelaide. July. Available from: <http://www.awri.com.au/wp-content/uploads/mcp_fact_sheet.pdf>.

AWRI (Australian Wine Research Institute), 2011a. *Botrytis* and Disease Pressures from Vintage 2011. Australian Wine Research Institute, Adelaide. Technical Review, August 17. Available from: <http://www.awri.com.au/>.

AWRI (Australian Wine Research Institute), 2011b. Managing *Botrytis* Infected Fruit. Australian Wine Research Institute, Adelaide. Fact Sheets. April, 4. Available from: <http://www.awri.com.au/wp-content/uploads/managing_botrytis_infected_fruit_fact_sheet.pdf>.

AWRI (Australian Wine Research Institute), 2012a. Agrochemicals Registered for Use in Australian Viticulture. Australian Wine Research Institute, Adelaide. Available from: <http://www.awri.com.au/wp-content/uploads/agrochemical_booklet.pdf>.

AWRI (Australian Wine Research Institute), 2012b. Maximum Residue Limits. Australian Wine Research Institute, Adelaide. Available from: <http://www.awri.com.au/industry_support/viticulture/agrochemicals/mrls/>.

Bakker, J., Clarke, R.J., 2011. Wine tasting procedures and overall wine flavour, Wine Flavour Chemistry. second ed. Wiley-Blackwell, Oxford.

Barnett, J.A., 2000. A history of research on yeasts 2: Louis Pasteur and his contemporaries, 1850–1880. Yeast. 16, 755–771.

Bartoshuk, L.M., 1989. Taste. Robust across the age span? Ann. N.Y. Acad. Sci. 561, 65–75.

Bartoshuk, L.M., 1993. Biological basis of food perception and acceptance. Food Qual. Pref. 4, 21–32.

Bartoshuk, L.M., Duffy, V.B., Lucchina, L.A., Prutkin, J., Fast, K., 1998. PROP (6-*n*-propylthiouracil) supertasters and the saltiness of NaCl. Ann. N.Y. Acad. Sci. 855, 793–796.

Bartoshuk, L.M., Duffy, V.B., Chapo, A.K., Fast, K., Yiee, J.H., Hoffman, H.J., et al., 2004. From psychophysics to the clinic: missteps and advances. Food Qual. Pref. 15, 617–632.

Bearzatto, G., 1986. Wine fining with silica sol. Aust. N.Z. Wine Ind. J. 1, 39–40.

Beauchamp, G., 2009. Sensory and receptor responses to umami: an overview of pioneering work. Am. J. Clin. Nutr. 90, 723S–727S.

Bell, S.-J., Henschke, P.A., 2005. Implications of nitrogen nutrition for grapes, fermentation and wine. Aust. J. Grape Wine Res. 11, 242–295.

Benedict, S.R., 1928. The determination of blood sugar. II. J. Biol. Chem. 76, 457–470.

Berg, H.W., Keefer, R.M., 1958. Analytical determination of tartrate stability in wine. I. Potassium bitartrate. Am. J. Enol. Vitic. 9, 180–193.

Berthels, N.J., Otero, R.R.C., Bauer, F.F., Thevelein, J.M., Pretorius, I.S., 2004. Discrepancy in glucose and fructose utilisation during fermentation by *Saccharomyces cerevisiae* wine yeast strains. FEMS Yeast Res. 4, 683–689.

Bisson, L.F., 1999. Stuck and sluggish fermentations. Am. J. Enol. Vitic. 50, 107–119.

Bisson, L.F., Butzke, C.E., 2000. Diagnosis and rectification of stuck and sluggish fermentations. Am. J. Enol. Vitic. 51, 168–177.

Black, C., Francis, L., Henschke, P.A., Capone, D.L., Anderson, S., Day, M., et al., 2012. Aged Riesling and the Development of TDN. Australian Wine Research Institute, Adelaide. September/October 2012. Available from: <http://www.awri.com.au/wp-content/uploads/Sept-Oct-2012-AWRI-Report.pdf>.

Blouin, J., Guimberteau, G., 2000. Maturation et Matutité des Raisins. Éditions Féret, Bordeaux.

Boulton, R., 1979. The heat transfer characteristics of wine fermenters. Am. J. Enol. Vitic. 30, 152–156.

Boulton, R., 2001. The copigmentation of anthocyanins and its role in the color of red wine: a critical review. Am. J. Enol. Vitic. 52, 67–87.

Boulton, R.B., 1983. The Conductivity Method for Evaluating the Potassium Bitartrate Stability of Wines (with Edits). Enology Briefs. Department of Enology and Viticulture, University of California, Cooperative Extension, Davis, CA.

Boulton, R.B., Singleton, V.L., Bisson, L.F., Kunkee, R.E., 1996. Principles and Practices of Winemaking. Chapman & Hall, New York.

Boyes, S., Strübi, P., Dawes, H., 1997. Measurement of protein content in fruit juices, wine and plant extracts in the presence of endogenous organic compounds. Food Sci. Technol. Leb. 30, 778–785.

Bramley, R.G.V., Proffitt, A.P.B., Hinze, C.J., Pearse, B., Hamilton, R.P., 2005. Generating Benefits from Precision Viticulture Through

Selective Harvesting. Precision Agriculture '05. Wageningen Academic Publishers, Wageningen.

Buck, L., 1993. Identification and analysis of a multigene family encoding odorant receptors: implications for mechanisms underlying olfactory information processing. Chem. Senses. 18, 203–208.

Bureau, S.M., Baumes, R.L., Razungles, A.J., 2000. Effects of vine or bunch shading on the glycosylated flavor precursors in grapes of *Vitis vinifera* L. cv. Syrah. J. Agric. Food Chem. 48, 1290–1297.

Burgos, P., Malcolm, A., Baigent, M., Doedens, J., Considine, J.A., 2008. Calibrating Vine Biomass and Productivity for cv Chardonnay: Plant Cell Density by Precision Digital Imagery, Ground Penetrating Radar and Physical Measurements. University of Western Australia, Specterra & Baigent Geosciences, Perth.

Butzke, C.E., Bisson, L., 1997. Ethyl Carbamate: Preventative Action Manual. Department of Enology and Viticulture, University of California, Davis, Davis, CA.

Calhelha, R.C., Andrade, J.V., Ferreira, I.C., Estevinho, L.M., 2006. Toxicity effects of fungicide residues on the wine-producing process. Food Microbiol. 23, 393–398.

Capone, D., 2012. The origin of eucalyptol in wine (1-8 cineole). Australian Cabernet Symposium. Australian Wine Research Institute, Penola, South Australia.

Castillo-Muñoz, N., Gómez-Alonso, S., García-Romero, E., Hermosín-Gutiérrez, I., 2007. Flavonol profiles of *Vitis vinifera* red grapes and their single-cultivar wines. J. Agric. Food Chem. 55, 992–1002.

Chapman, D.M., Roby, G., Ebeler, S.E., Guinard, J.-X., Matthews, M.A., 2005. Sensory attributes of Cabernet Sauvignon wines made from vines with different water status. Aust. J. Grape Wine Res. 11, 339–347.

Chapman, J., Baker, P., Wills, S., 2001. Winery Wastewater Handbook. Winetitles, Adelaide.

Chatonnet, P., Dubourdieu, D., 1998. Comparative study of the characteristics of American white oak (*Quercus alba*) and European oak (*Quercus petraea* and *Q. robur*) for production of barrels used in barrel aging of wines. Am. J. Enol. Vitic. 49, 79–85.

Chuine, I., Yiou, P., Viovy, N., Seguin, B., Daux, V., Ladurie, E.L., 2004. Historical phenology: grape ripening as a past climate indicator. Nature. 432, 289–290.

Clingeleffer, P., 2001. Crop development, crop estimation and crop control to secure quality and production of major wine grape varieties: a national approach. Victoria/Adelaide; Commonwealth Scientific and Industrial Research Organisation and Department of Natural Resources and Environment; CSH 96/1. September. 159. Available from: <http://www.gwrdc.com.au/completed_projects/>.

Cochran, W.G., 1962. Design and analysis of sampling. In: Snedecor, G.W. (Ed.), Statistical Methods. Iowa State University Press, Ames, IA.

Cohen, S.D., Tarara, J.M., Gambetta, G.A., Matthews, M.A., Kennedy, J.A., 2012. Impact of diurnal temperature variation on grape berry development, proanthocyanidin accumulation, and the expression of flavonoid pathway genes. J. Exp. Bot. 63, 2655–2665.

Compton, S., Jones, C., 1985. Mechanism of dye response and interference in the Bradford protein assay. Anal. Biochem. 151, 369–374.

Coombe, B.G., Bishop, G.R., 1980. Development of the grape berry. II. Changes in diameter and deformability during veraison. Aust. J. Agric. Res. 31, 499–509.

Coombe, B.G., Iland, P.G., 2004. Grape berry development and winegrape quality. In: Dry, P.R., Coombe, B.G. (Eds.), Australian Viticulture, second ed. Winetitles, Adelaide.

Cortell, J.M., Kennedy, J.A., 2006. The effect of shading on accumulation of flavonoid compounds in (*Vitis vinifera* L.) Pinot noir fruit and extraction in a model system. J. Agric. Food Chem. 54, 8510–8520.

Cozzolino, D., Cowey, G., Lattey, K., Godden, P., Cynkar, W., Dambergs, R., et al., 2008. Relationship between wine scores and visible-near-infrared spectra of Australian red wines. Anal. Bioanal. Chem. 391, 975–981.

Cozzolino, D., Cynkar, W., Shah, N., Smith, P., 2011. Technical solutions for analysis of grape juice, must, and wine: the role of infrared spectroscopy and chemometrics. Anal. Bioanal. Chem. 401, 1475–1484.

Čuš, F., Raspor, P., 2008. The effect of pyrimethanil on the growth of wine yeasts. Lett. Appl. Microbiol. 47, 54–59.

Daniel, M.A., Capone, D.L., Sefton, M.A., Elsey, G.M., 2009. Riesling acetal is a precursor to 1,1,6-trimethyl-1,2-dihydronaphthalene (TDN) in wine. Aust. J. Grape Wine Res. 15, 93–96.

Danilewicz, J.C., 2007. Interaction of sulfur dioxide, polyphenols, and oxygen in a wine-model system: central role of iron and copper. Am. J. Enol. Vitic. 58, 53–60.

Danilewicz, J.C., 2012. Review of oxidative processes in wine and value of reduction potentials in enology. Am. J. Enol. Vitic. 63, 1–10.

Devatine, A., Gerbaud, V., Gabas, N., Blouin, J., 2002. Prediction and mastering of wine acidity and tartaric precipitations: the MEXTAR (R) software tool. J. Int. Sci. Vigne Vin. 36, 77–91.

Downey, M.O., Harvey, J.S., Robinson, S.P., 2003. Analysis of tannins in seeds and skins of Shiraz grapes throughout berry development. Aust. J. Grape Wine Res. 9, 15–27.

Drinkine, J., Glories, Y., Saucier, C., 2005. (+)-Catechin–aldehyde condensations: competition between acetaldehyde and glyoxylic acid. J. Agric. Food Chem. 53, 7552–7558.

Drysdale, G., Fleet, G., 1989. The growth and survival of acetic acid bacteria in wines at different concentrations of oxygen. Am. J. Enol. Vitic. 40, 99–105.

Duffy, V.B., Bartoshuk, L.M., 2000. Food acceptance and genetic variation in taste. J. Am. Diet. Assoc. 100, 647–655.

Dukes, B.C., Butzke, C.E., 1998. Rapid determination of primary amino acids in grape juice using an *o*-phthaldialdehyde/*N*-acetyl-L-cysteine spectrophotometric assay. Am. J. Enol. Vitic. 49, 125–134.

Edwards, C.G., Haag, K.M., Semon, M.J., Rodriguez, A.V., Mills, J.M., 1999a. Evaluation of processing methods to control the growth of *Lactobacillus kunkeei*, a micro-organism implicated in sluggish alcoholic fermentations of grape musts. S. Afr. J. Enol. Vitic. 20, 11–18.

Edwards, C.G., Reynolds, A.G., Rodriguez, A.V., Semon, M.J., Mills, J. M., 1999b. Implication of acetic acid in the induction of slow/stuck grape juice fermentations and inhibition of yeast by *Lactobacillus* sp. Am. J. Enol. Vitic. 50, 204–210.

Etiévant, P.X., 1991. Wine. In: Maarse, H. (Ed.), Volatile Compounds in Food. Marcel Dekker, New York.

Farnet, A.M., Qasemian, L., Guiral, D., Ferré, E., 2010. A modified method based on arsenomolybdate complex to quantify cellulase activities: application to litters. Pedobiologia. 53, 159–160.

Ferrari, G., 2002. Influence of must nitrogen composition on wine and spirit quality and relation with aromatic composition and defects. J. Int. Sci. Vigne Vin. 36, 1–10.

Ferreira, A.M., Climaco, M.C., Faia, A.M., 2001. The role of non-*Saccharomyces* species in releasing glycosidic bound fraction of grape aroma components: a preliminary study. J. Appl. Microbiol. 91, 67–71.

Ferreira, V., López, R., Cacho, J.F., 2000. Quantitative determination of the odorants of young red wines from different grape varieties. J. Sci. Food Agric. 80, 1659–1667.

Fielding, P.J., Harris, J.M., Lucas, M.I., Cook, P.A., 1986. Implications for the assessment of crystalline style activity in bivalves when using the Bernfeld and Nelson–Somogyi assays for reducing sugars. J. Exp. Mar. Biol. Ecol. 101, 269–284.

Firestein, S., 2001. How the olfactory system makes sense of scents. Nature. 413, 211–218.

Flamini, R., De Luca, G., Di Stefano, R., 2002. Changes in carbonyl compounds in Chardonnay and Cabernet Sauvignon wines as a consequence of malolactic fermentation. Vitis. 41, 107–112.

Flecknoe-Brown, A., 2004. Controlled permeability, moulded wine tanks—new developments in polymer catalyst technology. Aust. N.Z. Grapegrower Winemaker. Winetitles, Adelaide.

Fleet, G.H. (Ed.), 1993. Wine Microbiology and Biotechnology. Harwood Academic Publishers, Chur.

Fleet, G., Heard, G., 1993. Yeasts: growth during fermentation. In: Fleet, G. (Ed.), Wine Microbiology and Biotechnology. Harwood Academic Publishers, Chur.

Fleet, G.H., 2001. Alcoholic beverages. In: Moir, C.J. (Ed.), Spoilage of Processed Foods: Causes and Diagnosis. Australian Institute of Food Science and Technology, Waterloo, NSW.

Forde, C.G., Cox, A., Williams, E.R., Boss, P.K., 2011. Associations between the sensory attributes and volatile composition of Cabernet Sauvignon wines and the volatile composition of the grapes used for their production. J. Agric. Food Chem. 59, 2573–2583.

Francis, I.L., Newton, J.L., 2005. Determining wine aroma from compositional data. Aust. J. Grape Wine Res. 11, 114–126.

Francis, I.L., Tate, M.E., Williams, P.J., 1996. The effect of hydrolysis conditions on the aroma released from Semillon grape glycosides. Aust. J. Grape Wine Res. 2, 70–76.

de Freitas, V.A.P., Glories, Y., 1999. Concentration and compositional changes of procyanidins in grape seeds and skin of white *Vitis vinifera* varieties. J. Sci. Food Agric. 79, 1601–1606.

de Freitas, V.A.P., Glories, Y., Laguerre, M., 1998. Incidence of molecular structure in oxidation of grape seed procyanidins. J. Agric. Food Chem. 46, 376–382.

de Freitas, V.A.P., Glories, Y., Monique, A., 2000. Developmental changes of procyanidins in grapes of red *Vitis vinifera* varieties and their composition in respective wines. Am. J. Enol. Vitic. 51, 397–403.

FSA (Food Standards Australia New Zealand), 2007. Copper Citrate as a Processing Aid for Wine. Food Standards Australia, Sydney; 8 August. Available from: <http://www.foodstandards.gov.au/foodstandards/applications/applicationa562coppe3375.cfm>.

FSAC (FSA Consulting), 2006. Best Practice Guide for Water and Waste Management in the Queensland Wine Industry. FSA Consulting, Toowoomba. Available from: <http://www.fsaconsulting.net/fsa/docs/EPA_Wine.pdf>.

Fugelsang, K.C., 1997. Wine Microbiology. Chapman & Hall, New York.

Fulcrand, H., Cheynier, V., Oszmianski, J., Moutounet, M., 1997. An oxidized tartaric acid residue as a new bridge potentially competing with acetaldehyde in flavan-3-ol condensation. Phytochemistry. 46, 223–227.

Gladstones, J.S., 1965. The climate and soils of southwestern Australia in relation to vine growing. J. Aust. Inst. Agric. Sci. 31, 275–288.

Gladstones, J.S., 1992. Viticulture and Environment. Winetitles, Adelaide.

Godden, P., Francis, L., Field, J., Gishen, M., Coulter, A., Valente, P., et al., 2001. Wine bottle closures: physical characteristics and effect on composition and sensory properties of a Semillon wine 1. Performance up to 20 months post-bottling. Aust. J. Grape Wine Res. 7, 64–105.

Godden, P., Lattey, K., Gishen, M., Francis, L., Cowey, G., Holdstock, M., et al., 2005. Towards offering wine to the consumer in optimal condition—the wine, the closures and other packaging variables: a review of AWRI research examining the changes that occur in wine after bottling. Aust. N.Z. Wine Ind. J. 20, 20–30.

Gutierrez, A.V.L., 2003. Sensory descriptive analysis of red wines undergoing malolactic fermentation with oak chips. J. Food Sci. 68, 1075–1079.

Hamilton, R.P., Coombe, B.G., 1992. Harvesting of winegrapes. In: Coombe, B.G., Dry, P.R. (Eds.), Australian Viticulture, first ed. Winetitles, Adelaide.

Harbertson, J.F., Spayd, S., 2006. Measuring phenolics in the winery. Am. J. Enol. Vitic. 57, 280–288.

Hayes, J.E., Pickering, G.J., 2012. Wine expertise predicts taste phenotype. Am. J. Enol. Vitic. 63, 80–84.

Henschke, P., Jiranek, V., 1993. Yeasts—metabolism of nitrogen compounds. In: Fleet, G.H. (Ed.), Wine Microbiology and Biotechnology. Harwood Academic Publishers, Chur.

Henschke, P.A., Ough, C.S., 1991. Urea accumulation in fermenting grape juice. Am. J. Enol. Vitic. 42, 317–321.

Herderich, M.J., Smith, P.A., 2008. Analysis of grape and wine tannins: methods, applications and challenges. Aust. J. Grape Wine Res. 11, 205–214.

Hernández-Orte, P., Cacho, J.F., Ferreira, V., 2002. Relationship between varietal amino acid profile of grapes and wine aromatic composition. experiments with model solutions and chemometric study. J. Agric. Food Chem. 50, 2891–2899.

Hernández-Orte, P., Ibarz, M.J., Cacho, J., Ferreira, V., 2006. Addition of amino acids to grape juice of the Merlot variety: effect on amino acid uptake and aroma generation during alcoholic fermentation. Food Chem. 98, 300–310.

Hocking, A.D., Pitt, J.I., 2001. Moulds. In: Moir, C.J. (Ed.), Spoilage of Processed Foods: Causes and Diagnosis. Australian Institute of Food Science and Technology, Waterloo, NSW.

Howell, G., Vallesi, M., 2003. Don't Waste Your Time Measuring the pH of Your Wine Unless You do It Properly! Aust. N.Z. Grapegrower Winemaker. Winetitles, Adelaide.

Howell, K.S., Swiegers, J.H., Elsey, G.M., Siebert, T.E., Bartowsky, E.J., Fleet, G.H., et al., 2004. Variation in 4-mercapto-4-methyl-pentan-2-one release by *Saccharomyces cerevisiae* commercial wine strains. FEMS Microbiol. Lett. 240, 125–129.

Huglin, P., Schneider, C., 1998. Biologie et Écologie de la Vigne. second ed. Technique & Documentation. Payot-Laussana, Paris.

Iland, P.G., Bruer, N., Edwards, G., Weeks, S., Wilkes, E., 2004. Chemical Analysis of Grapes and Wine: Techniques and Concepts. Patrick Iland Wine Promotions, Cambelltown, SA.

Internal Revenue Service (IRS), 1955. Code of Federal Regulations. Internal Revenue Service, Washington, DC.

Intrieri, C., Poni, S., 2000. Physiological response of winegrape to management practices for successful mechanization of quality vineyards. Acta Hortic. 33–47.

Jackson, R.S., 2008. Wine Science: Principles, Practice, Perceptions. third ed. Academic Press, San Diego, CA.

Jeandet, P., Bessis, R., Sbaghi, M., Meunier, P., 1995. Production of the phytoalexin resveratrol by grapes as a response to *Botrytis* attack under natural conditions. J. Phytopathol. 143, 135–139.

Jin, Y.L., Speers, R.A., 1998. Flocculation of *Saccharomyces cerevisiae*. Food Res. Int. 31, 421–440.

Jones, G.V., Duff, A.A., Hall, A., Myers, J.W., 2010. Spatial analysis of climate in winegrape growing regions in the western United States. Am. J. Enol. Vitic. 61, 313–326.

Jones, R.S., Ough, C., 1985. Variations in the percent ethanol (v/v) per degrees Brix conversions of wines from different climatic regions. Am. J. Enol. Vitic. 36, 268–270.

Joseph, C.M.L., Kumar, G., Su, E., Bisson, L.F., 2007. Adhesion and biofilm production by wine isolates of *Brettanomyces bruxellensis*. Am. J. Enol. Vitic. 58, 373–378.

Josyln, M.A., 1955. Sulfur dioxide content of wine, I. Iodometric titration. Am. J. Enol. Vitic. 6, 1–10.

Joyeux, A., Lafon-Lafourcade, S., Ribéreau-Gayon, P., 1984. Evolution of acetic acid bacteria during fermentation and storage of wine. Appl. Environ. Microb. 48, 153–156.

Kennison, K.R., Gibberd, M.R., Pollnitz, A.P., Wilkinson, K.L., 2008. Smoke-derived taint in wine: the release of smoke-derived volatile phenols during fermentation of Merlot juice following grapevine exposure to smoke. J. Agric. Food Chem. 56, 7379–7383.

Kilmartin, P., 2009. The oxidation of red and white wines and its impact on wine aroma. Chem. N.Z. 73, 18–22.

Krstic, M., Whiting, J., Mollah, M., Clingeleffer, P., Degaris, P., Moulds, G., et al., 2002. Compendium of Wine Grape Specifications and Measurement. Cooperative Research Centre, Viticulture, Adelaide. Final (Research) Report. CRV 99/6. Available from: <http://www/gwrdc.com.au/completed_projects/>.

Kunkee, R.E., 1967. Malo-lactic fermentation. Adv. Appl. Microbiol. 9, 235–279.

Lasanta, C., Gómez, J., 2012. Tartrate stabilization of wines. Trends Food Sci. Technol. 28, 52–59.

Lee, J., Durst, R.W., Wrolstad, R.E., 2005. Determination of total monomeric anthocyanin pigment content of fruit juices, beverages, natural colorants, and wines by the pH differential method: collaborative study. J. AOAC Int. 88, 1269–1278.

Lee, T., Simpson, R., 1993. Microbiology and chemistry of cork taints in wine. In: Fleet, G. (Ed.), Wine Microbiology and Biotechnology. Harwood Academic Publishers, Chur.

Lindemann, B., 2001. Receptors and transduction in taste. Nature. 413, 219–225.

Lohnertz, O., Prior, B., Bleser, M., Linsenmeier, A., 2000. Influence of N-supply and soil management on the nitrogen composition of grapes. Acta Hortic. 55–64.

Luthi, H., 1957. Symbiotic problems relating to the bacterial deterioration of wines. Am. J. Enol. Vitic. 8, 167–181.

Maicas, S., Mateo, J.J., 2005. Hydrolysis of terpenyl glycosides in grape juice and other fruit juices: a review. Appl. Microbiol. Biotechnol. 67, 322–335.

Makhotkina, O., Herbst-Johnstone, M., Logan, G., du Toit, W., Kilmartin, P.A., 2013. Influence of sulfur dioxide additions at harvest on polyphenols, C6-compounds, and varietal thiols in Sauvignon blanc. Am. J. Enol. Vitic. 64, 203–213.

Margalit, Y., 1996. Winery Technology & Operations: A Handbook for Small Wineries. Wine Appreciation Guild, San Francisco, CA.

Margalit, Y., 2004. Concepts in Wine Chemistry. Wine Appreciation Guild, San Francisco, CA.

Marks, L.E., Stevens, J.C., Bartoshuk, L.M., Gent, J.F., Rifkin, B., Stone, V.K., 1988. Magnitude-matching: the measurement of taste and smell. Chem. Senses. 13, 63–87.

Martini, A., Ciani, M., Scorzetti, G., 1996. Direct enumeration and isolation of wine yeasts from grape surfaces. Am. J. Enol. Vitic. 47, 435–440.

Matthews, M.A., Ishii, R., Anderson, M.M., O'Mahony, M., 1990. Dependence of wine sensory attributes on vine water status. J. Sci. Food Agric. 51, 321–335.

Mattivi, F., Guzzon, R., Vrhovsek, U., Stefanini, M., Velasco, R., 2006. Metabolite profiling of grape: flavonols and anthocyanins. J. Agric. Food Chem. 54, 7692–7702.

Maujean, A., Vallée, D., Saucy, L., 1986. Determination de la sursaturation en bitartrate de potassium d'un vin. Quantification des effets colloides-protecteurs. Rev. Fr. Oenol. 104, 39–49.

Mazza, G., Francis, F.J., 1995. Anthocyanins in grapes and grape products. Crit. Rev. Food Sci. 35, 341–371.

Mazza, G., Miniati, E., 1993. Anthocyanins in Fruits, Vegetables and Grains. CRC Press, Boca Raton, FL.

McCloskey, L.P., 1976. An acetic acid assay for wine using enzymes. Am. J. Enol. Vitic. 27, 176–180.

Menashe, I., Man, D., Gilad, L., Gilad, Y., 2003. Different noses for different people. Nat. Genet. 34, 143–144.

Mercurio, M.D., Smith, P.A., 2008. Tannin quantification in red grapes and wine: comparison of polysaccharide- and protein-based tannin precipitation techniques and their ability to model wine astringency. J. Agric. Food Chem. 56, 5528–5537.

Mesquita, P.R., Pereira, M.A.P., Monteiro, S., Loureiro, V.B., Teixeira, A.R., Ferreira, R.B., 2001. Effect of wine composition on protein stability. Am. J. Enol. Vitic. 52, 324–330.

Miller, M., 2009. Malolactic Fermentation e-Book 2. Accuvin. Available from: <http://accuvin.com/Malolactic%20Fermentation.pdf>.

Mirabel, M., Saucier, C., Guerra, C., Glories, Y., 1999. Copigmentation in model wines solutions: occurrence and relation to wine aging. Am. J. Enol. Vitic. 50, 211–218.

Moio, L., Etievant, P.X., 1995. Ethyl anthranilate, ethyl cinnamate, 2,3-dihydrocinnamate, and methy lanthranilate: four important odorants identified in Pinot noir wines of Burgundy. Am. J. Enol. Vitic. 46, 392–398.

Moskowitz, H.R., 1971. The sweetness and pleasantness of sugars. Am. J. Psychol. 84, 387–405.

Mpelasoka, B.S., Schachtman, D.P., Treeby, M.T., Thomas, M.R., 2003. A review of potassium nutrition in grapevines with special emphasis on berry accumulation. Aust. J. Grape Wine Res. 9, 154–168.

Myles, S., Boyko, A.R., Owens, C.L., Brown, P.J., Grassi, F., Aradhya, M.K., et al., 2011. Genetic structure and domestication history of the grape. Proc. Natl. Acad. Sci. U.S.A. 108, 3530–3535.

NATA (National Association of Testing Authorities), 2010. User Checks and Maintenance of Laboratory Balances. National Association of Testing Authorities, Sydney. Technical Note 13; May. Available from: <http://www.nata.asn.au/>.

Noble, A.C., Arnold, R.A., Buechsenstein, J., Leach, E.J., Schmidt, J.O., Stern, P.M., 1987. Modification of a standardized system of wine aroma terminology. Am. J. Enol. Vitic. 38, 143–146.

NTWG (North Texas Winemakers Guild), 2007. Pearson's Square. North Texas Winemakers Guild. Available from: <http://www.northtexaswinemakers.org/downloads-software.html> (accessed 28.02.13.).

Nykänen, L., 1986. Formation and occurrence of flavor compounds in wine and distilled alcoholic beverages. Am. J. Enol. Vitic. 37, 84–96.

Oliveira, C.M., Ferreira, A.C.S., De Freitas, V., Silva, A.M.S., 2011. Oxidation mechanisms occurring in wines. Food Res. Int. 44, 1115–1126.

Olmo, H.P., 1956. A Survey of the Grape Industry in Western Australia. Department of Agriculture, Perth.

OIV (Organisation Internationale de la Vigne et du Vin), 2006a. Compendium of International Methods of Wine and Must Analysis, vol. 1. Organisation Internationale de la Vigne et du Vin, Paris.

OIV (Organisation Internationale de la Vigne et du Vin), 2006b. Compendium of International Methods of Wine and Must Analysis, vol. 2. Organisation Internationale de la Vigne et du Vin, Paris.

Otero, R.R.C., Iranzo, J.F.U., Briones-Perez, A.I., Potgieter, N., Villena, M.A., Pretorius, I.S., et al., 2003. Characterization of the β-glucosidase activity produced by enological strains of non-*Saccharomyces* yeasts. J. Food Sci. 68, 2564–2569.

Ough, C., Amerine, M., 1988. Methods for Analysis of Musts and Wines. second ed. John Wiley & Sons, New York.

Ough, C., Kriel, A., 1985. Ammonium concentrations of musts of different grape cultivars and vineyards in the Stellenbosch area. S. Afr. J. Enol. Vitic. 6, 7–11.

Ough, C.S., 1992. Winemaking Basics. Food Products Press (Haworth), New York.

Ough, C.S., Crowell, E., 1979. Pectic-enzyme treatment of white grapes: temperature, variety and skin contact time factors. Am. J. Enol. Vitic. 30, 22–27.

Park, S.K., Boulton, R.B., Noble, A.C., 2000. Formation of hydrogen sulfide and glutathione during fermentation of white grape musts. Am. J. Enol. Vitic. 51, 91–97.

Pedneault, K., Dubé, G., Turcotte, I., 2012. L' évaluation de la maturité du raisin par analyse sensorielle: un outil d'aide à la décision. Centre de Development Bioalimenaire du Québec. Quebec. August 2012. Available from: <http://www.agrireseau.qc.ca/petitsfruits/documents/Pedneault,%20Dub%C3%A9,%20Turcotte,%202012%20-%20Suivi%20de%20la%20maturit%C3%A9%20par%20analyse%20sensorielle%20du%20raisin.pdf>.

Pellenc, R., 2011. Selective-sorting harvesting machine and sorting chain including one such machine. United States patent application 12/674,125. US2011/0223684 A1.

Peng, Z., Duncan, B., Pocock, K.F., Sefton, M.A., 1998. The effect of ascorbic acid on oxidative browning of white wines and model wines. Aust. J. Grape Wine Res. 4, 127–135.

Perez, J., Kliewer, W.M., 1990. Effect of shading on bud necrosis and bud fruitfulness of Thompson Seedless grapevines. Am. J. Enol. Vitic. 41, 168–175.

Pérez-Coello, M.S., Sánchez, M.A., Garcia, E., Gonzaléz-Vinas, M.A., Sanz, J., Cabezudo, M.D., 2000. Fermentation of white wines in the presence of wood chips of American and French oak. J. Agric. Food Chem. 48, 885–889.

Peynaud, E., 1987. The Taste of Wine. Wine Appreciation Guild, San Francisco, CA.

Pineau, B., Barbe, J.-C., Van Leeuwen, C., Dubourdieu, D., 2007. Which impact for β-damascenone on red wines aroma? J. Agric. Food Chem. 55, 4103–4108.

Pocock, K.F., Waters, E.J., 1998. The effect of mechanical harvesting and transport of grapes, and juice oxidation, on the protein stability of wines. Aust. J. Grape Wine Res. 4, 136–139.

Possner, D., Kliewer, W.M., 1985. The localization of acids, sugars, potassium and calcium in developing grape berries. Vitis. 24, 229–240.

Puech, J.-L., Feuillat, F., Mosedale, J.R., 1999. The tannins of oak heartwood: structure, properties, and their influence on wine flavor. Am. J. Enol. Vitic. 50, 469–478.

Puppazoni, M., 2007. *Oenococcus oeni* is the only bacterial species needed for malolactic fermentation in Cabernet Sauvignon wine. B. Sc. Honours, University of Western Australia.

R Development Core Team, 2012. R: A Language and Environment for Statistical Computing. 2.15.2 ed. R Foundation for Statistical Computing, Vienna.

Rankine, B., Fornachon, J., Bridson, D., 1969. Diacetyl in Australian dry red wines and its significance in wine quality. Vitis. 8, 129–134.

Rankine, B.C., 1968. Formation of α-ketoglutaric acid by wine yeasts and its oenological significance. J. Sci. Food Agric. 19, 624–627.

Rankine, B.C., 1970. Alkalimetric determination of sulphur dioxide in wine. Aust. Wine Brew. Spirit Rev. 88, 40–44.

Rankine, B.C., 1989. Making Good Wine. Sun (Macmillan), Melbourne.

Rankine, B.C., 1990. Tasting and Enjoying Wine. Winetitles, Adelaide.

Rankine, B.C., Cellier, K.M., Boehm, E.W., 1962. Studies on grape variability and field sampling. Am. J. Enol. Vitic. 13, 58–72.

Razungles, A.J., Baumes, R.L., Dufour, C., Sznaper, C.N., Bayonove, C.L., 1998. Effect of sun exposure on carotenoids and C-13-*nor*-isoprenoid glycosides in Syrah berries (*Vitis vinifera* L.). Sci. Aliment. 18, 361–373.

Reed, D.R., Bartoshuk, L.M., Duffy, V., Marino, S., Price, R.A., 1995. Propylthiouracil tasting: determination of underlying threshold distributions using maximum likelihood. Chem. Senses. 20, 529–533.

Ribéreau-Gayon, P., 1964. Les composés phénoliques du raisin et du vin. III. Les tannins. Ann. Physiol. Vég. Institut national de la Recherche agonomique, Paris. p. 81.

Ribéreau-Gayon, P., 1974. The chemistry of red wine. In: Webb, A.D. (Ed.), Chemistry of Winemaking. American Chemical Society, Washington, DC.

Ribéreau-Gayon, P., Boidron, J.N., Terrier, A., 1975. Aroma of Muscat grape varieties. J. Agric. Food Chem. 23, 1042–1047.

Ribéreau-Gayon, P., Dubourdieu, D., Donèche, B., Lonvaud, A., 2006a. second ed. Handbook of Enology: Microbiology of Wine and Vinifications, vol. 1. John Wiley & Sons, New York.

Ribéreau-Gayon, P., Glories, Y., Maujean, A., Dubourdieu, D., 2006b. second ed. Handbook of Enology: The Chemistry of Wine, Stabilization and Treatments, vol. 2. John Wiley & Sons, New York.

Ristic, R., Downey, M.O., Iland, P.G., Bindon, K., Francis, I.L., Herderich, M., et al., 2007. Exclusion of sunlight from Shiraz grapes alters wine colour, tannin and sensory properties. Aust. J. Grape Wine Res. 13, 53–65.

Ritchey, J.G., Waterhouse, A.L., 1999. A standard red wine: monomeric phenolic analysis of commercial Cabernet Sauvignon wines. Am. J. Enol. Vitic. 50, 91–100.

Robinson, S., Jacobs, A., Dry, I., 1997. A class IV chitinase is highly expressed in grape berries during ripening. Plant Physiol. 114, 771–778.

Roessler, E.B., Amerine, M., 1958. Studies on grape sampling. Am. J. Enol. Vitic. 9, 139–145.

Roessler, E.B., Amerine, M.A., 1963. Further studies on field sampling of wine grapes. Am. J. Enol. Vitic. 14, 144–147.

Rojas, V., Gil, J.V., Piñaga, F., Manzanares, P., 2003. Acetate ester formation in wine by mixed cultures in laboratory fermentations. Int. J. Food Microbiol. 86, 181–188.

Roland, A., Schneider, R., Razungles, A., Cavelier, F., 2011. Varietal thiols in wine: discovery, analysis and applications. Chem. Rev. 111, 7355–7376.

Rosi, I., Fia, G., Canuti, V., 2003. Influence of different pH values and inoculation time on the growth and malolactic activity of a strain of *Oenococcus oeni*. Aust. J. Grape Wine Res. 9, 194–199.

Roujou de Boubée, D., Van Leeuwen, C., Dubourdieu, D., 2000. Organoleptic impact of 2-methoxy-3-isobutylpyrazine on red Bordeaux and Loire wines. Effect of environmental conditions on concentrations in grape during ripening. J. Agric. Food Chem. 48, 4830–4834.

Roujou de Boubée, D., Cumsille, A.M., Pons, M., Dubourdieu, D., 2002. Location of 2-methoxy-3-isobutylpyrazine in Cabernet Sauvignon grape bunches and its extractability during vinification. Am. J. Enol. Vitic. 53, 1–5.

Rousseau, J., 2001. Suivi de la maturité des raisins par analyse sensorielle descriptive des baies. Relation avec les profils sensoriels des vins et les attentes des consommateurs. Bull. O.I.V. 74, 719–728.

Rousseau, J., 2003. Quantified Descriptive Sensorial Analysis of Grapes: Principle–Protocol–Interpretation. Analyse Sensorielle Raisin—Version Anglaise IAS-15 Rev/01. Institut Coopératif du Vin, ICV/Vignes et Vins, Lattes.

Royal Society of Chemistry (RSC), 2013. Visual Elements Periodic Table. Royal Society of Chemistry, London. Available from: <www.rsc.org/Periodic-Table>.

Ryona, I., Pan, B.S., Intrigliolo, D.S., Lakso, A.N., Sacks, G.L., 2008. Effects of cluster light exposure on 3-isobutyl-2-methoxypyrazine accumulation and degradation patterns in red wine grapes (*Vitis vinifera* L. cv. Cabernet Franc). J. Agric. Food Chem. 56, 10838–10846.

Sacchi, K.L., Bisson, L.F., Adams, D.O., 2005. A review of the effect of winemaking techniques on phenolic extraction in red wines. Am. J. Enol. Vitic. 56, 197–206.

Sala, C., Busto, O., Guasch, J., Zamora, F., 2005. Contents of 3-alkyl-2-methoxypyrazines in musts and wines from *Vitis vinifera* variety Cabernet Sauvignon: influence of irrigation and plantation density. J. Sci. Food Agric. 85, 1131–1136.

Sarmento, M.R., Oliveira, J.C., Slatner, M., Boulton, R.B., 2000. Influence of intrinsic factors on conventional wine protein stability tests. Food Control. 11, 423–432.

Sarneckis, C.J., Dambergs, R.G., Jones, P., Mercurio, M.D., Herderich, M.J., Smith, P.A., 2006. Quantification of condensed tannins by precipitation with methyl cellulose: development and validation of an optimised tool for grape and wine analysis. Aust. J. Grape Wine Res. 12, 39–49.

Schinner, F., von Mersi, W., 1990. Xylanase-, CM-cellulase, and invertase activity in soil: an improved method. Soil Biol. Biochem. 22, 511–515.

Sefton, M., Simpson, R.F., 2005. Compounds causing cork taint and the factors affecting their transfer from natural cork closures to wine: a review. Aust. J. Grape Wine Res. 11, 226–240.

Shi, P., Zhang, J., 2006. Contrasting modes of evolution between vertebrate sweet/umami receptor genes and bitter receptor genes. Mol. Biol. Evol. 23, 292–300.

Shimizu, K., 1993. Killer yeasts. In: Fleet, G.H. (Ed.), Wine Microbiology and Biotechnology. Harwood Academic Publishers, Chur.

Singleton, V.L., 1974. Some aspects of the wooden container as a factor in wine maturation. In: Webb, A.D. (Ed.), Chemistry of Winemaking. American Chemical Society, Washington, DC, pp. 254–277.

Singleton, V.L., 1985. Caftaric acid disappearance and conversion to products of enzymic oxidation in grape must and wine. Am. J. Enol. Vitic. 36, 50–56.

Singleton, V.L., 1987. Oxygen with phenols and related reactions in musts, wines, and model systems: observations and practical implications. Am. J. Enol. Vitic. 38, 69–77.

Singleton, V.L., Trousdale, E.K., 1992. Anthocyanin–tannin interactions explaining differences in polymeric phenols between white and red wines. Am. J. Enol. Vitic. 43, 63–70.

Skouroumounis, G.K., Kwiatkowski, M.J., Francis, I.L., Oakey, H., Capone, D.L., Duncan, B., et al., 2005. The impact of closure type and storage conditions on the composition, colour and flavour properties of a Riesling and a wooded Chardonnay wine during five years' storage. Aust. J. Grape Wine Res. 11, 369–377.

Singleton, V.L., Webb, A.D. (Eds.), 1974. Chemistry of Winemaking. American Chemical Society, Washington, DC.

Smart, R., Robinson, M., 1991. Sunlight into Wine. Winetitles, Adelaide.

Somers, T.C., Evans, M.E., 1977. Spectral evaluation of young red wines: anthocyanin equilibria, total phenolics, free and molecular SO_2, 'chemical age'. J. Sci. Food Agric. 28, 279–287.

Somers, T.C., Evans, M.E., 1979. Grape pigment phenomena: interpretation of major colour losses during vinification. J. Sci. Food Agric. 30, 623–633.

Somers, T.C., Zeimelis, G., 1973. Direct determination of wine proteins. Am. J. Enol. Vitic. 24, 47–50.

Somers, T.C., Ziemelis, G., 1985. Spectral evaluation of total phenolics in *Vitis vinifera*: grapes and wines. J. Sci. Food Agric. 36, 1275–1284.

Somogyi, M., 1952. Notes on sugar determination. J. Biol. Chem. 195, 19–23.

Soubeyrand, V., Julien, A., Sablayrolles, J.M., 2006. Rehydration protocols for active dry yeast and the search for early indicators of yeast activity. Am. J. Enol. Vitic. 52, 474–480.

Speedy, B., 2012. Ray Beckwith OAM: 23rd February 1912–7th November 2012. The Australian. Available from: <http://www.theaustralian.com.au/news/nation/father-of-aussie-wine-marks-rare-vintage/story-e6frg6nf-1226278778984> (accessed 23.02.12.).

Spillman, P.J., Pollnitz, A.P., Liacopoulos, D., Skouroumounis, G.K., Sefton, M.A., 1997. Accumulation of vanillin during barrel-aging of white, red, and model wines. J. Agric. Food Chem. 45, 2584–2589.

Spillman, P.J., Sefton, M.A., Gawel, R., 2004. The contribution of volatile compounds derived during oak barrel maturation to the aroma of a Chardonnay and Cabernet Sauvignon wine. Aust. J. Grape Wine Res. 10, 227–235.

Sponholz, W., 1993. Wine spoilage by microorganisms. In: Fleet, G.H. (Ed.), Wine Microbiology and Biotechnology. Harwood Academic Publishers, Chur.

Stephens, P., 2003. Culture methods. In: McMeekin, T.A. (Ed.), Detecting Pathogens in Food. Woodhead Publishing, Cambridge.

Stone, H., Bleibaum, R., Thomas, H.A., 2012. Sensory Evaluation Practices. Academic Press, San Diego, CA.

Swiegers, J.H., Bartowsky, E.J., Henschke, P.A., Pretorius, I.S., 2005. Yeast and bacterial modulation of wine aroma and flavour. Aust. J. Grape Wine Res. 11, 139–173.

Tattersall, D.B., Van Heeswijck, R., Hoj, P.B., 1997. Identification and characterization of a fruit-specific, thaumatin-like protein that accumulates at very high levels in conjunction with the onset of sugar accumulation and berry softening in grapes. Plant Physiol. 114, 759–769.

Thompson, A., Taylor, B.N., 2008. Guide to the Use of the International System of Units (SI). National Institute of Standards and Technology, Gaithersburg, MD.

Timberlake, C.F., Bridle, P., 1977. Anthocyanins: colour augmentation with catechin and acetaldehyde. J. Sci. Food Agric. 28, 539–544.

du Toit, W.J., Marais, J., Pretorius, I.S., du Toit, M., 2006. Oxygen in must and wine: a review. S. Afr. J. Enol. Vitic. 27, 76–94.

Tominaga, T., Murat, M.-L., Dubourdieu, D., 1998. Development of a method for analyzing the volatile thiols involved in the characteristic aroma of wines made from *Vitis vinifera* L. cv. Sauvignon blanc. J. Agric. Food Chem. 46, 1044–1048.

Tominaga, T., Baltenweck-Guyot, R., Peyrot de Gachons, C., Dubourdieu, D., 2000. Contribution of volatile thiols to the aromas of white wines made from several *Vitis vinifera* grape varieties. Am. J. Enol. Vitic. 51, 178–181.

Treeby, M.T., Holzapfel, B.P., Pickering, G.J., Friedrich, C.J., 2000. Vineyard nitrogen supply and Shiraz grape and wine quality. Acta Hortic. 77–92.

Tryon, C.R., Edwards, P.A., Chisolm, M.G., 1988. Determination of the phenolic content of some French–American hybrid white wines using ultraviolet spectroscopy. Am. J. Enol. Vitic. 39, 5–10.

Ugliano, M., Kwiatkowski, M., Vidal, S.E., Capone, D., Siebert, T., Dieval, J.-B., et al., 2011. Evolution of 3-mercaptohexanol, hydrogen sulfide, and methyl mercaptan during bottle storage of Sauvignon blanc wines. Effect of glutathione, copper, oxygen exposure, and closure-derived oxygen. J. Agric. Food Chem. 59, 2564–2572.

Upreti, G.C., Wang, Y., Finn, A., Sharrock, A., Feisst, N., Davy, M., et al., 2012. U-2012: an improved Lowry protein assay, insensitive to sample color, offering reagent stability and enhanced sensitivity. Biotechniques. 52, 159–166.

Vèrette, E., Noble, A.C., Somers, T.C., 1988. Hydroxycinnamates of *Vitis vinifera*: sensory assessment in relation to bitterness in white wines. J. Sci. Food Agric. 45, 267–272.

Waterhouse, A.L., 2002. Wine phenolics. Ann. N.Y. Acad. Sci. 957, 21–36.

Waters, E.J., Shirley, N.J., Williams, P.J., 1996. Nuisance proteins of wine are grape pathogenesis-related proteins. J. Agric. Food Chem. 44, 3–5.

Waters, E.J., Alexander, G., Muhlack, R., Pocock, K.F., Colby, C., O'Neill, P.B., et al., 2005. Preventing protein haze in bottled white wine. Aust. J. Grape Wine Res. 11, 215–225.

Watts, D.A., Ough, C.S., Brown, W.D., 1981. Residual amounts of proteinaceous additives in table wines. J. Food Sci. 46, 681–683, 687.

Weber, F., Greve, K., Durner, D., Fischer, U., Winterhalter, P., 2013. Sensory and chemical characterization of phenolic polymers from red wine obtained by gel permeation chromatography. Am. J. Enol. Vitic. 64, 15–25.

Weiss, K.C., Bisson, L., 2001. Optimisation of the amido black assay for determination of the protein content of grape juices and wines. J. Sci. Food Agric. 81, 583–589.

Wenzel, K., Dittrich, H.H., Heimfarth, M., 1987. Die Zusammensetzung der Anthocyane in der Beeren verscheidener Rebsorten. Vitis. 26, 65–78.

White, M.A., Diffenbaugh, N.S., Jones, G.V., Pal, J.S., Giorgi, F., 2006. Extreme heat reduces and shifts United States premium wine production in the 21st century. Proc. Natl. Acad. Sci. U.S.A. 103, 11217–11222.

White, R.E., 2003. Soils for Fine Wines. Oxford University Press, Oxford.

Williams, L.A., 1982. Heat release in alcoholic fermentation: a critical reappraisal. Am. J. Enol. Vitic. 33, 149–153.

Wilson, J.E., 1998. Terroir: The Role of Geology, Climate, and Culture in the Making of French Wines. Mitchell Beazley, London.

Winemakers' Federation of Australia (WFA), 2008. Winery Code of Practice. Winemakers' Federation of Australia, Adelaide. Available from: <http://www.wfa.org.au/resources/4/PDF_Resources/FC_Environmental_Winery_Code.pdf>.

Winkler, A.J., 1963. General Viticulture. Jacaranda Press, Brisbane.

Winter, E., Whiting, J., Rousseau, J., 2004. Winegrape Berry Assessment in Australia. Winetitles, Adelaide.

Wong, O.S., Sternson, L.A., Schowen, R.L., 1985. Reaction of *o*-phthalaldehyde with alanine and thiols: kinetics and mechanism. J. Am. Chem. Soc. 107, 6421–6422.

Wright, J.D., Bean, V.E., Aguilera, J., 2008. NIST Calibration Services for Hydrometers. National Institute of Standards and Technology. Special Publications, Gaithersburg, MD. 250–278. 34. Available from: <http://www.nist.gov/pml/div685/grp02/upload/250_78.pdf>.

Wu, Y.C., Koch, W.F., Durst, R.A., 1988. Standardization of pH Measurements. SP260-53.pdf. US Department of Commerce, National Institute of Standards and Technology, Gaithersburg, MD.

Zoecklein, B.W., 1995. Wine Analysis and Production. Chapman & Hall, New York.

Appendix

List of Abbreviations and Symbols

ε (epsilon)	extinction coefficient (spectrophotometry)
ρ (rho)	density = m/*v* (kg/m^3)
A	measured absorbance of a solution
3MH	3-mercaptohexan-1-ol
3MHA	acetate of 3-mercaptohexan-1-ol
4-MMP	4-mercapto-4-methylpentan-2-one
ACS	acetyl coenzyme A synthase
AGA	*Acetobacter–Gluconobacter* agar (culture medium for lactic acid bacteria)
An	anthocyanin
AR	analytical grade reagent
ASA	Australian Standards Association
ASTM	American Society for Testing Materials (now ASTM International)
ATP	adenosine triphosphate (nucleotide)
AU	absorbance units
AWBC	Wine Australia (formerly Australian Wine and Brandy Corporation)
AWRI	Australian Wine Research Institute
° B	° Brix
° Bé	° Baumé
BP	boiling point
BSA	bovine serum albumen
BSM	British Standard Milk
c	concentration (mol/L)
ca	circa (approximately)
CCP	critical control point
cfu	colony-forming unit (microbiological term relevant to counting colonies: a single cell or a clump of cells may form a single colony)
CIP	clean-in-place
CIPFF	clean-in-place flat-face
CL	confidence limit: standard deviation of the mean × Student's *t* for 95% probability (or as stated)
CO_2	carbon dioxide
CoA	coenzyme A
conc.	concentration
cP	centipoise, a measure of viscosity (mPa.s)
CS	citrate synthase
CuS	copper sulfide
$CuSO_4$	copper sulfate
CV	coefficient of variation
DAP	diammonium phosphate, $(NH_4)_2PO_4$
DF	dilution factor (fraction of original sample in the analysis volume)
DI	distilled or deionized water
dp	degree of polymerization (number of units linked together)
EtOH	ethanol
EU	European (Economic) Union
FAN	fermentable amino nitrogen (syn. YAN)
FSA	Food Standards Australia New Zealand
fSO_2	free sulfur dioxide (mg/L)
FTIR	Fourier transform infrared (spectroscopy)
g	gram (kg × 10^{-3})
g_n	acceleration due to gravity (m/s^2)
G2	a culture medium for lactic acid bacteria
GDD	growing degree days
GLC	gas–liquid chromatography
GOT	glutamate-oxaloacetate transaminase
h	hour
H_2O_2	hydrogen peroxide
H_2S	hydrogen sulfide
HACCP	hazard analysis and critical control point
Hg	mercury (usually refers here to a mercury vapor light source, ultraviolet)
HK	hexokinase
hL	hectoliter (100 L)
HPLC	high-pressure liquid chromatography
HS	hydrogen sulfide
IBMP	2-methoxy-3-isobutylpyrazine
IEC	International Electrotechnical Commission
ISO	Organization for Standardization
IUPAC	International Union of Pure and Applied Chemistry
j/wt	juice as a fraction of berry weight
$K_2S_2O_5$	potassium metabisulfite
KHT	potassium hydrogen tartrate
LR	laboratory reagent grade
m	mass (kg)
m/m	mass per unit mass (kg/kg)
M	molar mass (g/mol)
MDH	L-malate dehydrogenase
MEA	malt extract agar
MF	microscope factor

min	minute
mL	milliliter
mm	millimeter
MOG	material other than grapes
MRL	maximum residue limit
MRS	De Man, Rogosa, Sharpe
MSDS	material safety data sheet
MYGP	malt extract, yeast extract, glucose, peptone
NAD	nicotinamide adenine dinucleotide
NADH	nicotinamide adenine dinucleotide, reduced form
NADPH	nicotinamide adenine dinucleotide phosphate, reduced form
NaOH	sodium hydroxide
NATA	National Association of Testing Authorities
NDVI	normalized difference vegetation index
NH_3	ammonia
NH_4^+	ammonium ion
NIR	near-infrared
NIST	National Institute of Science and Technology
NTU	nephalometric turbidity unit
OD_{280}	optical density at 280 nm
OH&S	occupational health and safety
OIV	Organisation Internationale de la Vigne et du Vin
OPA	*o*-phthaldialdehyde
PCD	plant cell density
PGI	phosphoglucose isomerase
PMS	potassium metabisulfite
pp	per person
PP	pyrophosphate
PPO	polyphenol oxidase
pr	pair
PROP	6-*n*-propylthiouracil (a bitter compound)
PVPP	polyvinylpolypyrrolidone
QA	quality assurance
QDSA	quantitative descriptive sensory analysis
RCF	relative centrifugal force
R&D	research and development
RT	room temperature (ca 20–25°C)
s	second
SCBA	self-contained breathing apparatus
SD	standard deviation of a sample
SG	specific gravity (relative density) = density of substance/density water at 4°C)
SI	Système international
SO_2	sulfur dioxide
SOP	standard operating procedure
SR	settling rate
Std	standard
STP	standard (laboratory) temperature and pressure
TA	titratable acidity (here in tartartic acid equivalents, g/L)
TCA	2,4,6-trichloroanisole (primary cork taint)
TCA	tricarboxylic acid
TDN	1,1,6-trimethyl-1,2-dihydronaphthalene
TLC	thin-layer chromatography
tSO_2	total sulfur dioxide (mg/L)
tsp	teaspoon
TSS	total soluble solids
TTB	Alcohol and Tobacco Tax and Trade Bureau, US Department of the Treasury
UCD	University of California, Davis
UV	ultraviolet
v	volume (L)
VA	volatile acidity (principally acetic acid)
VSP	vertically shoot positioned
v/v	volume (L) per volume (L)
WFA	Winemakers' Federation of Australia
WLNA	WL nutrient agar
w/v	weight (mass, kg) per volume (L)
w/w	weight (kg) per unit weight (kg)
YAN	yeast-assimilable nitrogen
YEPD	yeast extract, peptone, dextrose

1. CALCULATORS

1.1 Introduction

The calculator provided on the companion website for this book are an aid to ensure consistency and the worksheets are protected to prevent inadvertent changes to the formulae. See http://booksite.elsevier.com/9780124080812. If your procedure differs and requires a change to the formulae then you will need to 'unprotect' the worksheet, make the changes and 'protect' again to minimize the risk of mistakes and false values (under the 'Review' menu). Teachers and tutors, managers would be well advised to use a password for added protection.

Note also that the Titratable acidity worksheet is linked to the Field Sample worksheet to minimize transpositional errors and the need for double-entry. The link assumes data is in a common order—so beware or delete the links by 'unprotecting' the sheet and deleting the link. The link is there for demonstration purposes. Links have value in routine analyses but without the added protection of a single 'key' as in a database, there is always the risk that the data transferred between worksheets may be mismatched.

Take note of the comments for cells with a red triangle in the top right hand corner because it contains important information regarding data for that cell.

Caution: The worksheets are designed using only functions available within Excel®. They have limited checks for mistaken entries and assume that data in linked worksheets is aligned. The functions are not hidden so that the user may see the approach and adopt and adapt as suits their own needs, especially at the student level. It is essential, therefore, that the user adopt a cautious approach to the results obtained with these worksheets and checks carefully the entry values. The underlying functions are intended to support teaching rather than commercial operations and to encourage a comprehensive approach to record keeping. For commercial purposes a database background would be more appropriate (e.g. Microsoft Access® or Filemaker Pro®, mySQL...). Such an approach should adopt multiple error checking for range and consistency and should align data by a common sample number (key) across all worksheets.

The worksheets should be checked against your own particular operating procedures as they have been designed for the protocols in this text. It is always good practice to check the assumptions and methodologies to ensure those provided suit your particular circumstances and reagents.

1.2 Protocol List

1 Field sampling.
2 Example record sheet for white wine cultivars.
3 Example record sheet for red wine cultivars.
4 Titratable acidity (Tartaric acid or H_2SO_4 equivalents).
5 Sulphur dioxide by aspiration.
6 Sulphur dioxide by titration.
7 Free amino nitrogen.
8 Ammonium by enzyme assay.
9 Reducing sugars by enzyme assay.
10 Malic acid by enzyme assay.
11 Acetic acid by enzyme assay.
12 Volatile acidity by distillation.
13 Potassium (sodium) by flame photometry.
14 Color in red grapes.
15 Color in red wine.
16 Phenolics and white grapes and wine.
17 Total phenolics by colorimetry.
18 Tannins.
19 Alcohol by ebulliometry.

2. MATERIALS LIST FOR WINERY OPERATIONS

TABLE A1 Equipment List

Item	Research	Teaching	Semi-Commercial
Individual ferment volume (L)	10 or 50	5	350–500
Picking buckets (18–22 L)	1/5	1/2	30
Certified balance	150 kg	30 kg	500 kg
Open fermenters (red wine, L)	20 or 100	10	600
Closed fermenters (red wine, L)	10 or 50	5	1000
Topping wine container (red wine, L)	2 or 5	2	20
Closed fermenters (white wine, L)	10 or 50	5	300–500
Silicone bungs (single hole)	1/fermenter	1/fermenter	–
Silicone bungs (solid)	1/fermenter	1/fermenter	–
Airlocks	1/fermenter	1/fermenter	1/fermenter/barrel
Oeno-tannin (red wine—optional)	50 mg/L	50 mg/L	50 mg/L
Pointed fruit secateurs	1 pr pp	1 pr pp	1 pr pp
Potassium metabisulfite (kg)	1	1	2
1 tonne electric forklift (optional)—no fumes allowed!	–	–	1
Destemmer–crusher (small)	1	1	1
Sorting table/conveyor	Optional	Optional	Optional
Plungers	SS paint mixer	SS paint mixer	SS commercial plunger
Press (L)	80/160	20/80/160	80/160
Transfer pump and food-grade hosing and fittings	–	–	1 set
Source of hot water (80°C)	√	√	√
Industrial high-pressure water cleaner (preferably steam)	√	√	√

(Continued)

TABLE A1 (Continued)

Item	Research	Teaching	Semi-Commercial
Cleaning agents	√	√	√
Industrial waste bin for disposing of marc	√	√	√
Racking	Siphon–racking tube	Siphon–racking tube	Racking plate (unless tank has a racking valve)
Yeast	200 mg/L	200 mg/L	200 mg/L
Yeast nutrients	√	√	√
DAP	150 mg/L	150 mg/L	150 mg/L
Tartaric acid	5 g/L	5 g/L	5 g/L
Baumé hydrometer	√	√	√
° Brix meter	√	√	√
Bunch sample press	√	√	√
Food-grade thermometers	√	√	√
Oak barrel or staves	–	–	225-L barrel or oak alternatives
Malolactic culture	60 mg/L	60 mg/L	60 mg/L
Food-grade bentonite	1.5 g/L	1.5 g/L	1.5 g/L
Citric acid	2 g/L	2 g/L	2 g/L
Copper sulfate or citrate	2 mg/L	2 mg/L	2 mg/L
In-line filters and bottling equipment	√	√	√
Bottles and closures (per ferment, 750 mL)	10 or 48	6	300
Carboy/bottle rinsing ball-head	√	√	√
Barrel rinsing ball-head			√
Labels	√	√	√
Labeling machine	√	√	√
Screw-capping machine (optional)	√	√	√
Corking machine	√	√	√
Capsule or foil applicator	√	√	√
Food-grade nitrogen & pressure regulator	√	√	√
Dry ice block maker (Snowpac® or equivalent)	√	√	√
Food-grade CO_2 cylinder with bottom siphon	√	√	√
CO_2 gas detector/alarm (if fermenting in an enclosed space)	√	√	√
Miscellaneous	Food-grade buckets	Food-grade buckets	Food-grade buckets
Miscellaneous	Cleaning tools	Cleaning tools	Cleaning tools
Air-tire wheeled trolley	√	√	–
Sack trolley	√	√	√
Fermentation room	0–25°C	0–25°C	0–25°C
Corrosion-proof shelving	√	√	–

(Continued)

TABLE A1 (Continued)

Item	Research	Teaching	Semi-Commercial
Barrel room (optional)	–	–	If $T < 20°C$
Bottled wine storage room	√	√	√
Fining agents—various	e.g. casein, PVPP, albumen	e.g. casein, PVPP, albumen	e.g. casein, PVPP, albumen

Note: Values are suggested and should be evaluated against individual circumstances.
DAP = diammonium phosphate; CO_2 = carbon dioxide; pr = pair; pp = per person; PVPP = polyvinylpolypyrrolidone.

3. OPERATION PROCEDURES FOR AN 80-L IDRO WATER BAG PRESS

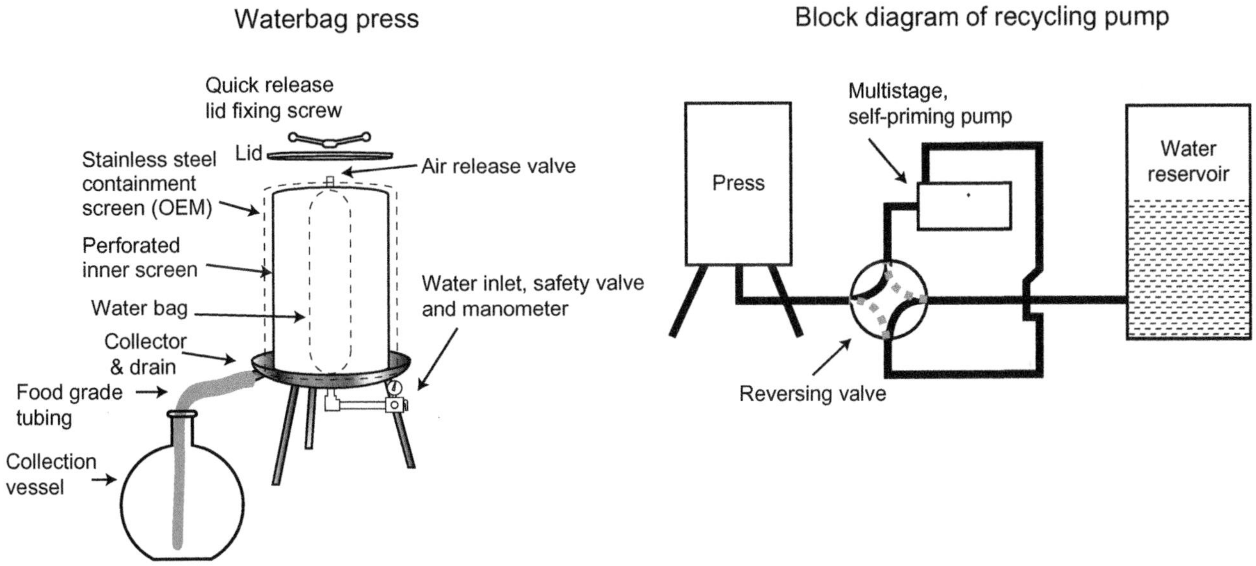

FIGURE A1 Water-bag press: 80-L capacity water-bag press modified by the addition of an external stainless steel containment screen (dotted line), and attached to a recycling pressure pump and reservoir as a source of water (not shown). *To minimize oxidation, direct the drain into a vessel under a cover of carbon dioxide. In order to achieve this, the press will need to be placed on a stand.* Block diagram of pump: Arrangement of a pump (preferably self-priming, multistage) to enable rapid filling and emptying of the water-bag press. The key item is a reversing valve, either a single unit (four-way) as shown or two separate, linked three-way valves.

3.1 Safety Issues

- Electrical due to operating in a wet area and using an extension lead (if using an electrical pump). The system must be used in conjunction with an earth leakage protection safety circuit. Care must be taken to avoid open connections being exposed to water.
- Tripping due to placement of the extension lead, which must be arranged and secured to minimize this risk.
- Cuts and abrasions: the stainless steel mesh filter and containment screen are sharp and represent a risk unless handled carefully, preferably with gloves.

3.2 Operation—Brief

3.2.1 Start

- Connect hoses between pump and press and reservoir.
- Remove outer stainless steel shell.
- Undo screw on lid and remove lid.
- Release air valve at top.
- Open tap on reservoir and turn reversing valve handle (s) on pump to the filling position.
- If necessary, use an appropriate tool to open priming outlet on pump and allow all air to escape, close valve.

- Switch pump on at pump and wait until all air has been displaced from press bladder, close air valve on press.
- Reverse valve handle(s) to 'empty" position and collapse the bladder (don't over-do); switch pump off at pump.

3.2.2 Cycle

- Fill press with whole bunches or crushed–destemmed fruit (white varieties) or mascerate from a red wine ferment. If bunches or destemmed berries, add liquid $H_2S_2O_5$ now and at each pressing—be sure to calculate the required amount of usually a 1 g/100 mL solution —half at first press and a quarter at each subsequent pressing.
- Replace lid (take care to ensure it fits within the screen) and screw down.
- Place stainless steel barrier in place to ensure all wine is directed to the lower reservoir (more important with white wine than red, but reduces spray!).
- Turn valve handle(s) on pump to 'fill' position and switch pump on.
- Wait till pressure comes up to ca 220 kPa (meter on pump) or first juice is released (white cultivars) and then set timer to 2 minutes; adjust valve handle(s) to maintain pressure at about 220 kPa by turning one or both (if two) a little toward the 'empty' position (i.e. allow some recycling to reduce the pressure).
- After 2 minutes, turn valve(s) on pump to 'empty' and remove the lid from the press.
- By hand, move all bunches, berries or macerate, working around the press so that all is loosened and cleared from the external mesh.
- Repeat from step 2 twice (but check the quality of the pressings at each stage). Keep separate if necessary.
- Remove lid and lift mesh, pushing the top slightly to one side, to enable one to get fingers underneath, and scrape remaining material off the press into a bin for discarding.
- If changing material, wash the mesh and the press before starting the next set of cycles.

3.2.3 Cleanup

- Dismantle the press and hot water or steam clean all components. Allow to dry and reassemble.

Index

Note: Page numbers followed by "*f*" and "*t*" refer to figures and tables respectively.

Y

Printed and bound by CPI Group (UK) Ltd, Croydon, CR0 4YY
03/10/2024
01040326-0019